DEVELOPMENTS IN ENVIRONMENTAL ECONOMICS

VOLUME 4

International Environmental Economics

Theories, Models and Applications to Climate Change, International Trade and Acidification

DEVELOPMENTS IN ENVIRONMENTAL ECONOMICS

Titles in this series:

1. Economics of Environmental Conservation
   by C.A. Tisdell
2. Macroeconomic Analysis of Environmental Policy
   by E.C. van Ierland
3. Macro-Environmental Policy: Principles and Design
   by G. Huppes
4. International Environmental Economics
   Theories, Models and Applications to Climate Change, International Trade and Acidification
   edited by E.C. van Ierland

DEVELOPMENTS IN ENVIRONMENTAL ECONOMICS

VOLUME 4

International Environmental Economics
Theories, Models and Applications to Climate Change, International Trade and Acidification

edited by

Ekko C. van Ierland

Department of General Economics
Wageningen Agricultural University
P.O. Box 8130
6700 EW Wageningen
The Netherlands

1994
ELSEVIER
Amsterdam • Lausanne • New York • Oxford • Shannon • Tokyo

ISBN 0-444-82042-6

Published by:
Elsevier Science B.V.
P.O. Box 211
1000 AE Amsterdam
The Netherlands

Printed in The Netherlands on acid-free paper

Preface

International economic and political relations are complex by their very nature. International environmental economics is even more complex, since it combines the analysis of international trade, transboundary pollution and international negotiations on protecting the global environment.

This book deals particularly with these topics, focusing on climate change, international trade and acidification. Its scope is very interesting, not only because it is an intellectual challenge to analyse these issues, but also because international policy making in this field is developing rapidly and urgently needs economic and institutional analysis in order to reach efficient and effective solutions.

When selecting and editing the various chapters of this volume - the idea for which was born during a lovely summer holiday in Sweden's marvellous nature - I tried to include contributions that show interesting and fresh approaches towards analysing and solving the economic and environmental issues that will occupy us in the coming decades.

My personal feeling is that we are living in a beautiful world that is badly suffering from a rapidly expanding economic system which - by the nature of the decision making mechanism - pays little attention to protecting the environment in its attempt to increase per capita income for a rapidly expanding population. The cumulative character of most environmental problems means that they will become more serious over time unless very drastic reductions in emissions are achieved. In this respect we are facing a match between economic growth and technological progress, and the task of environmental economists is to provide the analysis that leads to proper incentives for economic agents to give priority to technological progress and environmental protection. For this purpose we should continue to build the international institutions that make it possible to reach international agreements on emissions reduction strategies, while at the same time providing opportunities to developing nations to improve their living standards.

The chapters of this book improve our understanding of the complex questions that we are facing in this respect. I hope that the analysis will contribute to formulating international agreements that will protect the global environment in the long run.

The editing of this volume could never have been accomplished without the assistance of a large number of persons. First of all Remco Kok has been very helpful, both in assisting in the editing work and the word processing. Robbie Vogelzang did a perfect job in preparing the manuscript for publishing. Also, I would like to thank Ms Adri Kooijman-Timmer, Ms Jos Michel and Ms Bea Prijn for assisting in the word processing of the various chapters. Finally, I am grateful to Ms J. Burrough-Boenisch for correcting the manuscripts of Chapters 1 and 14.

Ekko van Ierland
Wageningen, June 1994

CONTENTS

Part III Acidification and Tradeable Permits

Amann, Markus	*International Institute for Applied Systems Analysis, Schlossplatz 1, Laxenburg A-2361, Austria.*
Boer, Ineke de	*Wageningen Agricultural University, Department of General Economics, P.O. Box 8130, 6700 EW Wageningen, The Netherlands.*
Bovenberg, Lans	*OCFEB, Erasmus University Rotterdam, P.O. Box 1738, 3000 DR Rotterdam, The Netherlands.*
Conrad, Klaus	*University of Mannheim, Seminargebäude A5, D-68131 Mannheim, Germany.*
Eyckmans, Johan	*Catholic University Leuven, Center for Economic Studies, E. van Evenstraat 2B, B-3000 Leuven, Belgium.*
Hendrix, Eligius	*Wageningen Agricultural University, Department of Mathematics, P.O. Box 8130, 6700 EW, Wageningen, The Netherlands.*
Hordijk, Leen	*Center for Environment and Climate Research, P.O. Box 9101, 6700 HB Wageningen, The Netherlands.*
Ierland, Ekko van	*Wageningen Agricultural University, Department of General Economics, P.O. Box 8130, 6700 EW Wageningen, The Netherlands.*
Kennedy, Peter	*University of Victoria, P.O. Box 3050, Victoria, B.C. Canada V8W 3P5.*
Klaassen, Ger	*International Institute for Applied Systems Analysis, Schlossplatz 1, Laxenburg A-2361, Austria.*
Kolstad, Charles	*University of California, Department of Economics, Santa Barbara, California, CA 93106-9210, USA.*
Kruitwagen, Sonja	*Wageningen Agricultural University, Department of General Economics, P.O. Box 8130, 6700 EW Wageningen, The Netherlands.*

Kverndokk, Snorre	*Central Bureau of Statistics, P.O. Box 8131 Dep, N-0033 Oslo, Norway.*
Mäler, Karl-Göran	*The Beijer Institute, Box 50005, 10405 Stockholm, Sweden.*
Mooij, Ruud de	*OCFEB, Erasmus University Rotterdam, P.O. Box 1738, 3000 DR Rotterdam, The Netherlands.*
Ping-Cheng Li	*Tam Kang University, Department of Industrial Economics, Tamsh, Taipei Hsien, Taiwan 25137, Republic of China.*
Proost, Stef	*Catholic University Leuven, Center for Economic Studies, E. van Evenstraat 2B, B-3000 Leuven, Belgium.*
Rose, Adam	*The Pennsylvania State University, Department of Mineral Economics, University Park, 221 Eric A. Walker Building, PA 16802-5010, USA.*
Schokkaert, Erik	*Catholic University Leuven, Center for Economic Studies, E. van Evenstraat 2B, B-3000 Leuven, Belgium.*
Schöpp, Wolfgang	*International Institute for Applied Systems Analysis, Schlossplatz 1, Laxenburg A-2361, Austria.*
Stevens, Brandt	*The Pennsylvania State University, Department of Mineral Economics, University Park, 221 Eric A. Walker Building, PA 16802-5010, USA.*
Tietenberg, Tom	*Colby College, Department of Economics, Miller Library 234, Waterville, Maine 04902, USA.*
Zylicz, Tomasz	*Warsaw University, Economics Department, Dluga 44/50, 00-241 Warsaw, Poland.*

Chapter 1 INTRODUCTION TO INTERNATIONAL ENVIRONMENTAL ECONOMICS

Ekko C. van Ierland

ABSTRACT

Nations and individuals have international relations, not only through international trade and foreign policies, but also because of global changes in climate and environment, and because of several types of transboundary pollution at the continental level. Previously the theory of international economic relations could confine itself to the theory of international trade, including theories of absolute and comparative advantage, theories of unequal exchange and the analysis of international monetary relations. Nowadays a large body of literature and theories is available on international environmental economics, i.e. the economics of trans-boundary pollution, the economics of global climatic change, transboundary pollution, and the effects of national and international environmental policies on competitiveness and international trade. This book deals in particular with these topics, which are rapidly rising on the international political agenda. Section I deals with the economics of climatic change. Section II contains a number of chapters that deal with International trade and the environment. Section III is devoted to the problems of transboundary air pollution, in particular to the case of acidification at the European level and the theoretical aspects related to it. The chapters in all three sections pay attention to the theoretical aspects, the application of the theories to empirical analysis, and they discuss the policy implications, sometimes as considered from the *national* point of view, in other cases as considered from the *international* point of view or for some *coalition of countries*.

In this introductory chapter the main issues in international environmental economics are shortly discussed. The second part of the chapter consists of summaries of the remaining chapters of the book.

1.1 THE EVOLUTION OF CONCERN ABOUT THE ENVIRONMENT

Concern for the environment initially focused on *local* problems of water pollution, environmental quality in inner cities and management of domestic waste. Rapidly thereafter attention turned to *transboundary* water pollution in river basins, the pollution of oceans and coastal seas and transboundary air pollution, in particular acidification and smog formation. With the discovery of the depletion of the ozone layer, the *global* environmental problems became manifest, and in recent decades global warming and climatic change have become the focus of scientific and political attention. National environmental problems could be solved more easily if international competitiveness were not one of the most essential aspects of economic policy making. The increasing costs of environmental protection threaten to damage the competitiveness of any individual country that tries to take the lead in implementing strict environmental policies. This implies that even if the environmental problems are fully national in character, international trade relations will cause international relations to become dominant. In particular, these international trade relations may distort the correct levels of environmental policy, in terms of overall welfare.

Achieving optimal levels of pollution abatement in the case of transboundary air pollution is even more complex, since each country is faced with the fact that reducing its own emissions will largely result in reducing of environmental damage in neighbouring countries, particularly for relative small countries. This provides little incentive for strict environmental policies and, paradoxically, it provides disincentives: the more country A abates, the smaller is the incentive for country B to do the same, since A's actions already reduce the damage in country B.

However, it easily can be demonstrated that countries could benefit from cooperation if they decided to jointly minimize the total sum of abatement and damage costs for all countries. The potential gains could be distributed over participating countries, generally providing incentives to reach international agreements for abatement policies.

For transboundary pollution at the continental scale, such as acidification and water pollution, international agreements for a group of most affected countries are a possible option. For transboundary pollution at the global scale, like stratospheric ozone, climate change, and pollution of oceans, global agreements are required to arrive at economically efficient and desirable outcomes.

1.2 INCOME DISTRIBUTION

Environmental policies not only have an impact on the quality of our environment, they also affect income distribution in various ways: personal income distribution, distribution of income over wages and profits, distribution amongst countries and the intertemporal distribution, i.e. income distribution over generations. This makes international environmental economics even more challenging. Various principles for income distribution can be applied and it can be analysed how the choice of these principles affects the outcome of our calculations on optimal (first best or second best) environmental policies. This topic is becoming increasingly relevant now that rapid international communication and large disparities in international income distribution are emphasizing the need to improve living conditions, not only for the present generation but also for future generations, in the light of the rapidly growing world population. Increasing per capita production and technological change offer scope for a more egalitarian income distribution. However, concomitantly, rapid population growth is undermining the economic structure and the environment in many countries, widening the gap between poor and rich. The impact of environmental policies on income distribution is therefore an essential and often dominant element in international negotiations on the environment and international trade.

1.3 DEVELOPING COUNTRIES

Environmental problems in developed and developing countries have different backgrounds. In developed countries environmental problems are caused by high levels of production and consumption, a lack of incentives for proper environmental management and little support from the electorate for strict environmental policies. In developing countries it is not so much the level of industrial production that causes problems, but the lack of means to apply proper technologies and the need to provide the basic daily requirements for the growing population. This leads to deforestation, soil erosion, desertification and the pollution of water, air and soil. The rapid economic growth of developing countries in the coming decades will increase their share in the emissions of global pollutants such as carbon dioxide. It is, for example, expected that in a few decades more than fifty percent of the global emissions of greenhouse gases will come from developing countries. Clearly, emission reduction is not the first priority in these countries, where the top priority is to improve living standards by producing food and basic consumption commodities.

Given the outdated technologies that are applied in most developing countries, technology transfer and improvements in energy efficiency are key factors for reducing environmental problems in these countries at relatively low costs. It is not yet clear to what extent the joint implementation of environmental policy programmes might contribute to this development. There is certainly scope for joint implementation, since both developed and developing countries may benefit from the potential efficiency gains; however, the impact on income distribution and the distribution of the credits among the participating countries should be clearly specified beforehand, if joint implementation is to be an attractive and viable policy option.

1.4 INSTRUMENTS OF ENVIRONMENTAL POLICY

Which instruments should be deployed to improve the environment depends typically on the characteristics of the given problem , the local circumstances, and the technological and economic characteristics. The theoretical advantages of *economic instruments* for achieving efficient allocation of production factors are well known and clearly documented in the literature. Whereas the history of environmental policy making has relied heavily on *direct regulation* and detailed legislation, examples of economic instruments have also been introduced, often with great success, despite the associated practical problems, such as monitoring of emissions and the risk of non-compliance. Several deposit systems, charges and tradable discharge permits systems are already in operation, providing ample scope for further empirical analysis of the advantages and drawbacks of each system, in its specific circumstances.

For non-uniformly mixing pollutants, such as sulphur dioxide and nitrogen oxides, the application of economic instruments will be complex and should be accompanied by additional regulations in order to take into account the regional distribution of the emissions and the regional environmental impacts. An unrestricted system of tradable discharge permits for non-uniformly mixing pollutants will lead to a cost-effective reduction of the total amount of emissions, but ignores the regional distribution of impacts and emissions. In the US system for tradeable discharge permits for sulphur dioxide, these shortcomings are compensated by simultaneously setting environmental quality standards at regional level.

Tradeable discharge permits can more easily be applied for uniformly mixing pollutants, such as carbon dioxide. In that case the main policy objective is to restrict the total level of emissions, irrespective of the regional distribution of the emissions. If this is the case, a system of tradeable permits will lead to a first best cost effective solution, after the initial allocation of the discharge permits, either on the basis of grandfathering or by means of an auction. Speculating about whether a system of tradeable discharge permits for carbon dioxide will ever be introduced is unrewarding, because of the difficulty of establishing the initial distribution of the permits, which will clearly determine the impact of the policy measures on income distribution over nations. Although, economists may be in favour of introducing the system from the point of view of efficiency, it will be a difficult - if not impossible - for the international negotiators to implement such a system. It is more likely international agreements by some coalitions will gradually be made to reduce carbon dioxide emissions by a certain percentage, whilst other countries will be reluctant to participate in such agreements, unless they are compensated via side payments or favours in other areas of international relations. An alternative instrument, the global carbon tax, seems more likely to be introduced in the future, although, again, some countries can be expected to be reluctant to participate in an agreement. The advantage of a carbon tax system is that it can be introduced gradually, step by step. Furthermore, it can be so arranged that the national governments will receive the national revenues of the tax, which makes it more likely that the system will be introduced. However, we may expect strong resistance from energy-intensive industries.

1.5 THE STRUCTURE OF THE BOOK

This book is centred around three main themes. Part I "Climate Change" focuses on the economics of uniformly mixing pollutants leading to global warming. Part II "International Trade and the Environment" deals with the international context in which environmental policies have to be formulated. This part contains a number of studies that relate environmental policy to international trade policies. Finally Part III "Acidification and Tradable Discharge Permits" deals with non-uniformly mixing pollutants for which the application of economic instruments becomes even more complicated, particularly if synergetic effects are taken into account.

1.5.1 Climate change

In Part I, Chapter 2, Johan Eyckmans, Stef Proost and Erik Schokkaert from Belgium present a profound analysis of equity and efficiency implications of three different international carbon agreements under full participation. First, they introduce the command and control solution where emission abatement targets are specified for each country. Secondly, they analyse a *uniform* international carbon tax with redistribution of the receipts. Thirdly, a minimal national carbon tax coordinating individual countries' carbon policies is discussed. The full participation constraint in all three approaches requires agreements to be designed such that every individual nation is at least as well off in terms of utility as under the noncooperative outcome. The analysis is illustrated by a two-period carbon emissions simulation model with twelve regions. The analysis by Eyckmans, Proost and Schokkaert suggests that efficiency losses in the case of the second-best tax instruments need not be too substantial compared to a first-best world with possibilities for redistributing lump sums.

Chapter 3, by Snorre Kverndokk from Norway, deals with coalitions and side payments in international carbon dioxide treaties. Based on the experience from the 1992 Rio conference and on the theory of self-enforcing agreements to restrict pollution among sovereign countries he shows that full participation will probably not be the outcome of an international treaty on reducing carbon dioxide emissions. Kverndokk analyses what the optimal policy is for a given commitment of cooperation by a defined group of countries. What is the global loss of partial cooperation compared to full cooperation (social optimum), and how is the optimal abatement level affected by the number of countries committed to cooperate? The analysis assumes that a group of OECD countries have committed themselves to cooperating on the global warming problem. The coalition chooses emission levels and offers the non-cooperating countries transfers if they restrict their emissions. The abatement and side payments are chosen so that the coalition's intertemporal utility function is maximized. The chapter demonstrates that side payments are an effective policy instrument for extending a limited treaty. The study is based on data provided by OECD (Burniaux et al., 1992), and uses the temperature equations of Peck and Teisberg (1992). Whereas the chapter is much more detailed in analysing coalitions and side payments for various groups of countries in the world, it is limited to the extent that man-made capital formation does not play such a prominent role as in the DICE model (Nordhaus, 1992).

Chapter 4, by Charles Kolstad from the University of California, pays attention to uncertainty and learning in global warming. Instead of dealing with given and fixed scenarios for temperature change and resulting damages, Kolstad deals with uncertainty about the severity of climate change and assumes that in the coming decades further information on the exact causes and nature of climate change will become available. He assumes that a learning process is taking place and analyses the implications of this learning process for emission reductions today and in the future. The empirical analysis is based on the Nordhaus (1992) DICE model, to which Kolstad explicitly adds the uncertainty and learning aspects. The best paper in the fifth annual conference of the European Association of Environmental and Resource Economists in Dublin, June 1994, by Ulph and Ulph (1994) is a further theoretical elaboration of the questions raised by Kolstad. Ulph and Ulph conclude that strong conclusions on the impact of learning on the optimal reduction of current emissions cannot be made purely on the basis of theoretical analysis: the empirical relations determine whether or not current emission reductions will be larger in the case of learning.

Adam Rose, Brandt Stevens and Ping-Cheng Li, in Chapter 5, study the implications of global marketable permits for carbon dioxide for energy demand in the USA. The study has three components: 1) a global political economy module to determine the efficient levels of carbon dioxide reduction for all countries and the equilibrium permit price, 2) an abatement strategy module to determine the direct costs and permit transactions by each sector of the US economy and 3) a general equilibrium module to determine the broader effects on the US economy and feedback effects on energy industries. A preliminary version of the model is applied to analyse the short-run impacts on US energy industries of the country's optimal mitigation strategy. The abatement allocations are based on various international equity principles that were discussed in Rose (1992). The study is a very interesting example of a study that relates international environmental economics to decision making that should take place at the national level, which will obviously be based on the implications for the domestic economy. It illustrates that national impacts play an important role in international decision making. The authors conclude "our results indicate that general equilibrium effects may be helpful in fine-tuning global warming policy to minimize the negative effect on the US economy. Still it appears that global warming mitigation strategies will have a slightly negative impact on the US economy as a whole and a very strongly negative impact on the energy sectors even under the so-called *no regrets* strategy".

However, so far the benefits of mitigation policies have not been included. If they are, it leads the authors to the following conclusions "our analysis indicates that the projected total benefits of carbon dioxide mitigation will exceed the costs in the US under most permit allocation rules. There is considerable uncertainty over the gross benefits, but it is still unfortunate that national accounts omit such environmental factors. Still these additional considerations are unlikely to offset the negative impacts on the energy industries".

Chapter 6, by Tom Tietenberg, concludes part I by discussing implementation issues for globally tradeable carbon entitlements. Whereas the advantages of tradeable permit systems are widely discussed in papers and textbooks, the implementation issues received little attention, except for emphasizing all kinds of complications that are related to a system of tradeable permits. In his contribution, which was originally prepared for the United Nations Conference on Trade and Development Secretariat, Tom Tietenberg explores in detail the implementation issues associated with crafting a workable international tradeable entitlements system for greenhouse gas emissions. The analysis focuses on the certification, monitoring, and enforcement functions associated with the system. To avoid the drawbacks of establishing the initial permit allocation for a long period in the future, special attention is given to assuring that the system serves current purposes, while facilitating further development into a more mature, comprehensive system. The chapter discusses in detail the various stages of arriving at a full system of marketable permits. It pays attention to joint implementation and how it is related to marketable permits. Of course, monitoring and enforcement are also broadly covered. Chapter 6 is of particular importance, since it translates theoretical economic insights into actual policy making and clearly sketches the institutional framework required to enable the system to work in practice.

1.5.2 International Trade and Environment

Part II "International Trade and Environment" starts with Chapter 7, written by Ineke de Boer, one of the younger researchers who have contributed to this volume. She presents a "Brief Survey of Current Issues on Trade and Environment" and deals, *inter alia*, with the following questions:

- What is the impact of trade on national and international welfare, including the environment?
- What impact might (regional) trade liberalization have on the welfare of countries both within and outside the region in question if the environmental impacts are considered?
- What is the impact of domestic unilateral environmental politics on international competitiveness and on patterns of production, consumption and trade?
- What is the impact of full or partial agreements on international trade or environmental policy on welfare?
- Is it possible to use trade policies as an incentive to enhance full cooperation in international environmental policy?

Typical topics that are discussed are the impacts of trade on national and international welfare, the externalities of the traded goods (unidirectional or reciprocal), the market structure and the impact of trade policies, especially trade liberalization on welfare. Furthermore, the impact of environmental policy on international competitiveness is analysed in terms of national and international environmental policies. Compensation and penalties for unidirectional environmental problems are discussed, as are unilateral actions with reciprocal environmental problems. Next international cooperation on the environment is covered, paying attention to game theory approaches and interconnected games that are particularly useful if it for unidirectional environmental problems, where it is extremely difficult to reach an agreement, because of the lack of incentives for the source countries to cooperate in the absence of clearly defined property rights.

Klaus Conrad from Germany contributed Chapter 8 on "Emission Taxes and International Market Share Rivalry", analysing strategic behaviour and rent capturing aspects of environmental policy. The study focuses on the implications of environmental policy at the national level for international trade relations and economic welfare. Based on work by Brander and Spencer (1983, 1985), Mintz and Tulkens (1986), and Krutilla (1991), it is illustrated how environmental policy can be used to optimize national welfare in a strategic way, taking into account the impact on international trade. The chapter clearly explains why there is a tendency for environmental taxes in an international setting to be lower than would be expected on the basis of pigovian taxation.

In Chapter 9 Peter Kennedy from Canada examines the strategic incentives to distort pollution taxes in open economies with imperfect competition. He analyses the topic in the context of the debate over the North American Free Trade Agreement (NAFTA), where opponents of the deal argued that the United States and Canada would be forced to weaken their environmental standards in an attempt to compete with Mexico. Proponents of NAFTA counterargued that freer trade would make Mexicans richer and thus more demanding of higher environmental quality. The framework of analysis has three main features: (i) trade is driven by imperfect competition among producers; (ii) the tax choice game between competing countries is modelled explicitly, and (iii) the model allows for transboundary air pollution. The chapter deals with equilibria with and without subsidies, and with restricted subsidies. Kennedy's results confirm the conclusions of Klaus Conrad, "the pollution tax is set below marginal domestic damage in the post-agreement equilibrium. This is clearly inefficient: efficiency requires that the tax be set equal to global marginal damage. This inefficiency reflects both the usual transboundary externality, and the strategic distortion of the pollution tax in response to the trade agreement".

Chapter 10, the last chapter in part II, deals with pollution taxes in an open economy under full competition, in particular focusing on the "double dividend hypothesis", which aims at creating employment by reducing taxation on labour and, concomitantly improving environmental quality by raising pollution taxes. Lans Bovenberg and Ruud de Mooij provide an extensive general equilibrium model, showing that a simple double dividend effect is not present in an open economy under full competition. Improving the environment results either in lower levels of employment or in changes in income distribution. Given the assumptions of the model it is illustrated that competition in the international markets precludes the double dividend hypothesis from applying to small open economies.

1.5.3 Acidification and Tradeable Permits

Part I dealt with a uniformly mixing pollutant at the global level. Part III is complementary, insofar as it deals with non-uniformly mixing pollutants, such as the acidifying compounds sulphur dioxide, nitrogen oxides and ammonia. For acidification, efficiency is not defined as minimizing the total quantity of emissions. On the contrary, efficiency requires that the location of the emissions and the regional impacts of deposition are taken into account. Obviously, this complicates the problem of finding optimal solutions, in particular if a first best command and control solution is not feasible for political reasons. All the chapters in this part deal with the question of how to arrive at solutions that approximate the first best solution for reaching specific environmental quality standards. However, in each chapter this question is tackled from a different perspective, analysing different alternative policy instruments or combinations of them.

In Chapter 11, Markus Amann, Leen Hordijk, Ger Klaassen and Wolfgang Schöpp compare the command-and-control strategy of the "best available technology concept" (BAT concept) and the critical loads approach. The BAT concept calls for uniform implementation of certain emission standards throughout the region. In the critical loads approach a scenario in which the gap between current depositions and the critical loads is everywhere reduced by at least 50 percent is analysed. The chapter demonstrates that the BAT concept does not result in a cost-effective solution. However, a regional zoning of BAT standards could produce results which are very close to the cost-minimal solution. In the negotiations on the second sulphur protocol that was signed in June 1994, the gap closure scenario played an important role, but using a 60 percent gap closure, not a 50 percent gap closure as in this study.

In Chapter 12 Tomasz Zylics from Warsaw University, Poland, explores the question how economic instruments could be applied in the case where several sources of environmental impact affect several economic quality indicators in a non-necessarily linear way. The chapter illustrates that there is a potential for cost savings if various sources of environmental impact are regulated in a coordinated way. However, economic instruments could achieve a least cost solution only under very restrictive assumptions. In particular, it is pointed out that some of the known synergistic interactions - implying non-convexity of the constraint function delineating the permit market (or determining a tax) - clearly violate these conditions. Zylics therefore argues that the applicability of economic instruments is restricted to simple linear cases. This topic of synergism of pollutants is particularly interesting for studying the interactions between acidifying compounds (in particular NOx) and ozone precursors (VOCs and NOx). An interesting conclusion is that a combination of a tradeable permit system and a pollution tax could be applied in the non-linear case. However, the rigid-convexity assumptions and the complexity of the trading mechanism in the non-linear case suggest that in many instances synergetic pollutants have to be handled by means of direct regulations.

Chapter 13, by Ger Klaassen of the International Institute of Applied Systems Analysis (IIASA), Austria, suggests using tradeable discharge permits for acidification by means of an exchange rate for permits. Given the importance of the location of the emissions and the regional distribution of the impacts of acidification (related to large variations in soil sensitivity and sensitivity of ecosystems over Europe) straightforward trade in emission permits will not be cost-effective for achieving critical loads. Based on the IIASA Rains model a routine has been developed to simulate bilateral sequential emission trading for sulphur dioxide in Europe. It is shown that generally this system will not achieve the optimal solution, because other necessary conditions for an optimal solution are neglected. The analysis reveals how complicated trade systems for non-uniformly mixing pollutants will have to be to achieve specified critical loads cost-effectively.

An alternative approach for tradeable discharge permits for acidification is presented in Chapter 14 by Sonja Kruitwagen, Eligius Hendrix and Ekko van Ierland. Instead of introducing exchange rates in emissions trading, the quantities of permits that can be traded by each individual country are restricted by a trade coordinating institution. In the case of 'guided bilateral trade' the optimal solution can be approximated by sequential bilateral trade, provided that these restricitions are applied. However, the paper also illustrates that a number of non-profitable trade transactions are required in order to fully reach the first best cost-effective solution. Guided bilateral trade clearly has advantages over non-guided trade, since it contributes both to reducing costs and improving environmental quality at a large number of receptor points. It is interesting to compare the systems of guided bilateral trade with the actual trade system for sulphur permits in the USA, where additional ambient air quality constraints have also been set to avoid that permits being concentrated in heavily polluted areas after trade.

The book concludes with a contribution by Karl-Göran Mäler of the Beijer Institute, Sweden, on the dynamics of the acid rain game. The chapter clearly shows that in the case of acidification we should consider the stock pollutant character of acid depositions in a dynamic framework, which allows repeated games, including trigger strategies for non-compliance and cheating. Using this approach it becomes possible to take into account to what extent the buffer capacity of the soil has already been used and how much time is left to reduce emissions (and simultaneously depositions) to avoid exhausting the soil's buffer capacities. This dynamic approach has not yet been fully applied in empirical research and clearly further research is needed to analyse these dynamic aspects in more detail in order to answer the question of how long current reduction strategies can be continued and how ecological catastrophies can be avoided when nature's carrying capacity is depleted over time.

1.6 CONCLUSION

Climate change, environmental aspects of international trade and acidification are very interesting and also very serious test cases for international environmental policy making. The various chapters in this book clearly reflect the considerations and the various aspects that are related to these topics of international environmental economics. The coming decades will reveal which parts of the analyses prove to be the most relevant for actual policy making. At least we can say that economists are clearly presenting the theoretical and practical options and implications of the various policy alternatives. Hopefully, this book will enhance our understanding of the issues at stake, and will contribute to reaching international agreement on policies to protect the global environment.

REFERENCES

Barrett et al. (1992) *Combating Global Warming: A Global System of Tradeable Carbon Emission Entitlements*, Unctad, Geneva, Switzerland.

Brander, J. A. and B.J. Spencer (1983) Strategic commitment with R&D: The symmetric case. *Bell Journal of Economics*, 14, 225-235.

Brander, J. A. and B.J. Spencer (1985) Export subsidies and international market share rivalry. *Journal of International Economics*, 18, 83-100.

Burniaux, J.M., et al. (1992) *Green - A Multi-sector, Multi-region Dynamic General Equilibrium Model for Quantifying the Costs of Curbing CO_2 Emissions: A Technical Manual.* Working Paper No. 116, Economics Department, OECD, Paris.

Krutilla, K. (1991) Environmental regulation in an open economy. *Journal of Environmental Economics and Management 20*, 127-142.

Mintz, J. and M. Tulkens (1986) Commodity tax competition between member states of a federation: Equilibrium and efficiency. *Journal of Public Economics* 29, 133-172.

Nordhaus, W.D. (1992) *The DICE model: Background and Structure of a Dynamic Integrated Climate-Economy Model of the Economics of Global Warming.* Cowles Foundation Discussion Paper 1009, New Haven, CT.

Peck, S. and T.J. Teisberg (1992) CETA: A Model for Carbon Emission Trajectory Assessment. *The Energy Journal*, 13:55-77.

Rose, A., 1992, Equity considerations of Tradeable Carbon Entitlements, in: Barrett et al., 1992, *Combating Global Warming: A Global System of Tradeable Carbon Emission Entitlements.*

Ulph, A. and D. Ulph (1994) *Global Warming: Why Irreversibility May Not Require Lower Current Emissions of Greenhouse Gases*. Paper presented at the fifth Annual Conference of the European Association of Environmental and Natural Resource Economists, Dublin.

PART I

CLIMATE CHANGE

Chapter 2

A COMPARISON OF THREE INTERNATIONAL AGREEMENTS ON CARBON EMISSION ABATEMENT

Johan Eyckmans, Stef Proost, Erik Schokkaert[1]

ABSTRACT

This paper examines equity and efficiency implications of three different international carbon abatement agreements under full participation: (i) the command-and-control solution where emission abatement targets are specified for each country, (ii) a uniform international carbon tax with redistribution of the receipts and (iii) a minimal national carbon tax coordinating individual countries' carbon policies. The full participation constraint requires agreements to be designed such that every individual region is at least as well off in terms of utility than under the noncooperative outcome. We consider these participation requirements as necessary conditions for the international carbon abatement agreements to be acceptable in the absence of an international authority. It is shown that Lagrange multipliers associated to the participation constraints are to be added to the social welfare weights and do not change the form of the optimal tax rules. The analysis is illustrated by a two-period carbon emission simulation model with twelve regions. The empirical part suggests that efficiency losses in the case of the second-best tax instruments need not to be too substantive compared to a first-best world with lump sum redistribution possibilities.

2.1 INTRODUCTION

In recent years there has been a growing interest in modeling international negotiations on environmental problems. As is well known, the need for international coordination of environmental policies follows from the fact that the noncooperative equilibrium will often be socially nonoptimal. A prominent example is offered by the greenhouse or global warming problem. In this context attention has been focused mainly on carbon dioxide (CO_2) emissions because CO_2 is believed to be the principal contributor to the enhanced greenhouse effect (IPCC, 1990). We will assume throughout the paper that an international

1 This chapter benefited from valuable comments by Charles Kolstad, Ekko van Ierland, Remco Kok and Pierre van Mouche. All remaining errors and flaws remain ours.

scientific and political consensus emerges on the need to curb CO_2 emissions in order to safeguard future generations of the negative repercussions of enhanced greenhouse warming.

The purpose of this chapter is twofold. First, we will analyze analytically three possible types of international greenhouse agreements: (i) an agreement specifying specific emission reduction targets for all participants (the command-and-control approach), (ii) an international uniform carbon tax scheme with redistribution of the receipts (the international uniform carbon tax), and (iii) a less ambitious instrument of harmonized national carbon taxes (the minimal carbon tax approach). It is assumed that an effective international lump sum side-payment mechanism is not politically feasible. In order to get a better insight into the merits and drawbacks of the agreements, we maximize a weighted sum of the utilities of the different countries. The weights can be interpreted both in ethical and power terms.

The noncooperative Nash-Cournot equilibrium concept will be used here to define threat points in negotiations. We assume that countries are only willing to join an agreement if it is in their own self-interest; we are looking for Pareto improvements compared to the noncooperative Nash-Cournot outcome. It is shown that these participation constraints, requiring every acceptable agreement to be a Pareto improvement vis à vis the threat points, can be incorporated into the analysis by an appropriate reinterpretation of the social marginal valuation of income.

Secondly, we present a simple empirical model of the world economy incorporating the best available information on costs and benefits of emission reduction for 12 regions. The empirical model is used to illustrate the results of the analytical section, but also to provide approximations where the analytical framework cannot yield clear-cut results. The empirical results are conditional upon the assumptions on the functional forms of cost, benefit and utility functions, but provide some relevant additional insight in the differences between the three policy instruments. The main conclusion is that efficiency losses of the second-best tax instruments need not be so important as we might guess from the analytical section.

The importance of participation constraints has been stressed by Barrett (1992) in the context of a bargaining model. The idea for an international carbon tax with redistribution of the receipts and for a minimal national tax agreement have been proposed in Hoel (1992*a*) and Hoel (1992*b*). In Hoel (1992*c*) participation constraints are explicitly taken into account for the case of uniform reduction of emissions. Our analysis is more general because it relaxes the small country assumption and introduces explicitly participation constraints as well as nonlinear utility functions. In general, our approach resembles the benefit framework presented by Welsch (1993) but we stay closer to the public good provision literature than to the cost sharing literature. International carbon tax solutions have been tested empirically by means of applied general equilibrium models like the OECD-GREEN model (Burniaux e.a., 1992) and Whalley and Wigle (1991). We use a simpler simulation model, which is calibrated to results obtained by Nordhaus (1991*a*) and Nordhaus (1991*b*) and Burniaux e.a. (1992), and explicitly optimize the tax and redistribution parameters instead of simulating some particular solution.

2.2 EMISSION ABATEMENT AS A PUBLIC GOOD

We will describe briefly the interdependence between economic production, emissions of an atmospheric pollutant and costs and benefits of emission abatement. The voluntary provision or noncooperative equilibrium will be characterized.

2.2.1 Costs and Benefits of Emission Abatement

Consider a world $N=\{1,2,...,n\}$ of n countries. Economic activity in country i causes emissions of carbon dioxide, or conversely, CO_2 emissions can be regarded as an input in economic production of country i. From the scientific greenhouse debate we know that global warming is essentially a dynamic pollution problem. It is the accumulated stock of the pollutant, and not the flow of emissions, that causes environmental damage. However, in order to concentrate on the international aspect of the problem, we will compress the dynamics into a simple two-period model. Suppose that in period zero all countries decide individually on their emission level. These decisions determine their income level in period zero, and the concentration level and associated environmental damage which accrues in period one. In this simplified model we can write, without loss of generality, country i's national income Y_i as a function of its emissions E_i: $Y_i = Y_i(E_i)$. It is assumed that national income increases in emissions but at a declining rate, i.e. Y_i is increasing and strictly concave in E_i. Environmental damage D_i on the other hand is a function of the total world emission level: $D_i = D_i(E)$ with $E = \sum_{j\in N} E_j$. The time interval between period zero and one is defined long enough to allow for the environmental repercussions to emerge. Environmental damage is assumed increasing and convex in the global emission level. Now we will translate the emission framework into an emission abatement model.

Define individual emission abatement R_i as the difference between a fixed reference emission level $\overline{E}_i$ and the new emission level E_i: $R_i = \overline{E}_i - E_i$. By choosing its emission level E_i in period zero a country determines uniquely its emission reduction R_i and vice versa. The choice of the reference situation will turn out to be of crucial importance for determining bargaining positions of individual countries. In our approach some fixed base year will serve as reference situation[2]. Notice that always $R_i \leq \overline{E}_i$ because a country cannot abate more than the emission level in the reference year. Emissions are considered as an input in economic production and cannot be negative. Abatement effort R_i can however be negative if country i increases its emissions relative to the reference period. For many developing countries environmental concern has a lower priority than economic growth and they might well decide to emit more pollutants than in the reference period. Other regions might benefit instead of suffer from the increased stock of the pollutant. E.g., it is believed that regions like Canada or Siberia could gain initially from moderate global warming because more agricultural land would become available to them. These countries might therefore prefer even higher concentration levels E of the pollutant. For a comprehensive description of the upper and lower bounds on emission abatement, see Welsch (1993). He defines the lower bound on abatement as $\underline{R}_i = \overline{E}_i - \overline{z}_i$ where $\overline{z}_i$ is the maximal economically sustainable emission level for country i which refers to a capacity constraint on economic production. Hence, individual emission abatement R_i is defined over the interval $[\underline{R}_i, \overline{E}_i] \subset \Re$ and the lower bound $\underline{R}_i$ can be negative.

On the one hand, reducing emissions results in a lower national income for country i because emissions are an input in production. We will denote the forgone income as $C_i(R_i) = Y_i(\overline{E}_i) - Y_i(E_i) = Y_i(\overline{E}_i) - Y_i(\overline{E}_i - R_i)$, the economic cost of emission reduction. The cost functions $C_i(R_i)$ are increasing and convex in R_i because of the assumptions concerning $Y_i(E_i)$. We further assume that $C_i(R_i)$ is twice continuously differentiable over the interval $[\underline{R}_i, \overline{E}_i]$.

On the other hand, lower emissions will lead to a lower stock of the atmospheric pollutant and hence to less environmental damage. The avoided damage can be regarded as the benefit of emission abatement : $B_i(R) = D_i(\overline{E}) - D_i(E)$ where $R = \sum_{j \in N} R_j$ stands for total emission abatement, $\overline{E} = \sum_{j \in N} \overline{E}_j$ for total emissions in the reference period, and $E = \sum_{j \in N} E_j$ for total emissions of CO_2 in period zero. Using the properties of the environmental damage functions $D_i(E)$ it is easy to show that benefit functions $B_i(R)$ are increasing and concave in R. Also $B_i(R)$ is assumed twice continuously differentiable over the interval $[\underline{R}, \overline{E}]$ where $\underline{R} = \sum_{j \in N} \underline{R}_j$ stands for the global lower bound on emission reduction.

---

2 Our assumptions imply that environmental concern was only recently introduced in countries' decision making process and that countries did not care about the environment and simply maximized national income before the reference period. Other authors consider the appropriate reference situation to be a noncooperative Nash-Cournot equilibrium, allowing them to calibrate their reaction functions, cfr. Mäler (1989).

It is clear that in this setting total emission abatement R is an international public good. No country can be excluded from consuming the benefits of world-wide abatement efforts R because emissions mix directly and uniformly in the atmosphere. The fact that country i enjoys benefits of total emission reduction R does not prevent all other countries from doing the same. Provision of the public good is organized in a decentralized way: every country can contribute to R by reducing its individual emissions R_i at cost $C_i(R_i)$. There is no world authority which can provide the public good and there is no collective technology to produce emission abatement. It is a stylized fact in the literature that under these circumstances voluntary provision of such a public good will in general be too low compared to the social optimum see, e.g., Cornes and Sandler (1986). In what follows we will characterize the voluntary provision equilibrium and contrast it with socially optimal outcomes under different assumptions concerning the available policy instruments.

2.2.2 Voluntary Provision of Emission Control

Preferences in a Two-Period Model

Each country's preferences are defined over per capita income in both periods and we assume the same functional form for the utility function for all countries[3]:

$$U_i = U(y_i^0(R_i), y_i^1(R_i, R_{-i})) \equiv U\left(\frac{Y_i^0(R_i)}{n_i^0}, \frac{Y_i^1(R_i, R_{-i})}{n_i^1}\right) \quad \forall i \in N \qquad (1)$$

where (Y_i^0, Y_i^1) are total income levels for country i in period zero and one respectively and lower-case letters denote per capita variables and (n_i^0, n_i^1) the exogenous population levels in the two periods. The utility functions $U_i(y_i^0, y_i^1)$ are twice continuously (partially) differentiable and bounded over $\Re_+^2$. We assume that every country decides in period zero on its individual contribution R_i to the public good and incurs a corresponding economic cost $C_i(R_i)$. In period one it enjoys benefits $B_i(R)$ of joint emission reduction efforts R. Income in both periods is defined as:

$$Y_i^0(R_i) = \overline{Y}_i^0 - C_i(R_i) \quad \forall i \in N \qquad (2a)$$

$$Y_i^1(R_i, R_{-i}) = \overline{Y}_i^1 + B_i(R_i + R_{-i}) \quad \forall i \in N \qquad (2b)$$

3 This assumption is not necessary for the analytical results but it is however quite useful for the empirical illustration.

where $R_{-i} \equiv \sum_{j \neq i} R_j = R - R_i$ is the emission abatement by all other countries but i, and $\bar{Y}_i^0$ and $\bar{Y}_i^1$ are baseline income levels resulting from straightforward income maximization without environmental concern. By introducing benefits only in period one we want to stress the time lag between abatement and resulting slowing of global warming. The formulation in (2) can easily be generalized to include benefit terms in period zero and cost terms in period one.

Nash-Cournot Noncooperative or Voluntary Provision Equilibrium

Consider first the case of nonagreement or noncooperation. Under the assumption of Nash-behaviour on parts of all participants every country chooses its abatement level R_i to maximize (1) taking as given the abatement effort R_{-i} of the rest of the world[4].

$$\underset{R_i}{maximize} \ U\left(\frac{\bar{Y}_i^0 - C_i(R_i)}{n_i^0}, \frac{\bar{Y}_i^1 + B_i(R_i + R_{-i})}{n_i^1}\right) \qquad \underline{R}_i \le R_i \le \bar{E}_i \qquad \forall i \in N \qquad (P.1)$$

We assume that the constraint $R_i \le \bar{E}_i$ is never binding because it is extremely costly to abate the last unit of carbon dioxide emissions. We also assume that R_i can only be strictly positive because marginal abatement costs tend to zero for R_i approaching zero[5]. Define MC_i^{NC} and MB_i^{NC} as the marginal cost and corresponding marginal benefit of the emission reduction vector $R^{NC} = (R_i^{NC}, R_{-i}^{NC})$. The first-order conditions for an interior solution are given by[6] (the superscript *NC* refers to the noncooperative equilibrium):

$$-\left(\frac{\partial U}{\partial Y_i^0}\right)^{NC} MC_i^{NC} + \left(\frac{\partial U}{\partial Y_i^1}\right)^{NC} MB_i^{NC} = 0 \qquad \forall i \in N \qquad (3)$$

---

4 Cornes and Sandler (1984) analyze the provision of public goods under non-Nash behavioural assumptions.

5 According to Welsch (1993) however, the latter assumption on abatement cost functions is inappropriate because it corresponds to zero marginal productivity of emissions in the reference period. Therefore, he incorporates the constraint $\underline{R}_i \le R_i$ in the maximization (P.1).

6 Welsch (1993) gives a more complete characterization of the noncooperative equilibrium with Kuhn-Tucker conditions for the case where the constraint $\underline{R}_i \le R_i$ is binding. If a country's minimal marginal cost, which occurs at $\underline{R}_i$, exceeds the sum of marginal benefits, it is optimal for that country to choose abatement level $R_i = \underline{R}_i$.

which can be written as[7]:

$$MC_i^{NC} = \theta_i^{NC} MB_i^{NC} \qquad \forall i \in N \tag{4}$$

where θ_i^{NC} stands for the marginal rate of intertemporal substitution between income in the first and second period, evaluated in the noncooperative equilibrium:

$$\theta_i^{NC} = \frac{(\partial U / \partial Y_i^1)^{NC}}{(\partial U / \partial Y_i^0)^{NC}} \qquad \forall i \in N \tag{5}$$

According to (4) all countries abate emission up to the point where their marginal abatement costs equal their private discounted reduction benefits[8]. Solving (3) w.r.t. R_i yields the reaction curve $R_i = \mathcal{R}_i(R_{-i})$ which defines the optimal strategy for country i, given abatement level R_{-i} of the rest of the world. The slope of the reaction curve is given by (totally differentiating (4) and rearranging):

$$\frac{dR_i}{dR_{-i}} = \frac{(\partial\theta_i/\partial R_{-i})MB_i + \theta_i B_i''}{C_i'' - (\partial\theta_i/\partial R_i)MB_i - \theta_i B_i''} \qquad \forall i \in N \tag{6}$$

Because θ_i is decreasing in both R_i and R_{-i}, see Appendix (A.1), the reaction curves are downward sloping.

The Nash-Cournot noncooperative equilibrium is found by solving simultaneously the system of n first-order conditions in (3). Since:

(i) $\forall i \in N$, the strategy spaces $S_i = [0, \overline{E}_i] \subset \Re$ of the noncooperative game are compact and convex,

---

7 The population terms n_i^0, n_i^1 vanish in (3) because marginal utility of income is expressed in terms of income levels in stead of per capita income.

8 Second-order condition for a maximum of (P.1) can be shown to be (we omitted the superscript *NC* but all functions are evaluated in the noncooperative Nash-Cournot equilibrium, and C_i'' and B_i'' stand for the second-order derivatives of cost and benefit functions):

$$-C_i'' + \left(\frac{\partial\theta_i}{\partial R_i} + \frac{\partial\theta_i}{\partial R_{-i}}\right) MB_i + \theta_i B_i'' \le 0 \qquad \forall i \in N$$

In appendix (A.1) we show that for the usual assumptions on the utility function, the effects of changes in R_i and R_{-i} on θ_i are negative. Hence, the second-order condition is satisfied for all abatement vectors $R = (R_i, R_{-i})$.

(ii) $\forall i \in N$, the utility functions $U_i(y_i^0, y_i^1)$ are twice continuously (partially) differentiable and bounded over $\Re^2_{++}$

(iii) $\forall i \in N$, $U_i(y_i^0, y_i^1)$ are concave in R_i, cf. the second-order condition,

there exists at least one Nash-Cournot noncooperative equilibrium, see Friedman (1986).

Uniqueness of the noncooperative equilibrium requires the Jacobian of the system of first-order conditions (3) to be negative quasi-definite $\forall (R_1, R_2, \ldots, R_n) \in S_1 \times S_2 \times \ldots \times S_n$. This imposes restrictions on the relative magnitudes of second-order derivatives of utility, cost and benefit functions for all i in N. We will assume in the rest of the text that these conditions hold, and hence that the noncooperative equilibrium is unique. For later reference we define the utility level of region i in the Nash-Cournot equilibrium as U_i^{NC}.

2.3 THREE DIFFERENT TYPES OF INTERNATIONAL CARBON AGREEMENTS

2.3.1 A Welfare Maximization Framework with Participation Constraints

In order to investigate the negotiation process, we consider a weighted sum of the utility functions of all participating countries. This formulation can be interpreted as the policy problem faced by a hypothetical world planner (or a greenhouse fund administrator) maximizing a generalized social welfare function, but as argued extensively in Eyckmans, Proost and Schokkaert (1993), it can be interpreted in a bargaining context as well:

$$\underset{\omega}{maximize} \quad \sum_{i \in N} \lambda_i U(y_i^0(R_i), y_i^1(R_i, R_{-i})) \qquad \text{with } \sum_{i \in N} \lambda_i = 1 \quad (7)$$

where ω stands for the set of available policy instruments. The weights λ_i can be interpreted as *ethical* weights in a social welfare maximization framework, or as power weights in a *bargaining* context. It is also possible to start from a given proposal on the distribution of abatement efforts or tax scheme to compute the implicit ethical or power weights used by the negotiators, the so-called *inverse optimum* approach, see Eyckmans, Proost and Schokkaert (1993). In addition, if the utility locus is strictly convex, varying the weights λ_i describes the complete *Pareto-frontier*.

As noted before, noncooperative or voluntary provision of the public good is not Pareto optimal because individual countries only take into account private benefits of abatement when they determine their contribution, see among others Cornes and Sandler (1986). They do not consider the spill over or external effects of their abatement decisions on other countries. A frequently suggested solution is to impose a corrective externality tax on the polluting activity to confront every country with the correct "price" for its pollution. In the context of global environmental problems however, a strong world authority which could impose such tax solutions does not exist. One can only hope for some voluntary agreement and cooperation emerging between the polluters. Countries will only be willing to sign such an agreement if it is in their self-interest, or alternatively, if they are better off under the agreement than when they play their noncooperative Nash-Cournot strategy. The noncooperative Nash-Cournot equilibrium defined in Section 2 acts as a threat point in negotiations. For the three types of agreement that we consider in this paper, we will impose explicitly that the outcomes are Pareto improvements over the noncooperative voluntary provision equilibrium. Ideally, the complete treatment of voluntary cooperation requires an inquiry into intermediate coalition formation and coalition stability. In this context interesting work is provided by Bauer (1992), Hoel (1993), Kverndokk (1994) and Petrakis and Xepapadeas (1993). We will however simplify the analysis by assuming that countries can only choose between joining a complete agreement including all nations, or no agreement at all. Formally, we require for every agreement that all countries are at least as well off than when they play their Nash strategy:

$$U(y_i^0(R_i), y_i^1(R_i, R_{-i})) \geq U_i^{NC} \qquad \forall i \in N \tag{8}$$

with U_i^{NC} defined as in section 2. We will call this condition the *participation constraint* of country i.

2.3.2 Command-and-control Optimum

Command-and-control Optimum without Lump Sum Transfers

Suppose that all countries in N voluntary start negotiating on an emission abatement vector $\boldsymbol{R}=(R_1, R_2, \ldots, R_n)$ and that it is considered politically unfeasible to design an effective international side-payment scheme. Hence, differentiating the emission abatement efforts is the only instrument available to the negotiators to overcome all participation constraints and to do better than the noncooperative outcome. This is in fact the type of international environmental agreement that we often observe in reality. Existing agreements, the Montreal Protocol on the Protection of the Ozone Layer as the most prominent example, usually stipulate emission abatement efforts for all participants (e.g., equal percentage reduction targets) without possibility for substantial international income redistribution. In our framework, this amounts to the problem of finding an emission abatement vector $\boldsymbol{R}=(R_1, R_2, \ldots, R_n)$ which maximizes the objective function (7) and respects at the same time all participation constraints (8). This vector is defined as the solution to the problem:

$$\underset{(R_1,\ldots,R_n)}{\text{maximize}} \quad \sum_{i\in N} \lambda_i U\left(\frac{\overline{Y}_i^0 - C_i(R_i)}{n_i^0}, \frac{\overline{Y}_i^1 + B_i(R_i + R_{-i})}{n_i^1}\right) \tag{P.2}$$

$$\text{subject to:} \qquad U(y_i^0(R_i), y_i^1(R_i, R_{-i})) - U_i^{NC} \geq 0 \qquad \forall i \in N$$

The corresponding Lagrangian function is defined as:

$$L(R_1,\ldots,R_n,\pi_1,\ldots,\pi_n) = \sum_{i\in N} \lambda_i U(y_i^0(R_i), y_i^1(R_i, R_{-i})) +$$

$$\sum_{i\in N} \pi_i [U(y_i^0(R_i), y_i^1(R_i, R_{-i})) - U_i^{NC}]$$

The first-order conditions for an interior solution for $\boldsymbol{R}=(R_1, R_2, \ldots, R_n)$ are given by (superscript *CC* denotes Command-and-Control):

$$\tilde{\lambda}_i^{CC} MC_i^{CC} - \sum_{j\in N} \tilde{\lambda}_j^{CC} \theta_j^{CC} MB_j^{CC} = 0 \qquad \forall i \in N$$

$$\text{with} \quad \tilde{\lambda}_i \equiv (\lambda_i + \pi_i^{CC})\left(\frac{\partial U}{\partial Y_i^0}\right)^{CC} \quad \text{and} \quad \theta_i^{CC} \equiv \frac{(\partial U/\partial Y_i^1)^{CC}}{(\partial U/\partial Y_i^0)^{CC}} \tag{9}$$

The shadow prices π_i^{CC} of the participation constraints are determined by the following Kuhn-Tucker conditions:

$$U(y_i^0(R_i^{CC}), y_i^1(R_i^{CC}, R_{-i}^{CC})) - U_i^{NC} \geq 0 \quad ; \quad \pi_i^{CC} \geq 0 \quad ;$$

$$\pi_i^{CC}[U(y_i^0(R_i^{CC}), y_i^1(R_i^{CC}, R_{-i}^{CC})) - U_i^{NC}] = 0 \qquad \forall i \in N \tag{10}$$

θ_i^{CC} stands for the marginal rate of intertemporal substitution between income in period zero and one, evaluated in the Command-and-Control equilibrium. $\tilde{\lambda}_i^{CC}$ should be interpreted as the marginal social value to country *i* of an additional unit of period zero income evaluated in the Command-and-Control optimum. It is a function of three parameters: first, the weight λ_i of country *i* in the objective function, secondly, the shadow price of its participation constraint π_i^{CC}, and thirdly, its marginal utility of period zero income. In the case where for country *i* the participation constraint is binding, the social welfare weight is corrected by the shadow price π_i^{CC}. These shadow prices are a measure of the power a country can derive from the threat to return to its noncooperative Nash-Cournot strategy and are an outcome of the maximization exercise. The π_i^{CC} are therefore a function of the weights λ_i, the utility level in the noncooperative threat point U_i^{NC}, and of the available policy instruments. It is interesting to notice that introducing full participation constraints does not alter the general *form* of the first-order conditions.

Their effect works through the new definition of the social marginal valuation of income $\tilde{\lambda}_i$. If lump sum redistribution is impossible, the first-order conditions for a social optimum require that for all *i* weighted marginal costs are equal to the sum of weighted marginal benefits. Countries with a high ethical/power weight λ_i in the objective function, or with a strong bargaining position (high π), or with high marginal utility of income (i.e. relatively poor regions), are required to abate less than other participants.

Command-and-control Optimum with Lump Sum Transfers

It is illuminating to compare the first-order conditions (9) with corresponding conditions for a scenario with lump sum side-payments. The latter scenario refers to a first-best world and will serve as a bench-mark in what follows. Assume that an effective international side-payment mechanism can be implemented, for instance through intermediation of an international greenhouse fund. The administrators of this fund can vary emission abatement efforts R_i and use also lump sum transfers in order to achieve more abatement than under noncooperation, and respecting at the same time the participation constraints. It can be shown that the optimal transfers should be such that they equalize the marginal social valuation $\tilde{\lambda}_i$ of income across all nations, see Eyckmans, Proost and Schokkaert (1993). Substituting this condition for $\tilde{\lambda}_i$ in (9) results in (superscript *FB* denotes First-Best optimum):

$$MC_i^{FB} = \sum_{j \in N} \theta_j^{FB} MB_j^{FB} \qquad \forall i \in N \tag{11}$$

where θ_i^{FB} is defined like in (5) but evaluated in the first-best optimum. Condition (11) has two important implications. First, marginal costs should be equalized across all nations to obtain *cost-efficiency*, i.e. achieving the global emission reduction target at the lowest cost possible. Secondly, (11) requires that for all *i* in N marginal costs should equal the sum of all discounted marginal benefits. The latter condition assures an optimal provision of the public good emission reduction *R*.

For an agreement fixing directly emission abatement levels for all countries we can conclude in general:

(i) participation constraints can analytically be incorporated in an easy way through an appropriate redefinition of the social marginal valuation of income,

(ii) the participation constraints do not preclude cost efficiency and an optimal supply of global emission reduction in a first-best world with lump sum transfers,

(iii) in a second-best world without lump sum transfers, efficiency and distributional considerations cannot be separated.

2.3.3 An International Uniform Carbon Tax with Redistribution of the Receipts

Countries' Individual Behaviour under the International Carbon Tax

In contrast with an agreement fixing individual abatement efforts for all countries, consider an international carbon tax scheme and assume that all nations face the same uniform tax rate t per unit of carbon emitted. The receipts of this tax system are redistributed among the participants according to a priori fixed shares $\beta_i \geq 0 \quad \forall i \in N$ with $\sum_{j \in N} \beta_j = 1$. We will assume that taxing and redistributing occurs in period zero. This international carbon tax agreement can be seen as a two-stage problem. At the lower level, individual countries choose their optimal abatement effort level R_i taking as given the tax-redistribution scheme $(t, \beta_1, \ldots, \beta_n)$ and the reactions of all other nations R_{-i}. At the upper level, some world-wide greenhouse authority, or international negotiators, determine the optimal tax rate t and redistribution shares β_i to maximize social welfare and taking into account the individual participation constraints. We will first focus on the problem faced by the participating countries, and then we turn to the international level. Period zero income is now defined as:

$$Y_i^0(R_i) = \overline{Y}_i^0 - C_i(R_i) - t(\overline{E}_i - R_i) + t\beta_i(\overline{E} - R) \qquad \forall i \in N \qquad (2a')$$

and period one income is still defined according to (2.b). Here, $t(\overline{E}_i - R_i) = tE_i$ stands for country i's carbon tax payment to the Greenhouse fund, and $t(\overline{E} - R) = tE$ is the total tax revenue collected of which country i can claim its share β_i. Assuming selfish behaviour on part of all participants, maximization problem (P.1) changes to:

$$\underset{R_i}{maximize} \; U\left(\frac{\overline{Y}_i^0 - C_i(R_i) - t(\overline{E}_i - R_i) + t\beta_i(\overline{E} - R_i - R_{-i})}{n_i^0}, \frac{\overline{Y}_i^1 + B_i(R_i + R_{-i})}{n_i^1}\right) \qquad \forall i \in N$$

(P.3)

For a given carbon taxation scheme $(t, \beta_1, \ldots, \beta_n)$ the first-order condition for a noncooperative optimal strategy for country i is defined by (where superscript *int* denotes international carbon tax solution and we focus on interior solutions only):

$$MC_i^{int} - t(1 - \beta_i) = \theta_i^{int} MB_i^{int} \qquad \forall i \in N \qquad (12)$$

θ_i^{int} stands for the marginal rate of intertemporal substitution between income in both periods evaluated under the international tax scenario. Compared to condition (4) all countries equate *net* marginal costs to discounted marginal benefits. Equation (12) implicitly defines the reaction curve of country i: $R_i = \mathcal{R}(t, \beta_1, \ldots, \beta_n, R_{-i})$. The properties of this noncooperative supply function of emission abatement are found by totally differentiating (12):

$$\left.\frac{\partial \mathcal{R}_i}{\partial t}\right|_{d\beta = dR_{-i} = 0} = \frac{1-\beta_i}{M_i} > 0 \tag{13a}$$

$$\left.\frac{\partial \mathcal{R}_i}{\partial \beta_i}\right|_{dt = dR_{-i} = 0} = \frac{-t}{M_i} < 0 \tag{13b}$$

$$\left.\frac{\partial \mathcal{R}_i}{\partial R_{-i}}\right|_{d\beta = dt = 0} = \frac{(\partial \theta_i / \partial R_{-i}) MB_i + \theta_i B_i''}{M_i} \leq 0 \tag{13c}$$

$$\text{with} \qquad M_i = C_i'' - (\partial \theta_i / \partial R_i) MB_i - \theta_i B_i'' \geq 0$$

These partial derivatives describe partial effects of changes in the tax rate, the own redistribution share and emission abatement by all others on the emission reduction effort of country i. The effects correspond to shifts of country i's *Nash-Cournot reaction curve.* A higher carbon tax rate leads ceteris paribus to an increase in the supply of emission abatement by country i, whereas an increase in its share of the receipts lowers its optimal effort level. Furthermore, the higher the supply of abatement efforts by the other participants, the lower one's own effort[9]. The effects in (13) do not describe the shifts of the Nash-Cournot equilibrium in response to changes in tax rate and redistribution parameters. It is an interesting exercise to determine the total effects of changes in the tax parameters, i.e. the shifts of the *Nash-Cournot equilibrium*, for a two-region scenario borrowing the methodology from, e.g., Boadway e.a. (1989). In appendix (A.2) we show that for two regions:

$$\frac{dR}{dt} = \frac{d(R_i + R_{-i})}{dt} \geq 0 \quad ; \quad \frac{dR_i}{d\beta_i} \leq 0 \qquad \forall i \in N$$

implying that an increase in the uniform tax rate raises the supply of the public good, and that countries receiving a higher share in tax receipts lower their own emission reduction effort. However, the incidence of an increase in the tax rate on individual emission reduction is not uniquely determined, it might be positive or negative. It can also be shown that a transfer of tax receipts from region 2 towards region 1 increases total emission reduction

9 The signs in (13*a*) and (13*b*) depend upon the assumptions that $t > 0$ and $\beta_i < 1, \quad \forall i \in N$.

R if region 1 is characterized by a steeper marginal abatement cost function than region 2. For more than 2 regions the analysis does not yield clear-cut results any more because of multiple interdependences. Only empirical simulations can help in that case.

Optimal Uniform Carbon Tax Rate and Redistribution Shares

Let us turn to the international level. Negotiators (or Greenhouse fund administrators) will try to optimize the carbon tax rate t and redistribution shares β_i taking into account the countries individual behaviour described by first-order conditions (12) and participation constraints (8). He has no lump sum redistribution instrument at his disposal. The optimal international carbon tax scheme solves the following program:

$$\underset{(t,\beta_1,\ldots,\beta_n)}{\text{maximize}} \quad \sum_{i\in N}\lambda_i U\left(\frac{\overline{Y}_i^0 - C_i(R_i) - t(\overline{E}_i - R_i) + t\beta_i(\overline{E} - R_i - R_{-i})}{n_i^0}, \frac{\overline{Y}_i^1 + B_i(R_i + R_{-i})}{n_i^1}\right)$$

subject to: (P.4)

$$U(y_i^0(R_i), y_i^1(R_i, R_{-i})) - U_i^{NC} \geq 0 \qquad \forall i \in N$$

$$\beta_i \geq 0 \quad \forall i \in N \quad , \quad t \geq 0 \quad , \quad \sum_{j\in N}\beta_j = 1$$

where R_i is defined by country i's reaction curve: $R_i = \mathcal{R}(t, \beta_1 \ldots \beta_n, R_{-i})$ following from the behavioural relationship in (12). R_{-i} stands again for the reduction effort by all other regions than i. Let π_i be the Lagrange multipliers associated with the participation constraints in (P.4), and μ the multiplier of the requirement that all β_i sum to unity. The corresponding Lagrangian function is defined as:

$$\mathcal{L}(t, \beta_1, \ldots, \beta_n, \pi_1, \ldots, \pi_n, \mu,) = \sum_{i\in N}(\lambda_i + \pi_i)U(y_i^0(R_i), y_i^1(R_i, R_{-i})) - \sum_{i\in N}\pi_i U_i^{NC} + \mu\left[1 - \sum_{i\in N}\beta_i\right]$$

The first-order conditions for an optimal set of redistribution parameters $(\beta_1, \ldots, \beta_n)$ are given by (we drop the superscript *int* but all quantities in (14) are evaluated in the international carbon tax equilibrium and we focus on interior solutions only):

$$\tilde{\lambda}_i\left\{(\overline{E} - R)t + [(1-\beta_i)t - MC_i]\left(\frac{dR_i}{d\beta_i}\right)\right\} + \left(\frac{dR_i}{d\beta_i}\right)\left\{-\sum_{j\neq i}\tilde{\lambda}_j t\beta_j + \sum_{j\in N}\tilde{\lambda}_j\theta_j MB_j\right\} = \mu \qquad \forall i \in N \qquad (14)$$

where θ_i is again the marginal rate of intertemporal substitution between income in period zero and one, and $\tilde{\lambda}_i$ denotes the marginal social valuation of income as defined in (10). As in the previous section the shadow price of the constraint is added to country i's weight λ_i in the social welfare (or power) function if the participation constraint is binding for country i . Substituting individual behaviour relationships (12) yields:

$$\tilde{\lambda}_i(\bar{E} - R)t \quad + \quad \left(\frac{dR_i}{d\beta_i}\right)\sum_{j \neq i} \tilde{\lambda}_j(\theta_j MB_j - \beta_j t) \quad = \quad \mu \qquad \forall i \in N \qquad (15)$$

Similarly we can derive the first order condition for an optimal international carbon tax rate t:

$$\sum_j \tilde{\lambda}_j \left\{ (\bar{E} - R)\beta_j - (\bar{E}_j - R_j) + (\theta_j MB_j - \beta_j t)\left(\sum_{k \neq j} \frac{dR_k}{dt}\right)\right\} \quad = \quad 0 \qquad (16)$$

Let us first look at the interpretation of (16). This condition states that the weighted sum of the effects of a marginal change in t on the different countries has to be equal to zero, where the weights are equal to the social marginal valuations of income (possibly incorporating the effect of the participation constraints). A change in t has a direct and an indirect effect on each country j. On the one hand an increase in the tax rate directly implies a change in its net tax payments (first two terms in (16)). On the other hand, all countries other than j will change their abatement efforts in response to a change in the tax rate t. These responses dR_k/dt are defined by the shifts of the noncooperative Nash-Cournot equilibrium and as we noted in the previous section, the sign of these effects cannot be determined a priori in a n-region model. The shift of the Nash-Cournot equilibrium has an indirect effect on region j consisting of a positive physical externality through $\theta_j MB_j$ and a negative pecuniary externality through the lower global tax revenue.

In the choice of the redistribution parameters there are also two effects to be considered. At the one hand a higher revenue share for country i (β_i) would directly increase its own tax reimbursement (first term in (15), where the monetary effect again is weighted by the marginal social valuation). But the change in β_i will also have an influence on the abatement effort of country i and, hence, on the welfare of all other countries (second term in (15)). In the optimum the effect on country i and on all other countries are exactly balanced. All the monetary effects are weighted by the appropriate marginal social value of income.

It is obvious that expressions (15)-(16) reflect a complicated mixture of redistributional and efficiency aspects. The importance of the distributional effects becomes clear by considering the special case where $\tilde{\lambda}_i = 1, \quad \forall i$. This situation could be reached if the agreement provides the possibility of lump sum redistribution. In all other cases it is a far-fetched assumption. (Remember that the vector $\tilde{\lambda}$ also captures the participation constraints[10]). If $\tilde{\lambda}_i$ is equal for all i, condition (16) implies immediately that $\beta_j t = \theta_j MB_j, \forall j \in N$[11] and (after summing over all j) $t = \sum_j \theta_j MB_j$. Hence the uniform carbon tax rate is equal to the sum of all marginal willingness-to-pay for emission reduction. Tax revenue is reimbursed to country i according to the share of this country in total willingness-to-pay. We are back in the first-best Pigovian tax solution[12].

$$t^{FB} = \sum_{j \in N} \theta_j^{FB} MB_j^{FB} \qquad \text{and} \qquad \beta_i^{FB} = \frac{\theta_i^{FB} MB_i^{FB}}{\sum_{j \in N} \theta_j^{FB} MB_j^{FB}} \qquad \forall i \in N \qquad (17)$$

2.3.4 A Minimal National Carbon Tax Solution

Countries' Individual Behaviour under the Minimal National Carbon Tax

It could be argued that an international tax system with redistribution of the receipts is a rather utopian policy instrument at the international level. It is indeed difficult to imagine that industrialized countries like the United States or the European Community would be willing to accept such loss of sovereignty. Therefore we consider in this section a less ambitious internationally coordinated tax instrument. Assume that all countries in the world agree upon a minimum national tax rate on carbon emissions. But in contrast to the previous section, national tax revenues are not collected by some international carbon fund. Every individual country can freely dispose of its own carbon tax receipts. Hoel (1992*a*) calls this a situation of "harmonized domestic taxes".

---

10 These participation constraints may become binding, even if the agreement reflects the maximization of a simple sum of monetary values (i.e., linear utility functions).

11 Assuming that $\sum_{k \neq j} dR_k / dt$ is nonzero for all k in N.

12 The same first-best Pigovian tax solution can be attained under the small-country assumption, see Hoel (1992*a*). This assumption stipulates that all participating countries are so small that they cannot influence significantly their reimbursement share by varying their abatement effort. Under this assumption expression (12) simplifies to:

$$MC_i^{int} - t = \theta_i^{int} MB_i^{int} \qquad \forall i \in N \qquad \text{and} \qquad \frac{dR_i}{d\beta_i} = 0$$

From (15) it then follows that all $\tilde{\lambda}_i$ are equal and we recover (17) immediately.

The agreement on a minimum national carbon tax would amount to the imposition of a lower limit for these national taxes. Call this lower limit t^{min}. It is obvious that the introduction of such a lower limit will only make a difference for countries that by themselves would impose a lower national tax rate than t^{min}. If each country implements the agreement honestly[13], its individual behaviour can therefore be described by the following two conditions:

$$MC_i^{min} = t^{min} \qquad \text{for} \qquad t^{min} > \theta_i^{min} MB_i^{min}$$

$$MC_i^{min} = \theta_i^{min} MB_i^{min} \qquad \text{for} \qquad t^{min} \leq \theta_i^{min} MB_i^{min} \tag{18}$$

where marginal costs, benefits and marginal rate of intertemporal substitution are defined as before and are evaluated in the minimal tax equilibrium. Hence, if the minimal tax rate is higher than the marginal benefit to country *i* in the minimal tax equilibrium, country *i* chooses an emission reduction effort that equalizes its marginal abatement costs to the tax rate t^{min}. If the tax rate is however smaller than its marginal benefit, country *i* continues to play its noncooperative strategy of equalizing marginal benefits and costs[14].

Optimal Minimal Carbon Tax Rate

The maximization problem describing the international agreement can then be written as:

$$\underset{t^{min}}{\text{maximize}} \quad \sum_{i \in N} \lambda_i U\left(\frac{\overline{Y}_i^0 - C_i(R_i)}{n_i^0}, \frac{\overline{Y}_i^1 + B_i(R_i + R_{-i})}{n_i^1} \right) \tag{P.5}$$

subject to:

$$U(y_i^0(R_i), y_i^1(R_i, R_{-i})) - U_i^{NC} \geq 0 \qquad \forall i \in N$$

$$t^{min} \geq 0$$

13 The assumption of honest behaviour is crucial here as monitoring of the domestic taxes will be very difficult, see Hoel (1992*a*). It can be expected that harmonization of national taxes will only be partly effective. Countries will try to cheat and will use countervailing measures to limit their real effort. This cheating is however difficult to monitor and we will only consider in this paper a national tax system with perfect monitoring.

14 Note, however that these individual optima need not coincide with the Nash-Cournot equilibrium of section 2 although for some regions the first-order conditions are of the same form. The imposition of a lower limit will influence the emissions of some other regions and, hence, the marginal benefit of emission reduction for country *i*.

Again, R_i is defined as the reaction curve of country i following from the behavioural relationship (18): $R_i = \mathcal{R}_i(t^{\min}, R_{-i})$. The first-order conditions for an optimal tax rate t^{min} do not yield much additional insight. Notice however that there does always exist a feasible solution to program (P.5) if all countries are characterized by nonnegative marginal costs in the noncooperative Nash-Cournot equilibrium. For $t^{\min} = \min\{MC_i^{NC} \mid i \in N\}$, i.e. a tax rate equal to the minimum of marginal costs in the noncooperative Nash-Cournot equilibrium, the individual behavioural relationships and all participation constraints are satisfied. This procedure mimics the noncooperative outcome.

Suppose the procedure works and it is even possible to increase the tax rate up to the point where $t^{\min} \geq \max\{MC_i^{NC} \mid i \in N\}$. From this point onwards, it is easy to see that the minimal tax solution yields a cost-efficient allocation of abatement efforts. If the participation constraints would not be binding, one could even increase the tax rate until $t^{\min} = \sum_{i \in N} \theta_i^{FB} MB_i(R_i^{FB} + R_{-i}^{FB})$. This would reproduce the first-best outcome (with lump sum side payments) described in (17). In general however, this point will not be reached because for some country the participation constraint will become binding before the first-best levels are attained. Only an empirical simulation can tell how far we can push the tax rate $t^{\min}$ before some region hits its participation constraint.

2.4 AN EMPIRICAL MODEL OF THE WORLD CARBON ECONOMY

2.4.1 Model Description and Noncooperative Nash-Cournot Equilibrium

Description of the Model

As was hinted above, an empirical simulation is indispensable to derive clear-cut results in this highly asymmetrical carbon emission game. We tried to use as well as possible the information available in the literature. For the details of the specifications of cost, benefit and utility functions, we refer to Eyckmans, Proost and Schokkaert (1993). In this text we describe only briefly the main features of the model.

We divided the world into 12 homogeneous blocks of countries. Table 2.1 summarizes some relevant information on those regions. Percentage shares in world carbon emissions (resulting from fossil fuel use), in world GDP and in world population are reported in the first three columns of Table 2.1. Also GDP per capita (\$ per capita), per capita carbon emissions (kg C per capita) and the ratio of emissions over GDP (tonnes of C per million \$ GDP) are shown. Emissions and GDP per capita are the highest in North America and the lowest in Rest of Asia (which includes India, Pakistan etc.). The industrialized world accounts for more than 70% of emissions and for two thirds of world GDP, although it only represent a quarter of world population. Especially former East European countries are characterized by a high ratio of emissions over GDP, indicating that they are inefficient carbon users.

Table 2.1. Participants in the greenhouse negotiations

	regional characteristics							noncooperative[15]		
	% emis.	% GDP	% pop.	GDP/ capita	emis./ capita	emis./ GDP	γ_i	r_i^{NC}	MC_i^{NC}	pay-off
N-America	26.45	23.10	5.32	20391	5.68	0.28	-0.0029	2.36	8.08	3.29
EC	14.59	17.83	6.59	12698	2.53	0.20	-0.0026	1.63	6.24	2.25
O-Europe	1.85	2.93	1.71	8046	1.23	0.15	-0.0020	0.20	1.03	1.51
Pacific	6.63	8.65	2.86	14200	2.64	0.19	-0.0008	0.24	3.03	2.66
ex-USSR	16.35	8.11	5.53	6876	3.37	0.49	-0.0069	1.93	2.84	1.20
E-Europe	4.75	2.68	2.33	5412	2.33	0.43	-0.0060	0.56	0.94	1.01
Middle East	2.98	2.32	2.31	4721	1.47	0.31	-0.0060	0.48	0.81	0.88
China	11.07	12.04	21.44	2636	0.59	0.22	-0.0194	7.85	4.21	0.40
Africa	2.99	3.73	12.27	1428	0.28	0.20	-0.0037	0.48	1.31	0.27
L-America	4.72	8.05	8.35	4523	0.64	0.14	-0.0027	0.75	2.82	0.84
D-Asia	3.37	5.33	7.47	3350	0.52	0.15	-0.0029	0.54	1.87	0.63
R-Asia	4.27	5.24	23.84	1031	0.21	0.20	-0.0038	0.70	1.83	0.19
World	100.0	100.0	100.0	4694	1.14	0.24		2.21		0.82

N(orth)-*America* = US and Canada
EC = European Community (including former Eastern Germany)
O(ther)-*Europe* = non-EC members plus Turkey
Pacific = Japan, Australia, New Zealand and Oceania
ex-USSR = all nation of former USSR
E(astern)-*Europe* = former East European countries except for ex-USSR and former East Germany
L(atin)-*America* = all countries of American continent excluding N-America
D(eveloping)-*Asia* = Hong-Kong, Indonesia, South Korea, Malaysia, Philippines, Singapore, Taiwan and Thailand
R(est)-*Asia* = other Asian countries.

Preferences of the different regions are represented by the well-known constant elasticity of marginal utility specification and we assume that all regions are in a steady state with respect to population and GDP. Also the marginal rate of intertemporal substitution θ_i is assumed to be constant and equal to 0.111 (for more details see Eyckmans, Proost and Schokkaert (1993)). Utility can then be represented by:

15 $r_i^{NC} = R_i^{NC}/\overline{E}_i$ is reported in percentage emission reduction compared to the base year 1990, MC_i^{NC} is given in \$ per tonne of carbon abated, and all pay-off figures are in \$ per capita. For more references on the data reported in Table 2.1, see Eyckmans, Proost and Schokkaert (1993).

$$U_i = \frac{k}{1-\epsilon}\left(\frac{\overline{Y}_i^0 - C_i(R_i) + \theta_i B_i(R_i + R_{-i}) + T_i}{n_i^0}\right)^{(1-\epsilon)} \qquad \forall i \in N \qquad (19)$$

where k is a normalization factor and T_i stands for country i's net carbon tax liabilities if any. The parameter ϵ determines the concavity of the utility function and can be interpreted as a measure of inequality aversion in a social welfare framework. For our purposes we put $\epsilon = 0.5$ throughout but this parameter can be the subject of a sensitivity analysis.

The abatement cost functions C_i were derived from the specification of Nordhaus (1991*a*) for the marginal cost of carbon emission abatement:

$$\ln\left(1 - \frac{R_i}{\overline{E}_i}\right) = \gamma_i MC_i \qquad \forall i \in N \qquad (20)$$

with R_i stands for emission reduction (million tonnes of C) and MC_i is the associated marginal cost of an additional unit of carbon abatement. The parameters γ_i were calibrated to simulation results reported in Burniaux e.a. (1992).

Abatement benefits are assumed linear in global emission reduction, hence marginal benefits are constant and are set equal to 35$ per tonne of carbon abated. This is the central estimate proposed by Ayres and Walter (1991). For the disaggregation of benefits over the different regions we adopted the simplest assumption: benefits are disaggregated according to the regions' shares in world GDP, cf. Table 2.1.

Noncooperative Nash-Cournot Solution

In the second part of Table 2.1 we report emission reduction, marginal abatement cost, and per capita pay-off[16]. The abatement vector results from solving the system of first-order conditions (4). The overall emission reduction of 2.21% compared to the base year 1990 is fairly low but there are important differences between individual effort levels. In general we can say that regions with high marginal benefits (i.e. with a high share in world GDP in our specification) and low marginal abatement cost (i.e. with high absolute value of γ_i in Table 2.1) will contribute more than other regions. N-America and EC for instance are characterized by approximately the same cost parameter γ_i, but N-America reduces its emissions more because it has a larger share in world GDP, and hence higher marginal benefits. China reduces its emissions by 7.85% because abatement is cheap and marginal benefits are substantial. Pacific and ex-USSR are facing approximately the same marginal

16 Per capita pay-off for country i is defined as: $(\theta_i B_i(R) - C_i(R_i))/n_i^0$.

abatement benefit (cf. their shares in world GDP in Table 2.1) but it is extremely costly to reduce emissions in the Pacific and therefore the latter region contributes less than ex-USSR to world emission reduction.

In Table 2.1 we also see large differences in marginal costs of abatement in the noncooperative equilibrium, which is obviously not cost-efficient. The same overall carbon emission reduction of 2.21% could be attained at a much lower cost. It should be noted that the Nash-Cournot solution depends upon the aggregation of countries into regions, and this aggregation is somewhat artificial in our application. Considering every country individually would generate an even lower overall emission abatement effort.

2.4.2 Command-and-Control Optimum

First, we draw attention to the last two columns of Table 2.2 which contain emission reduction levels and marginal costs for the case of a first-best world with a lump sum redistribution instrument. The vector of abatement efforts r^{FB} solves the expression (11) and can serve as a bench-mark for the interpretation of the other solutions. Average pay-off per capita amounts to 3.42$ per capita.

To illustrate the command-and-control equilibrium and carbon tax solutions we have to specify our overall objective function (7). To illustrate the applicability of the approach we chose two illustrations. The first is an *ethical* one, in which we weigh the utility level of the representative inhabitant of region i by the (relative) number of inhabitants: $\lambda_i = n_i^0/n, \forall i \in N$, where n is the total population of the world. This reflects a perfectly individualistic specification of the social welfare function where all human beings are of equal importance wherever they may live. This normative assumption may not correspond to the current situation in international organizations. The *power* of the different regions is perhaps better captured by weighting them in the objective function by their share in world GDP. This is our second illustrative exercise. In both cases we put the degree of inequality aversion equal to 0.5.

Table 2.2: Command-and-control optimum

	GDP-weights (without lump sum transfers)				population weights (without lump sum transfers)				First-Best (with lump sum)	
	r_i^{GDP}	MC_i^{GDP}	pay-off	π_i^{GDP}	r_i^{POP}	MC_i^{POP}	pay-off	π_i^{POP}	r_i^{FB}	MC_i^{FB}
N-America	8.10	28.55	20.25	0	15.60	57.29	3.29	2.74	9.84	35.00
EC	9.11	36.18	12.54	0	13.36	54.32	8.44	0	8.83	35.00
O-Europe	8.61	45.45	8.17	0	8.21	43.25	8.87	0	6.70	35.00
Pacific	2.63	34.22	17.42	0	4.38	57.45	16.14	0	2.69	35.00
ex-USSR	17.32	27.64	1.20	2.41	17.75	28.40	1.20	9.45	21.40	35.00
E-Europe	17.21	31.33	1.01	2.77	17.65	32.20	1.01	2.71	19.03	35.00
Mid. East	19.97	37.45	0.88	1.43	17.89	33.12	2.25	0	18.80	35.00
China	41.35	27.52	0.40	3.90	38.11	24.74	1.05	0	49.27	35.00
Africa	19.80	60.46	0.27	1.22	6.44	18.22	1.80	0	11.99	35.00
L-America	14.99	60.61	3.08	0	8.32	32.43	5.34	0	8.95	35.00
D-Asia	18.35	70.40	1.18	0	7.72	27.91	4.04	0	9.59	35.00
R-Asia	20.12	58.51	0.19	1.59	5.77	15.49	1.32	0	12.58	35.00
World	15.40		3.16	(10^{-5})	16.10		2.99	(10^{-7})	16.25	

Under the command-and-control regime without transfers, emission abatement levels are fixed directly and there is no lump sum redistribution instrument. Only differences in emission reduction efforts can be used to overcome the participation constraints. Table 2.2 reports abatement efforts, marginal costs, pay-offs and shadow prices of the participation constraints for the two interpretations of the objective function. Shadow prices of the participation constraints are normalized[17] and value "0" for the shadow price means that the constraint is not binding for that particular region.
In the *power* interpretation (left part of Table 2.2) we notice relatively high abatement efforts and hence marginal abatement costs for the developing regions. The OECD countries have such a predominant weight in the objective function that they can shift the reduction effort towards the developing countries, forcing the poorest regions stay on their participation constraints. This pattern of effort levels and marginal costs is completely reversed if we adopt the *ethical* interpretation of the objective function, cf. middle part of Table 2.2.

Notice that under both scenarios the optimal supply of emission abatement amounts to more than 15% which is close to the first-best level of provision. However, due to the large differences in marginal costs, the command-and-control solutions are cost-inefficient which is reflected in the relatively low average per capita pay-off. In conclusion a command-and-control solution which satisfies the participation constraints exists. The overall level of abatement is close to the first-best level but the distribution of efforts is far from cost-efficient. In the next section it will become clear that an international uniform carbon tax with redistribution of the receipts can do much better in terms of cost-efficiency and average per capita pay-off.

2.4.3 An International Uniform Carbon Tax with Redistribution of the Receipts

We now turn to the international uniform carbon tax solution with redistribution of the receipts. Results for both the ethical and power scenario are reported in Table 2.3. Remember that in the hypothetical first-best scenario with lump sum redistribution possibilities, the optimal tax rate amounts to 35$ (= total sum of marginal benefits which are assumed constant in our approach). The first-best optimal shares in total tax revenue are then given by the share of the regions in total marginal benefits, hence for our specification they are equal to the GDP-shares.

---

17 Shadow prices are normalized as: $\frac{\pi_i}{n_i^0}\frac{\partial U}{\partial Y_i^0}$ which gives the marginal effect on the objective function of a change in country i's threat point U_i^{NC}.

We start with the *power* interpretation of the objective function, i.e. GDP-weights. The vector β_i^{GDP} resembles the distribution of initial carbon emissions in Table 2.1. The same observation was already made in the literature on tradeable emission entitlements. In order to insure full participation in the permit agreement, the initial allocation of permits should not differ too much from the initial emission shares, this is the so-called grandfathering principle. The only exception to this observation in our simulation is N-America.

Comparing the redistribution shares β_i^{GDP} with the GDP shares in Table 2.1 (shares in first-best scenario), we notice that especially N-America, and to a lesser extent also ex-USSR, E-Europe and Middle East receive a higher share in total carbon tax revenue than in the hypothetical first-best outcome with lump sum redistribution. In the power interpretation N-America has a predominant role in the objective function, it can claim the largest share of tax receipts and it ends up with a pay-off far higher than in the noncooperative outcome. Notice also that N-America is the only region which can do strictly better than its noncooperative strategy. Because of its high share in tax revenue, it will contribute less to the public good of emission abatement and incurs a lower marginal abatement cost than in the first-best case.

Table 2.3. International carbon tax solution

	GDP-weights						population weights					
	β_i^{GDP}	r_i^{GDP}	MC_i^{GDP}	pay-off	T_i^{GDP}	π_i^{GDP}	β_i^{POP}	r_i^{POP}	MC_i^{POP}	pay-off	T_i^{POP}	π_i^{POP}
N-America	0.34	8.95	31.67	52.13	31.98	0	0.26	9.56	33.94	3.29	-15.62	0.173
EC	0.14	9.30	36.96	2.26	-10.95	0.016	0.14	9.17	36.43	2.25	-11.00	0.141
O-Europe	0.02	6.89	36.03	1.51	-8.06	0.012	0.02	6.77	35.40	1.51	-8.06	0.027
Pacific	0.06	2.80	36.35	2.66	-15.55	0.004	0.06	2.75	35.76	2.66	-15.52	0.068
ex-USSR	0.16	20.17	32.74	1.20	2.44	0.054	0.16	19.87	32.21	1.20	2.16	0.089
E-Europe	0.05	18.96	34.87	1.01	0.99	0.021	0.05	18.67	34.26	1.01	0.78	0.027
Mid. East	0.03	18.99	35.39	0.88	-0.84	0.020	0.03	18.69	34.78	0.88	-0.97	0.023
China	0.08	51.32	37.13	0.40	1.72	0.166	0.08	50.76	36.53	0.40	1.61	0.091
Africa	0.03	12.29	35.92	0.27	-1.10	0.087	0.03	12.09	35.30	0.27	-1.11	0.010
L-America	0.04	9.43	36.96	0.84	-4.28	0.066	0.04	9.28	36.36	0.84	-4.29	0.073
D-Asia	0.03	9.95	36.40	0.63	-3.07	0.061	0.03	9.79	35.79	0.63	-3.08	0.048
R-Asia	0.04	12.92	36.03	0.19	-0.76	0.153	0.12	11.80	32.70	11.10	10.08	0
World	1.00	16.18		3.41			1.00	16.12		3.42		
tax rate t				35.59						34.98		

with $T_i = (\beta_i tE - tE_i) / n_i^0$

Comparing world totals, it is remarkable how close this tax solution approximates the first-best outcome: 16.18% against 16.25% emission reduction and 3.41$ per capita against 3.42$ per capita pay-off. Also the distribution of marginal costs indicates little efficiency losses in the tax solution compared to the first-best outcome. Only N-America and ex-USSR have substantially lower marginal costs than the other regions.

For the *normative* interpretation, i.e., population weights, we notice in Table 2.1 that R-Asia has the largest share in world population. This fact is reflected in the distribution of net carbon tax liabilities T_i^{POP} and revenue shares β_i^{POP}: R-Asia can claim a relative high share of receipts compared to the first-best scenario. It is also the only region for which the participation constraint does not bind. Again the distribution of β_i^{POP} closely resembles the initial emission shares in Table 2.1, except for R-Asia itself. Little efficiency losses emerge because marginal costs are approximately equal for all regions.

Concluding, we can say that:

(i) If participation constraints are imposed, the optimal distribution of tax receipts (β_i) resembles closely the initial distribution of carbon emissions. A similar phenomenon has also been noticed in the literature on tradeable carbon emission rights.

(ii) Introducing participation constraints in the international uniform carbon tax scenario does <u>not</u> lead to an important efficiency loss. The tax-redistribution mechanism is apparently flexible enough to insure full cooperation and to attain more or less an equalization of marginal abatement costs.

(iii) The choice of weights λ_i and of inequality aversion ϵ in the objective function has a strong influence on the distribution of winners and losers in terms of pay-off. For the two interpretations of the objective function and for $\epsilon = 0.5$, there is only one unique winner compared to the noncooperative equilibrium.

(iv) Compared to the command-and-control solutions in Table 2.2, the international uniform tax performs much better in terms of cost-efficiency and average per capita pay-off.

2.4.4 A Minimal National Tax Solution

In Table 2.4 we report some simulation results for the minimum national tax model. The last column gives the tax revenue which remains in the individual regions. The solution is identical for the normative (population weights) and power (GDP weights) case. It turned out that the highest feasible tax rate which is consistent with the participation constraints amounts to 21.00$ per tonne of carbon. For all regions, this is higher than the marginal benefit under Nash-Cournot. The tax rate is therefore binding for all regions and yields a cost-efficient distribution of abatement efforts. However, the overall suppiy of carbon emission reduction is insufficient compared to the first-best bench-mark scenario. The tax rate cannot be increased beyond 21.00$ per tonne of carbon because this would activate the participation constraint for low cost countries. The constraint becomes binding first for China, then for ex-USSR followed by E-Europe.

Table 2.4. The minimum national tax solution

	R_i^{NAT}	MC_i^{NAT}	pay-off	T_i^{NAT}
N-America	6.03	21.00	14.46	111.99
EC	5.39	21.00	9.80	50.18
O-Europe	4.07	21.00	6.59	24.79
Pacific	1.63	21.00	12.10	54.60
ex-USSR	13.45	21.00	1.43	61.24
E-Europe	11.89	21.00	1.94	43.08
Middle East	11.75	21.00	2.39	27.29
China	33.45	21.00	0.40	8.23
Africa	7.38	21.00	1.05	5.41
L-America	5.47	21.00	3.63	12.79
D-Asia	5.87	21.00	2.65	10.18
R-Asia	7.75	21.00	0.75	3.96
World	10.39		2.95	

with $T_i^{nat} = t E_i^{nat}$

The overall efficiency of this instrument can be judged by comparing the average pay-off per capita (2.95$) with the one obtained in the first-best solution (3.42$) and in the international tax solution. Although the national tax solution is cost-efficient, it fails to attain a level of emission reduction which would be comparable to the abatement effort under the international tax scenario. The distribution of pay-offs is also different. Because the national tax solution lacks a redistribution mechanism, the resulting pay-off vector preserves to a large extent the existing welfare distribution. There is no longer a unique winner who claims the remaining part of the cooperative surplus. The major winners are those regions with a high valuation for climate change avoidance (reflected in GDP-shares in our specification) and high emission abatement costs: the OECD countries. The low cost regions which have to make relatively high abatement efforts do not gain anything compared to the situation where they would play their Nash-Cournot strategies.

2.5 CONCLUSIONS

It was the purpose of this paper to investigate the equity and efficiency properties of three types of international carbon agreements: a command-and-control solution without lump sum side-payments, and two carbon tax schemes, all under the requirement of full participation. Central in the analytical part was the explicit introduction of participation constraints in our framework. In section 2 we showed that participation constraints can be introduced easily in an optimal carbon taxation framework. Formally, the Lagrange multipliers of the participation constraints are simply added to the welfare weights in the marginal social valuation of income. These multipliers equal zero if the participation constraint is not binding.

In a direct regulation scenario (command-and-control) the availability of a lump sum redistribution instrument is crucial in determining the distributional and participation characteristics of a carbon emission treaty. Sidepayments allow to separate distributional from efficiency considerations. We considered two different taxation schemes: a uniform international carbon tax with redistribution of the receipts, and a minimum national tax agreement. The optimal international tax rate and reimbursement shares are determined taking into account optimizing behaviour by the individual regions. The resulting optimality conditions are easily interpreted.

Using a two-period simulation model, the noncooperative Nash-Cournot solution was characterized and yielded very modest overall carbon emission abatement of 2.2%. Mainly countries with high marginal abatement benefits or with cheap carbon abatement possibilities contributed to this result. The Command-and-control solution, assigning unique abatement levels to every region, results in more than 15% emission abatement compared to the base year 1990. However, the distribution of efforts is not cost-efficient because differences in abatement levels are used to overcome participation constraints.

It was shown that the first-best reference case, yielding some 16% of emission reduction, is reasonably well approximated by an international tax scheme with redistribution of the receipts. Of all three instruments considered, the international tax performs best in efficiency terms due to the powerful redistribution possibilities offered by the fixed tax revenue shares. The exact distribution of pay-offs depends upon the type of welfare (or power) function used in the negotiation process. For the minimum national tax case we assumed that all regions would implement at least the uniform tax rate. If this tax rate is lower than their marginal benefit of carbon emission abatement, they play their Nash-Cournot strategy. The simulations indicated that, although the minimum national tax solution is cost-efficient, it fails to attain the first-best bench-mark solution because of the participation constraints. Overall abatement amounts to 10% and marginal costs equal 21$ per tonne of carbon for all regions.

We are well aware that many aspects of the analysis require further investigation. First, concerning participation in international environmental agreements, the formation of coalitions should be addressed more adequately. Secondly, the two-period approach should be generalized by incorporating the stock dimension of atmospheric greenhouse gases, and by replacing the one-shot game by a dynamic emission game as in Hoel (1992*b*). Thirdly, more attention should be paid to monitoring and enforcement problems. However, our simple approach has the advantage that it allows to concentrate on the huge differences in the international welfare distribution and abatement costs and benefits. These distributional aspects surely play an important role in actual greenhouse negotiations.

REFERENCES

Ayres, R.U. and Walter, J. (1991) The Greenhouse Effect: Damages, Costs and Abatement, *Environmental and Resource Economics*, **1**, 237-270.

Barrett, S. (1992) *Conventions on Climate Change*. OECD, Paris.

Bauer, A. (1992), International Cooperation over Environmental Goods, mimeo. University of Munich, Germany.

Boadway, R., Pestieau, P., And Wildasin, D. (1989) Tax-transfer Policies and the Voluntary Provision of Public Goods, *Journal of Public Economics*, **39**, 157-176.

Burniaux, J.M. e.a. (1992) *The Costs of Reducing CO_2 Emissions: Evidence from GREEN*, OECD Working Paper. OECD, Paris.

Cornes, R. and Sandler, T. (1984) The Theory of Public Goods: Non-Nash Behaviour, *Journal of Public Economics*, **23**, 367-379.

Cornes, R. and Sandler, T. (1986), *The Theory of Externalities, Public Goods and Club Goods*. Cambridge University Press, Cambridge.

Eyckmans, J., Proost, S. and Schokkaert, E. (1993) Efficiency and Distribution in Greenhouse Negotiations, *Kyklos*, **46**, 363-397.

Friedman, J. (1986) *Game Theory with Applications to Economics*. Oxford University Press, Oxford.

Hoel, M. (1992a) Carbon Taxes - an International Tax or Harmonized Domestic Taxes, *European Economic Review*, **36**, 400-406.

Hoel, M. (1992b) Emission Taxes in a Dynamic Game of CO_2 Emissions, in: Pethig, R. (ed.), *Conflict and Cooperation in Managing Environmental Resources*. Springer, Berlin-Heidelberg.

Hoel, M. (1992c) International Environment Conventions: the Case of Uniform Reductions of Emissions, *Environmental and Resource Economics*, **2**, 141-159.

Hoel, M. (1993) Efficient Climate Policy in the Presence of Free Riders, mimeo. University of Oslo, Norway.

IPCC (1990) *Climate Change: the IPCC Scientific Assessment*. Cambridge University Press, Cambridge.

Kverndokk, S. (1994) Coalitions and Side Payments in International CO_2 treaties, Chapter 3 in this book.

Mäler, K.G. (1989) The Acid Rain Game, in: Folmer, H. and Van Ierland, E.C. (eds.), *Valuation Methods and Policy Making in Environmental Economics*. North-Holland, Amsterdam.

Nordhaus, W.D. (1991a) The Cost of Slowing Climate Change: a Survey, *The Energy Journal*, **12**, 37-65.

Nordhaus, W.D. (1991b) To Slow or Not to Slow: the Economics of the Greenhouse Effect, *The Economic Journal*, **101**, 920-937.

Petrakis, E, and Xepapadeas, A. (1993) Environmental Consciousness and Moral Hazard in International Agreements to Protect the Environment, mimeo. University of Crete, Greece.

Welsch, H. (1993) An Equilibrium Framework for Global Pollution Problems, *Journal of Environmental Economics and Management*, **25**, 64-79.

Whalley, J., and Wigle, R. (1991) Cutting CO_2 emissions: the Effect of Alternative Policy Approaches, *The Energy Journal*, **12**, 109-124.

APPENDICES

Appendix A.1

a. The effect of a change in R_i on the marginal rate of intertemporal substitution is given by: (U_i^0, U_i^1 denote the partial derivatives of the utility function w.r.t. country i's income in period zero and one respectively)

$$\frac{\partial \theta_i}{\partial R_i} = \frac{\partial}{\partial R_i}\left(\frac{U_i^1}{U_i^0}\right) = \frac{(\partial U_i^1/\partial R_i)U_i^0 - (\partial U_i^0/\partial R_i)U_i^1}{(U_i^0)^2}$$

$$= \frac{[U_i^{10}(-MC_i)+U_i^{11}MB_i]U_i^0 - [U_i^{00}(-MC_i)+U_i^{01}MB_i]U_i^1}{(U_i^0)^2}$$

$$= \frac{(U_i^{00}U_i^1 - U_i^{10}U_i^0)MC_i + (U_i^{11}U_i^0 - U_i^{01}U_i^1)MB_i}{(U_i^0)^2}$$

If $(U_i^{00}U_i^1 - U_i^{10}U_i^0)$ and $(U_i^{11}U_i^0 - U_i^{01}U_i^1)$ are both negative, the effect of a change of R_i is negative. These conditions on the first- and second-order derivatives of the utility function correspond to the requirements which insure normality in a two-good model in standard consumer theory.

b. The effect of a change in R_{-i} on the marginal rate of intertemporal substitution is given by:

$$\frac{\partial \theta_t}{\partial R_{-t}} = \frac{\partial}{\partial R_{-t}}\left(\frac{U_t^1}{U_t^0}\right) = \frac{(U_t^{11}U_t^0 - U_t^{01}U_t^1)MB_t}{(U_t^0)^2}$$

which is also negative under the same assumption on the utility function as in *a*.

Appendix A.2

For a two-region model, totally differentiating first-order condition (13) yields (assuming θ_t is constant and equal to unity, and $\beta_1 = \beta$ and $\beta_2 = 1-\beta$):

$$\begin{cases} C_1''dR_1 - (1-\beta)dt + t\,d\beta = B''_1(dR_1 + dR_2) \\ C_2''dR_2 - \beta dt - t\,d\beta = B''_2(dR_1 + dR_2) \end{cases}$$

$$\begin{pmatrix} C''_1 - B''_1 & -B''_1 \\ -B''_2 & C''_2 - B''_2 \end{pmatrix}\begin{pmatrix} dR_1 \\ dR_2 \end{pmatrix} = \begin{pmatrix} 1-\beta \\ \beta \end{pmatrix} dt + \begin{pmatrix} -t \\ t \end{pmatrix} d\beta$$

now define:

$$\text{determinant } A = |A| = \begin{vmatrix} C''_1 - B''_1 & -B''_1 \\ -B''_2 & C''_2 - B''_2 \end{vmatrix}$$

$$= C''_1C''_2 - C''_1B''_2 - C''_2B''_1 \geq 0$$

Hence, using Cramer's rule and assuming $|A| \neq 0$ we can determine the comparative statics of the system first-order conditions:

$$\frac{dR_1}{dt} = |A|^{-1}\begin{vmatrix} 1-\beta & -B''_1 \\ \beta & C''_2 - B''_2 \end{vmatrix} = \frac{(1-\beta)(C''_2 - B''_2) + \beta B''_1}{|A|}$$

$$\frac{dR_2}{dt} = |A|^{-1}\begin{vmatrix} C''_1 - B''_1 & 1-\beta \\ -B''_2 & \beta \end{vmatrix} = \frac{\beta(C''_1 - B''_1) + (1-\beta)B''_2}{|A|}$$

Summing both effects we get:

$$\frac{dR_1}{dt} + \frac{dR_2}{dt} = \frac{dR}{dt} = \frac{(1-\beta)C''_2 + \beta C''_1}{|A|} \geq 0$$

Hence, increasing in the tax rate raises the global emission reduction effort R.

$$\frac{dR_1}{d\beta} = |A|^{-1}\begin{vmatrix} -t & -B''_1 \\ t & C''_2 - B''_2 \end{vmatrix} = \frac{(B''_1 - C''_2 + B''_2)t}{|A|} \leq 0$$

$$\frac{dR_2}{d\beta} = |A|^{-1}\begin{vmatrix} C''_1 - B''_1 & -t \\ -B''_2 & t \end{vmatrix} = \frac{(C''_1 - B''_1 - B''_2)t}{|A|} \geq 0$$

Summing the effects we get:

$$\frac{dR_1}{d\beta} + \frac{dR_2}{d\beta} = \frac{dR}{d\beta} = \frac{(C''_1 - C''_2)t}{|A|}$$

A transfer of tax receipts towards a country with steeper marginal cost curve, raises the provision of the international public good of carbon emission abatement.

Chapter 3 COALITIONS AND SIDE PAYMENTS IN INTERNATIONAL CO_2 TREATIES

Snorre Kverndokk[1]

ABSTRACT

Most numerical studies analysing the costs and benefits of international CO_2 emissions abatement assume full participation by all countries and regions in the world. Based on the experience from the 1992 Rio conference on the one side, and the theory of self-enforcing agreements to restrict pollution among sovereign countries on the other, full participation will probably not be the outcome of an international treaty on reducing CO_2 emissions. In this study I focus on coalitions and side payments in international CO_2 treaties by answering questions such as: Given the commitment of cooperation by a defined group of countries, what is the optimal policy of the group? What is the global loss of partial cooperation compared to full cooperation (social optimum), and how is the optimal abatement level affected by the number of countries committed to cooperate? The framework of the analysis is as follows. A group of OECD countries have committed themselves to cooperate on the global warming problem. The coalition (or the cooperating countries) chooses emission levels and offers the non-cooperating countries transfers if they restrict their emissions. The abatement and side payments made by the coalition are chosen so that its intertemporal utility function is maximized. Compared to the social optimum, limited participation implies a significant global loss. However, compared to doing nothing, a treaty signed by a group of countries may be important. Side payments are an effective policy instrument for making a limited treaty significant.

3.1 INTRODUCTION

Abating CO_2 emissions has two aspects. First, CO_2 emissions are a by-product of production activities that create goods for final demand. Thus, emitting CO_2 creates income. Second, accumulation of CO_2 and other greenhouse gases (GHGs) in the

[1] I am grateful to Asbjørn Torvanger and Michael Hoel for helping me develop the model and for valuable comments, to Kjell Arne Brekke, Samuel Fankhauser, Jan Fuglestvedt, Rolf Golombek, Stef Proost, Ekko van Ierland and Haakon Vennemo for comments on earlier drafts, to J. Oliveira Martins (OECD) for providing me with data, and to Ted Gartner for editing assistance. This chapter was financed by the Norwegian Research Council (Economy and Ecology).

atmosphere may have negative effects in the future in the form of climate change. The negative external effects from CO_2 accumulation may harm all countries in the world and not only the emitting country. Thus, while the *benefits* from CO_2 emissions are immediate and directed towards the emitting country, the *damage costs* will be worldwide and continue as long as the CO_2 remains in the atmosphere.

The two aspects of CO_2 emissions abatement, i.e., the costs and the benefits, have been studied numerically in several papers (see Nordhaus (1991a,b, 1993a,b), Peck and Teisberg (1992, 1993a,b), Eyckmans *et al.* (1992) and Cline (1992a,b)). However, these studies assume full cooperation by all countries in the world in an international CO_2 treaty, which means that they calculate the socially efficient abatement of CO_2 emissions.

Another strand of the literature concentrates on the possibility of limited participation in the future CO_2 treaty (see Barrett (1990, 1991) and Carraro and Siniscalco (1992, 1993)). These papers discuss, among other things, whether self-enforcing agreements to restrict pollution among sovereign countries are possible, which means whether a coalition of cooperating countries can be stable. A coalition is stable if no country, whether it is a part of the coalition or not, wants to change its action given the actions of the other countries. According to numerical and theoretical analyses, only a small number of countries are likely to join a stable coalition. However, it is possible that the cooperating countries can expand the stable coalition through self-financed utility transfers. They are interested in providing incentives for emissions reductions in the non-cooperating countries, and side payments may be the right instrument to create this. The side payments could be carried out by linking environmental policy to other economic policies such as financial aid, trade policy, technology transfers, etc. Hence, such actions will make the treaty broader by compensating the gains from "free riding" (enjoying the benefits of reduced emissions without contributing to the reduction).

Full participation, as analysed by the first group of papers, may therefore not be the outcome of an international CO_2 treaty without introducing side payments. This is stressed by Fankhauser and Kverndokk (1992) who argue that a socially optimal treaty may only be achieved if side payments are offered to at least China and the countries of the former Soviet Union. They also conclude that the OECD countries are the only countries with an incentive for internal cooperation and unilateral reductions. The 1992 conference in Rio confirms the latter conclusion, and it also was the opening for side payments from the developed to the developing world as part of a future CO_2 treaty.

In this chapter I combine the two strands of literature by a numerical analysis of costs and benefits under constrained CO_2 treaties. By analysing different regimes of coalitions and side payments, I want to study problems such as:

- Given the commitment of cooperation by a defined group of countries, what is the optimal policy of the group (e.g., what are the side payments to other countries)?
- What is the global loss of partial cooperation compared to full cooperation (defined as the social optimum)?
- How is the optimal abatement level affected by the number of countries committed to cooperate?

The framework of such an analysis can be as follows. Following the arguments of Fankhauser and Kverndokk (1992), a group of OECD countries has committed itself to cooperation on the global warming problem. The problem is to find the optimal policy for this group. I assume that the group is fixed throughout the analysis, although I do not analyse the problem of stability (see above). Henceforth, the concept coalition is used only for this original group of cooperating countries.

The cooperating countries (the coalition) choose their emissions and offer the non-cooperating countries transfers if they restrict emissions to certain levels. The abatement in the non-cooperating countries will reduce accumulation of CO_2 in the atmosphere and therefore reduce the global warming problem. The cooperating countries face two trade-offs: (1) increasing current emissions increases current consumption possibilities but reduces prospective consumption due to future climate change; (2) emissions reductions by non-cooperating countries reduce climate change, which means higher prospective consumption for the cooperating countries, while giving side payments in itself reduces their current consumption possibilities. The abatement and side payments made by the coalition are chosen so that its intertemporal utility function is maximized. The constrained agreement is also compared to a treaty with full cooperation (the social optimum).

The chapter is organised in the following way. In section 3.2, the model is outlined. Section 3.3 describes the specific functions and data, while the simulation results are given in section 3.4. Some sensitivity analyses are provided in section 3.5, and conclusions are summarised in section 3.6.

3.2 THE MODEL

I is the set of cooperating countries (the coalition) and N is the set of all countries. The aim of the coalition is to find the optimal time path of consumption over the planning horizon of τ years. Thus, it trades income against damage from CO_2 emissions, as well as offers side payments to non-cooperating countries, to maximize the *intertemporal utility* function, W, formalised in equation (3.1). C(t) is the consumption of the coalition at time t, $U[\cdot]$ is the total utility of the coalition, and ρ is a discount rate representing pure time preferences.

$$W = \sum_{t=0}^{T} (1+\rho)^{-t} U[C(t)] \tag{3.1}$$

This optimization problem is subject to several constraints. Consider first the production of country n, n ∈ N, at time t, $P_n(t)$, defined in the following way:

$$P_n(t) = D_n[T(t)] \cdot F_n[e_n(t), z_n(t)] \tag{3.2}$$

$F_n[\cdot]$ is production in the absence of the climate-feedback effect. It is a function of domestic CO_2 emissions, $e_n(t)$, and a vector of exogenous technology variables at time t, $z_n(t)$.[2] This income function is increasing and concave in CO_2 emissions, *ceteris paribus,* and expresses the income to a country under different CO_2 emission levels.

The feedback from climate to production in country n, which is the damage from global warming, is represented multiplicatively by $D_n[\cdot]$:

$$0 \leq D_n[T(t)] \leq 1$$

$$D_n[T(0)] = 1$$

$$\frac{\partial D_n[T(t)]}{\partial T(t)} < 0$$

$$\frac{\partial^2 D_n[T(t)]}{\partial T(t)^2} \leq 0$$

The multiplicative damage function is decreasing and concave in T(t), which is the increase in the global mean temperature between preindustrial time (taken to be 1765) and time t. Hence I assume convexity of damages with respect to temperature, where 1 - $D_n[T(t)]$ is the relative loss in income in country n at temperature increase T(t). Thus, as a simplification, the various aspects of climate change are represented by this single variable. I set $T(t) \leq \bar{T}$, where the value of $\bar{T}$ is specified to avoid negative production, i.e., $D_n[\cdot] \geq 0$, and therefore $P_n(t) \geq 0$ for $T(t) \leq \bar{T}$. The model is specified such that there is no loss at time 0 (defined as 1990).

The relation between *temperature increase* and emissions of CO_2 can, in general, be described in the following way,

[2] In this model economic development is basically exogenous and is given by the vector of technology parameters (z). The only endogenous variables influencing the economic development in this model are CO_2 emissions from fossil fuel combustion (e) which enter as an input, and the increase in the global mean temperature (T) which determinates the feedback from the climate to production explained below. This means that savings and capital accumulation are not studied in the model.

$$\begin{aligned} T(t) &= f[e(0),...,e(t);t] \\ e(s) &= \sum_{n\in N} e_n(s), \quad s = 0,...,t \end{aligned} \tag{3.3}$$

which gives the temperature increase as a function of aggregated CO_2 emissions in previous and current periods.

To restrict the global CO_2 emissions, the coalition offers transfers to the non-cooperating countries if they are willing to limit their emissions. Based on Hoel (1993), the transfers offered are such that the non-cooperating countries obtain the same income with their imposed emissions and the side payments as they would with the right to choose emissions optimally. Thus, the optimal *side payment* to the non-cooperating country j is defined as follows:

$$\begin{aligned} D_j[T(t)] \cdot g_j[e_j(t),z_j(t)] &\equiv \{\max_x D_j[T(t)] \cdot F_j[x,z_j(t)]\} \\ &\quad - D_j[T(t)] \cdot F_j[e_j,z_j(t)] \\ &= D_j[T(t)] \cdot (\{\max_x F_j[x,z_j(t)]\} - F_j[e_j(t),z_j(t)]) \end{aligned} \tag{3.4}$$

In the absence of the greenhouse effect, $g_j[\cdot]$ is the production loss in country j from choosing $e_j(t)$ instead of the optimal emissions. $D_j[\cdot]g_j[\cdot]$ is hence the side payment, which must be offered to country j at time t, $j \notin I$, to make it indifferent (in the sense of equal income) between choosing e_j (and receiving the side payment) and the emissions maximizing production. Implicitly, country j is assumed to be small in the sense that it considers T(t) and therefore $D_j[T(t)]$ as exogenously given.[3]

Assuming that there is a constant ratio between consumption and production, the aggregated *consumption* for the coalition is specified in the following way:[4]

$$C(t) = \sum_{i\in I} P_i(t) - \sum_{j\notin I} D_j[T(t)] \cdot g_j[e_j(t),z_j(t)] \tag{3.5}$$

The object of the coalition is to maximize the present value of utility over the planning horizon. Hence, the maximization problem in equation (3.1), subject to

[3] With this specification of the side payments, all the efficiency gains from greenhouse policies go to the coalition. Other specifications where the gains are distributed over the various countries are of course also possible.

[4] The factor of proportionality between consumption and production, ζ, is in this study set equal to 1. However, this is irrelevant in this respect, since the assumption means that we can specify utility as a function of production, P, i.e., $u[C] = u[\zeta P] = \hat{u}[P]$.

equations (3.2) through (3.5), gives the optimal path of emissions and therefore consumption for this group of countries.

The first order conditions for this maximization problem, where l ∈ I, k ∉ I and 0 ≤ s ≤ τ, can be presented in the following way:

$$(1+\rho)^{-s} \cdot \frac{\partial U[.]}{\partial C(s)} \cdot \frac{\partial F[.]}{\partial e_l(s)} \cdot D_l[.]$$

$$- \sum_{t=s}^{\tau} \left[(1+\rho)^{-t} \cdot \frac{\partial U[.]}{\partial C(t)} \cdot \sum_{j \notin I} \left[\frac{\partial D_j[.]}{\partial T(t)} \cdot \frac{\partial T(t)}{\partial e_l(s)} \cdot g_j[.] \right] \right] \tag{3.6}$$

$$= -\sum_{t=s}^{\tau} \left[(1+\rho)^{-t} \cdot \frac{\partial U[.]}{\partial C(t)} \cdot \sum_{i \in I} \left[\frac{\partial D_i[.]}{\partial T(t)} \cdot \frac{\partial T(t)}{\partial e_l(s)} \cdot F_i[.] \right] \right]$$

The first part on the left side of equation (3.6) is the discounted marginal utility from an increase in the emissions from a country within the coalition due to higher production in period s. The second part shows the discounted utility gain over the planning horizon due to lower side payments. The increase in emissions involves higher damage and therefore lower side payments to the non-cooperating countries. The right side gives the discounted utility loss in the coalition over the planning horizon due to damage from the increased emissions. Thus, in the optimal state, the gains from higher production and lower side payments from emissions within the coalition should be balanced against current and future damage directed toward the coalition.

From the above optimality condition, we see that the emissions from the countries within the coalition are higher when side payments are given to other countries than they would be in the absence of side payments (the second part on the left side is zero). This is because the damage from global warming in the non-cooperating countries involves a gain for the coalition in terms of lower side payments, an irrelevant effect under unilateral actions.

Equation (3.7) gives the condition to determine the emissions from the non-cooperating countries. At the optimal solution, the discounted utility loss in the coalition due to damage from these emissions should be balanced against the utility gains from lower side payments. The first part on the right side of the equation represents the impacts from side payments due to higher production in the emitting country in period s, while the last part is the discounted utility gains from lower side payments due to higher damage in all non-cooperating countries caused by increased emissions.

$$\sum_{t=s}^{T}\left[(1+\rho)^{-t}\cdot\frac{\partial U[.]}{\partial C(t)}\cdot\sum_{i\in I}\left[\frac{\partial D_i[.]}{\partial T(t)}\cdot\frac{\partial T(t)}{\partial e_k(s)}\cdot F_i[.]\right]\right] =$$

$$(1+\rho)^{-s}\cdot\frac{\partial U[.]}{\partial C(s)}\cdot\frac{\partial g_k[.]}{\partial e_k(s)}\cdot D_k[.] \qquad (3.7)$$

$$+\sum_{t=s}^{T}\left[(1+\rho)^{-t}\cdot\frac{\partial U[.]}{\partial C(t)}\cdot\sum_{j\notin I}\left[\frac{\partial D_j[.]}{\partial T(t)}\cdot\frac{\partial T(t)}{\partial e_k(s)}\cdot g_j[.]\right]\right]$$

The *social optimum*, which is the optimal solution under full cooperation, is defined as the maximum of the global intertemporal utility function (1) subject to (2), (3), and $C(t) = \Sigma_n P_n(t)$, n,l $\in$ N. This gives the following optimum condition:

$$(1+\rho)^{-s}\cdot\frac{\partial U[.]}{\partial C(s)}\cdot\frac{\partial F[.]}{\partial e_l(s)}\cdot D_l[.] =$$

$$-\sum_{t=s}^{T}\left[(1+\rho)^{-t}\cdot\frac{\partial U[.]}{\partial C(t)}\cdot\sum_{n\in N}\left[\frac{\partial D_n[.]}{\partial T(t)}\cdot\frac{\partial T(t)}{\partial e_l(s)}\cdot F_n[.]\right]\right] \qquad (3.8)$$

Thus, in the optimal state, the utility gain from increased production in one country at time s should be balanced against the discounted utility loss from global warming in all countries over the rest of the planning horizon.

3.3 FUNCTIONAL FORMS AND DATA

3.3.1 Time periods and regions

A long planning horizon is justified by the long-term impacts of global warming. Impacts are only felt after some 30 to 50 years, but may then persist for as much as two or three centuries. Thus, the model is simulated for the period 1990 to 2230, giving a planning horizon of 240 years, where t = 0 refers to 1990.

I divide the world into seven blocks of countries derived from the OECD GREEN model (Burniaux *et al.* (1992a,b)).[5]

1. USA
2. EEC

[5] The GREEN model consists of 12 sectors.

3. Rest of OECD (ROECD)
4. The former Soviet Union (EX-USSR)
5. China
6. India
7. Rest of the World (REST)

According to this list, the OECD countries consist of the three first regions while regions 1 to 4 are industrialized countries. The last group is a mixture of both industrialized and developing countries.

3.3.2 The production module

In the present analysis, the production in a country is defined as its gross domestic product (GDP). The *income function* in equation (3.2), the F[•] function, is taken from Kverndokk (1993),

$$F_n[e_n(t),z_n(t)] = Y_n[e_n(t);t] = \hat{Y}_n(t) - G_n[e_n(t);t]$$

$$G_n[e_n(t);t] = \frac{q_n(t)\hat{e}_n(t)}{b_n(t)}\left[\frac{\hat{e}_n(t) - e_n(t)}{\hat{e}_n(t)}\right]^{b_n(t)}, \quad e_n(t) \leq \hat{e}_n(t) \tag{3.9}$$

where (the subscript is left out for simplicity):

- z = {b,q,Ŷ,ê},
- b = the constant elasticity of abatement costs with respect to abatement (ê - e),
- q = the shadow price of CO_2 when e = 0, i.e., the tax on CO_2 emissions which leads to a total substitution away from fossil fuels to non-fossil backstop technologies (the switch price of CO_2),
- Y = GDP in the absence of climate change,
- Ŷ = GDP in the absence of climate change and emissions constraints (BAU GDP without climate change),
- ê = CO_2 emissions from energy use in the absence of any emissions constraints (BAU CO_2), and
- G = abatement costs for abatement equal to ê - e.

The income function is concave in CO_2 emissions for b > 1 and q > 0. The condition $b \geq q\hat{e}/\hat{Y}$ ensures that GDP is non-negative. The data are mainly taken from the OECD's GREEN model (see Burniaux *et al.* (1992a,b)), and are given in the Appendix.

The exogenous BAU scenario (ê and Ŷ) is defined so that the emissions maximize the production in the absence of global warming, i.e., $G_n[\bullet] = 0$.

The *q-values* are based on Burniaux *et al.* (1992a,b), and they are assumed to decline due to an expected increase in fossil fuel prices, as well as falling costs of backstop technologies. However, the switch prices are constant from 2090 onwards. The *b-values* are calibrated such that the carbon taxes in the GREEN model (the Toronto scenario) equal the respective shadow prices of carbon in my model in 2015 at the specific emission levels. The exogenous paths of the b-parameters follow the paths for the autonomous energy-efficiency parameter (AEEI) in the GREEN model, which is a growth of 1% p.a. in all regions from 1990 to 2050. I assume a uniform growth of 0.5% p.a. until 2100 and zero growth in the parameter thereafter.

The annual *growth rates for BAU GDP* in the absence of climate change from 1990 to 2050 are based on Burniaux *et al.* (1992a). From 2051 onward, they build on the long-run growth rates in Manne and Richels (1992) and Cline (1992a), leading to an almost 21-fold multiple of gross world product (GWP) by 2230 if there are no feedbacks from global warming.[6]

The *BAU CO_2 growth rates* from fossil fuel use are based on GREEN to the middle of the next century and build on the long-run growth rates from Manne and Richels (1992) and Cline (1992a) thereafter. This gives an average annual growth rate of 0.4% for the period 2100-2230.

Under these assumptions, BAU CO_2 emissions from energy use increase to almost 60 billion ton carbon by 2230. The relative share of total emissions for different regions is shown in Figure 3.1. The share of OECD countries declines from 47% in 1990 to only 3% in 2230. This is due to a decline in emissions from 2111 onward derived from technology changes and backstop technologies, as well as a high increase in emissions from developing countries. The share of Chinese emissions increases from 10% in 1990 to 60% in 2230.

The *damage function* in equation (3.2) is specified in the following way, involving no damage in 1990 (period 0):

$$D_n[T(t)] = 1 - k_n \cdot \left[\frac{T(t) - T(0)}{\Lambda - T(0)} \right]^{\gamma}, \quad \gamma \geq 1 \qquad (3.10)$$

[6] To compare my estimate to other studies, Cline (1992a,b) assumes that the global production multiples 26-fold from 1990 to 2275 in the absence of global warming, while Nordhaus (1993a,b) assumes a 7 1/2-fold multiple over the same period. A high growth as assumed in this study may raise serious doubts of feasibility due to limits of the global resource and environmental base. The possibility of such a growth has been supported although not recommended by Cline (1992a), p. 287. Cline argues that lower per capita growth than in the past may be difficult to avoid, and that the per capita income projections do not seem high at least on the basis of the experience over the past century. However, the uncertainty in predicting production and technological changes over such a long time horizon should be emphasized.

Thus, k_n is the relative GDP loss in region n due to the climate-feedback effect at temperature increase Λ, where Λ is set to 2.5° C,[7] i.e., $1 - D_n[\cdot] = k_n$ when $T(t) = \Lambda$. For a similar additive specification of the damages, see Fankhauser and Kverndokk (1992).[8]

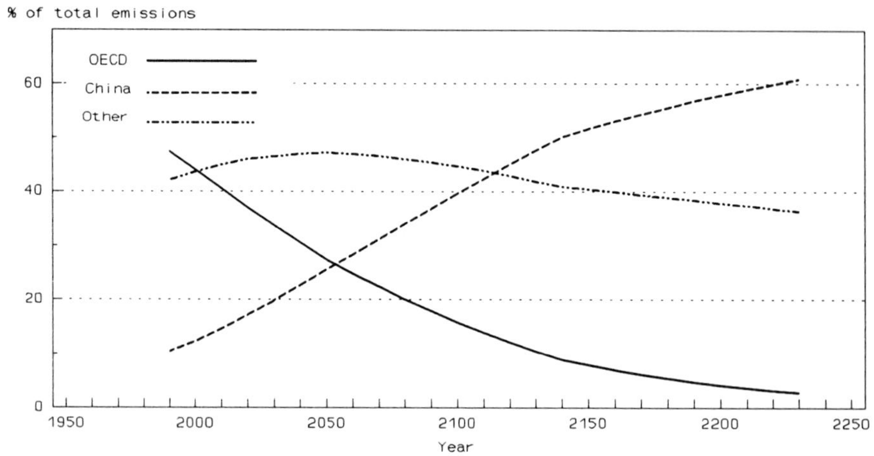

Figure 3.1 Relative CO_2 Emissions in BAU Scenario

Several studies, see, e.g., Nordhaus (1991a,b), Cline (1992a), and Fankhauser (1992), estimate that a temperature increase of 2.5° C since preindustrial time will cause a global damage in the interval of 1-2% of GWP. Thus, in this study I work with three different scenarios for total damage at 2.5° C actual global temperature increase: 1%, 2%, and 3% of GWP, named Scenario 1, 2, and 3, respectively (see Table 3.1).[9]

[7] This is IPCC's best guess for an equilibrium temperature change at a doubling of CO_2 equivalents in the atmosphere (see Houghton *et al.* (1990, 1992)).

[8] A multiplicative damage function like the one above may be a good description if most of the damage related to global warming comes from activity losses in GDP-producing activities like agriculture, while an additive damage function may best represent the damage if it reduces amenities not usually included in the GDP measure, like loss of wetlands due to rising sea levels, loss of bio-diversity, and so on. However, willingness to pay for such amenities increases with income or GDP, involving higher valuations in richer countries. Thus, damage should in some way be related to GDP. Fankhauser and Kverndokk (1992) solve this problem by employing an additive damage function, where the annual damage is assumed to grow proportionally with income.

[9] Much uncertainty exists on these figures. This implies that it is well possible that much higher percentages are relevant for damage estimates. Fankhauser (1994) illustrates this by providing damage probability distributions.

For parameter calibration of the *k-values*, I use the damage distribution pattern in Fankhauser (1992). *T(0)*, the global temperature increase from preindustrial time to 1990, is set equal to 0.6° C, based on Houghton *et al.* (1992), while γ is set equal to 1.3, which is the estimate of Cline (1992a).

Table 3.1 The Damage Scenarios

SCENARIO 1:	SCENARIO 2:	SCENARIO 3:
Total damage at T(t) = 2.5 °C is 1% of Gross World Product	Total damage at T(t) = 2.5 °C is 2% of Gross World Product	Total damage at T(t) = 2.5 °C is 3% of Gross World Product

3.3.2 The Climate Module

This model represents the 17 major GHGs, where only the CO_2 emissions from energy use are specified endogenously. Thus, I assume that the international agreement only concerns these emissions and that other emissions, including CO_2 emissions due to changes in land-use patterns, are not affected.

The climate module of the model consists of a stock equation, (3.11), and several temperature equations, (3.12) and (3.13).

The *stock equation* gives the atmospheric concentration of CO_2 at different emission levels. Emissions of GHGs increase the corresponding concentrations in the atmosphere. However, the observed concentrations are less than they would have been if all emissions since preindustrial time had been added to a constant preindustrial stock. For CO_2 for instance, this is due to the removal of carbon from the atmosphere into the oceans and the terrestrial biosphere.

One way of modelling this is as a depreciation model assuming an annual depreciation of the stock of GHGs based on the lifetime of the actual GHG. I have chosen this model for the accumulation of CO_2 in the atmosphere, where the preindustrial stock is assumed to be the equilibrium stock. This means that if all anthropogenic emissions were eliminated, the atmospheric concentration would approach the preindustrial level. However, I do not apply a constant depreciation rate, due to saturation of the carbon-sink capacity of the oceans (see Houghton *et al.* (1990, 1992)). This is shown in equation (3.11), where Q(t) is the atmospheric concentration of CO_2 at time t, Q^P is the preindustrial concentration, Z(t) is the difference between concentration at time t and preindustrial time, L(t) is the lifetime of CO_2, and X(t) is the global emissions (in PPM) at time t. The parameters are set according to

Houghton *et al.* (1990, 1992) and Peck and Teisberg (1992), and are specified in the Appendix.

$$Q(t) = Q^P + Z(t),$$
$$Z(t) = Z(t-1) \cdot \left(1 - \frac{1}{L(t)}\right) + X(t), \quad t > 0 \tag{3.11}$$

The *temperature equations* describe the reaction of temperature to changes in atmospheric GHG concentrations or radiative forcing, i.e., heating in watt per square metre (W m^{-2}), and consist of functions for potential and actual temperature increase. Potential temperature increase is the increase in the geophysical equilibrium. Before fully warming the earth's surface, the greenhouse effect must first heat up the oceans (ocean terminal lag). This may take as long as three decades or more, and the potential temperature increase is therefore the increase in temperature after full adjustment. Potential temperature increase due to CO_2 emissions, T_{CO2}^P, is specified according to Peck and Teisberg (1992). The A and B parameters in equation (3.12) are fit so that the increase is zero at the preindustrial concentration and 2.5° C for twice the preindustrial concentration of CO_2.[10]

$$T_{CO2}^P(t) = A \ln Q(t) - B,$$
$$T_{OGHG}^P(t) = \theta R(t), \tag{3.12}$$
$$T^P(t) = T_{CO2}^P(t) + T_{OGHG}^P(t)$$

The potential temperature increase due to emissions of the 16 other GHGs, T_{OGHG}^P, is the product of the total warming coefficient, θ (measured in °C/W m^{-2}),[11] and the radiative forcing above preindustrial level, R(t) (measured in W m^{-2}).[12] The time series for radiative forcing, where CFCs are assumed to be phased out by 2020, are taken from the calculations of Hoel and Isaksen (1993). The parameter specifications are given in the Appendix.

[10] Peck and Teisberg (1992) assume a temperature increase of 3°C for twice the preindustrial concentration of CO_2. However, 2.5°C corresponds to IPCC's best guess (Houghton *et al.* (1990, 1992)).

[11] Usually potential temperature increase is specified as the product of radiative forcing, R, the warming to be expected as the direct impact of radiative forcing before taking account of feedback, λ, and the feedback multiplier, β, i.e., $T^P = R\lambda\beta$ (see, e.g., Cline (1991, 1992a)). In my model, θ represents both the direct and indirect impact of radiative forcing, i.e., the product of λ and β.

[12] This is of course also true for CO_2 and is implicitly included in our specifications (11) and (12). The relation between radiative forcing and atmospheric concentration of CO_2 is usually assumed to be $R_{CO2} = 6.3 \ln [Q(t)/Q^P]$.

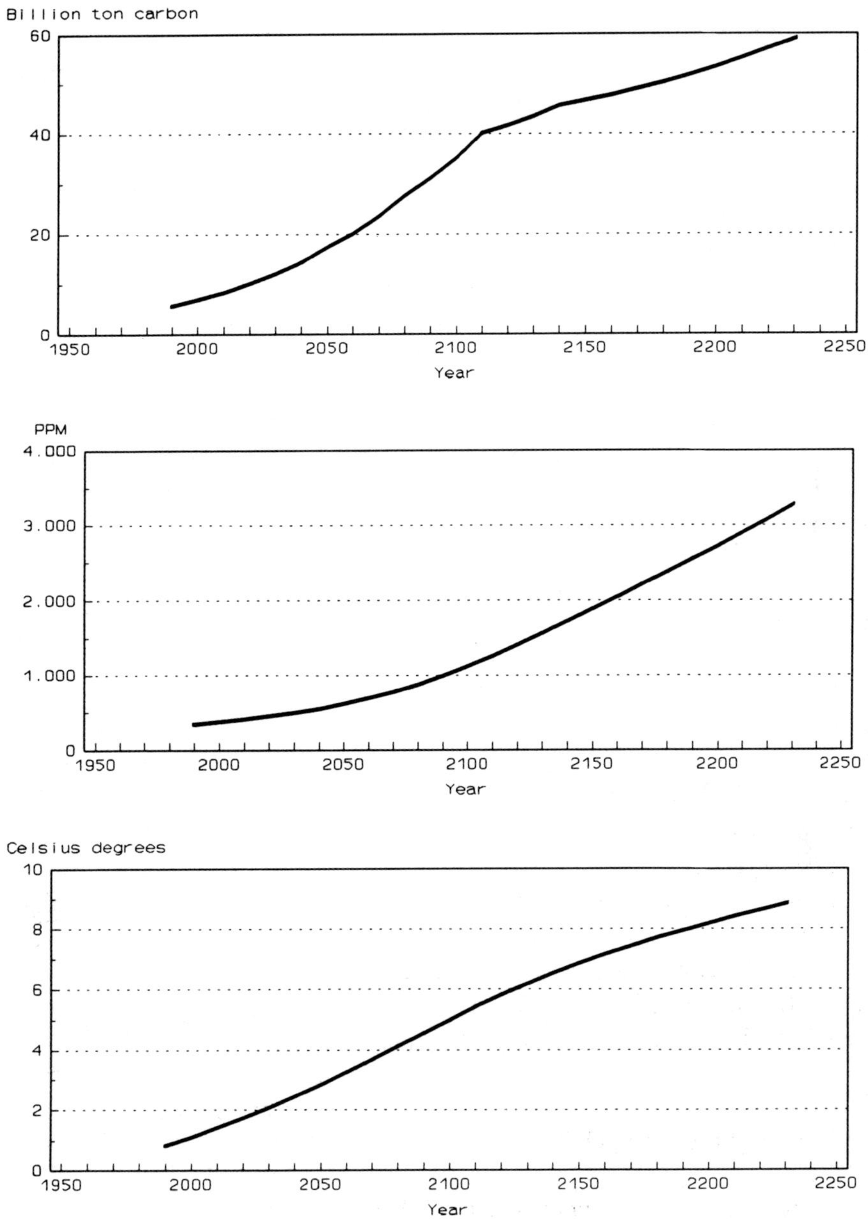

Figure 3.2 BAU Paths: CO_2 Emissions from Energy Use, Atmospheric CO_2 Concentration, and Actual Temperature Increase

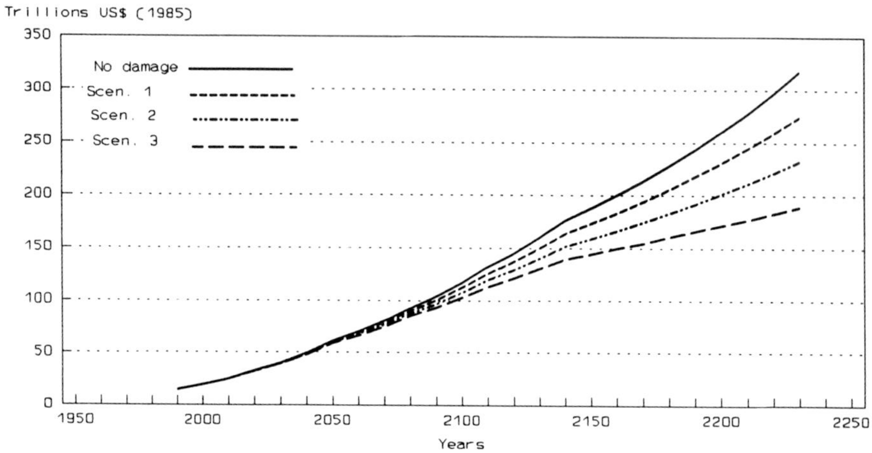

Figure 3.3 BAU GWP with Feedback from Global Warming

Based on Nordhaus (1991a,b), the increase in actual temperature (T) in each period is governed by a lagged adjustment process due to the thermal inertia of oceans as mentioned above, where the speed of adjustment is determined by the delay parameter α, which is set equal to 0.025.

$$T(t) = \alpha T^P(t) + (1 - \alpha)T(t - 1) \tag{3.13}$$

The development in endogenous CO_2 emissions, atmospheric CO_2 concentration, and actual temperature increase under the BAU scenario are shown in Figure 3.2. For a comparison to other studies, see Cline (1992a) for analysis in the very long term (2275), and Houghton *et al.* (1992) for analysis to the year 2100. Based on these comparisons, my estimates seem to be in the upper end.

To illustrate the importance of global warming in the long run, the development of GWP with feedbacks from the warmer climate, i.e., $\Sigma_n D_n[\cdot]\hat{Y}_n(\cdot)$, is shown in Figure 3.3. Global warming affects GWP relatively little until 2050. The effects in the 22nd century are high however. By 2230, more than 13% of potential GWP is lost under Scenario 1, and 40% is lost under Scenario 3.

3.3.3 The consumption module

I have chosen an isoelastic *utility function* where $1/\mu$ is the elasticity of substitution between consumption at two points in time.

$$U[C(t)] = \frac{1}{1-\mu}[C(t)^{1-\mu} - 1], \quad \mu \neq 1, \mu > 0$$
$$U[C(t)] = \ln C(t), \quad \mu = 1 \tag{3.14}$$

Compared to other studies, Nordhaus (1991a,b, 1993a,b) uses a μ value equal to 1, while Cline (1992a) proposes a value of 1.5 for the USA. In this model μ is set equal to 1.

The rate of pure time preference, ρ, is set equal to zero to avoid discrimination of future generations. This is discussed further in section 3.5.

3.4 RESULTS

In addition to the social optimum, simulations are run with two different coalitions. The first coalition consists of all OECD countries apart from USA, while in the second coalition, all OECD countries are assumed to sign the treaty.

With no transaction costs related to side payments, the coalition calculates and offers optimal side payments to all other countries. However, in the presence of *transaction costs*, the coalition may concentrate on compensating a limited group of countries. Thus to represent this, under a different regime, the coalition of EEC and ROECD offers side payments to the large emitters of USA, the countries of the former Soviet Union, China, and India, which are all among the top 10 emitters in the world, considering the former Soviet Union as one region.

The four different coalition regimes are summarised in Table 3.2. In Regime 1, the reference regime, the coalition of EEC and ROECD offers side payments to all other regions. The coalition is extended to all OECD countries in Regime 2, while in Regime 3 the original coalition (EEC and ROECD) does not offer side payments to REST due to transaction costs. The last regime is the social optimum.

Table 3.2 The Different Coalition Regimes

	REGIME 1	REGIME 2	REGIME 3	SOCIAL OPTIMUM
REGIONS IN THE COALITION	EEC ROECD	USA EEC ROECD	EEC ROECD	USA EEC ROECD EX-USSR China India REST
REGIONS OUTSIDE THE COALITION	USA EX-USSR China India REST	EX-USSR China India REST	USA EX-USSR China India	

The simulations were carried out for the time period 1990 to 2230 using the GAMS/MINOS system; see Brooke *et al.* (1992). However, I do not report results for the last 50 years to avoid extreme end effects caused by the relatively arbitrary time horizon. Due to the great uncertainties in analysis dealing with such a long time horizon, the results should only be considered illustrative estimates. In the graphs I only report the results for the "best guess" scenario, Scenario 2, but I nevertheless comment on all scenarios.

3.4.1 The Optimal Policy

Consider the results for Regime 1 and the social optimum; the results for Regimes 2 and 3 will be presented below. Figures 3.4 and 3.5 show the *optimal emission levels* under the different coalition regimes. The total emissions reductions compared to the BAU scenario are substantial, especially from the middle of the next century (see Figure 3.4), reaching 100% under the social optimum and 60-85% under the different scenarios of Regime 1. The high abatement levels are, among other things, due to low abatement costs[13] and high damage costs over time, as well as a low discount rate. The socially optimal abatement paths are high compared to the studies of Peck and Teisberg (1992) and Nordhaus (1993a,b). However, Nordhaus sets the rate of pure time preference equal to 3%, while I use a zero rate. Simulations with Nordhaus's DICE model employing lower rates of time preference give abatement paths within the range of this study (see Cline (1992b)).

The emission reductions in the regions outside the coalition are actually higher than the corresponding abatement in the coalition. Under Regime 1, Scenario 2, abatement is up to 72% in the coalition and 78% outside the coalition. The bulk of emissions reductions are taken in China and India (84% and 82%), among other things due to the relatively low marginal abatement costs in these countries using the GREEN data.

While the emissions reductions seem high compared to BAU emissions, this is not the impression comparing them to 1990 emissions, as shown in Figure 3.5. With limited participation, the emissions will increase compared to the 1990 level, which is not in accordance with recommendations from international conferences such as Toronto and Hamburg 1988. The first calling for emissions reductions of 20% of 1990 levels from the year 2005 onwards and the second recommending reductions of 30% by the year 2000.

[13] To illustrate this, in the model the abatement costs at 100% abatement, i.e., $G_n[0,t]$, are about 15% of BAU GDP in the absence of climate change ($\hat{Y}$) for the OECD countries in 1990. These costs are decreasing over time due to technological progress, and amount to about 1% within 2080 under the assumptions of the model. The corresponding abatement costs in the non-OECD countries are significantly higher; e.g., 85% in EX-USSR and 47% in China in 1990; however they also decrease over time to 7% and 5% respectively in 2080.

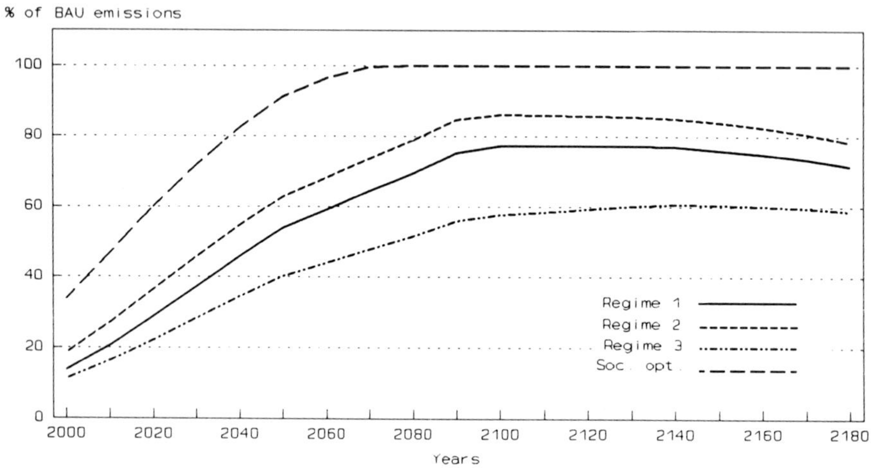

Figure 3.4 Global CO_2 Reductions relative to BAU, Scenario 2

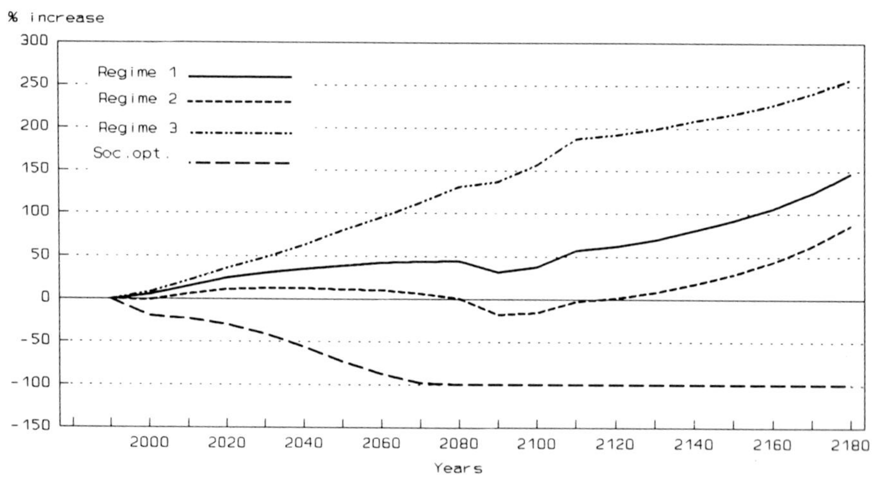

Figure 3.5 Global CO_2 Emissions Relative to 1990 Level, Scenario 2

The *optimal side payments* in per cent of GDP before action is taken (BAU GDP) are shown in Figures 3.6 and 3.7. The hump shape of the graphs can be explained by the abatement paths, with increasing abatement to about 2100 and lower abatement thereafter, as well as lower abatement costs and higher damage costs over time. The transfers are quite modest in most years, with the peak at 1.7% of the coalition GDP by 2100 under Regime 1, Scenario 3. Under the "best guess" scenario, Scenario 2, the peak is however 0.95%. In comparison, this is just above the development assistance disbursement target of 0.7% established by the United Nations, which however is not met by most industrialized countries. The transfers to the non-cooperating countries in terms of their GDP under Regime 1 are relatively smaller, owing to larger total GDP in this group.

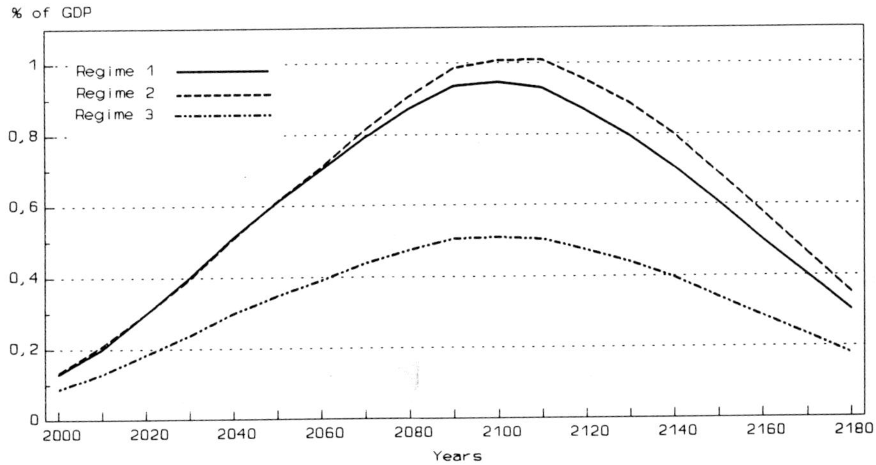

Figure 3.6 Compensation in % of GDP of Coalition, Scenario 2

Figure 3.8 shows the *actual temperature increase* under the different regimes. The abatement strategies have a minor effect on the temperature in the first decades of agreements. However, from the middle of the next century the differences are significant. But, even if the decline in prospective temperature increase is high under limited participation, there are big differences compared to the socially optimal paths.

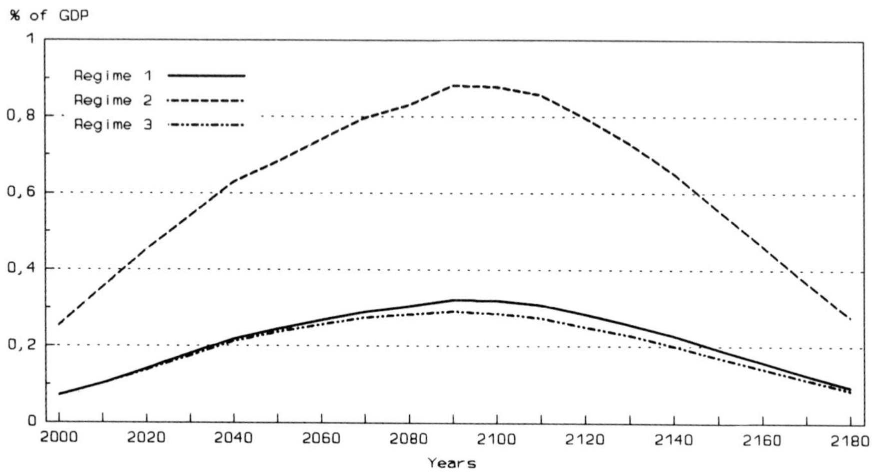

Figure 3.7 Compensation in % of GDP Outside Coalition, Scenario 2

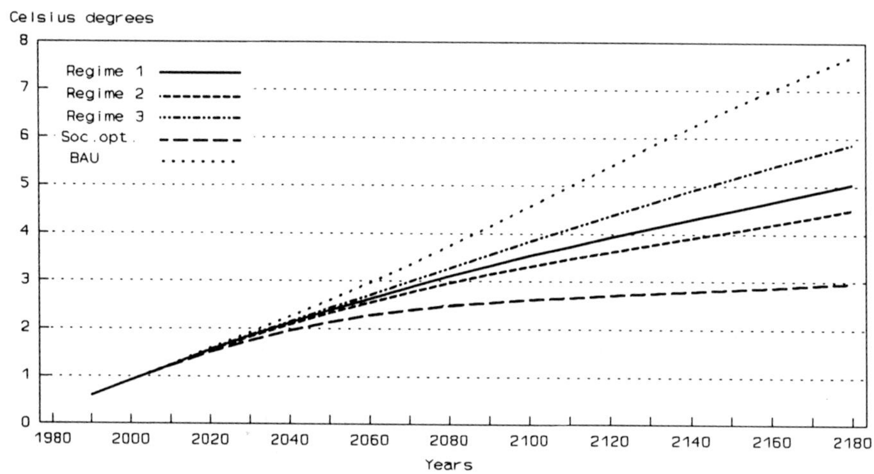

Figure 3.8 Actual Temperature Increase, Scenario 2

3.4.2 Gains from CO_2 Abatement

Figure 3.9 shows the *global GDP gains* compared to BAU GDP under the different coalition regimes. Until the middle/end of the next century, the global gains are negative — a loss of more than 3.5% at the most — with the highest losses under the

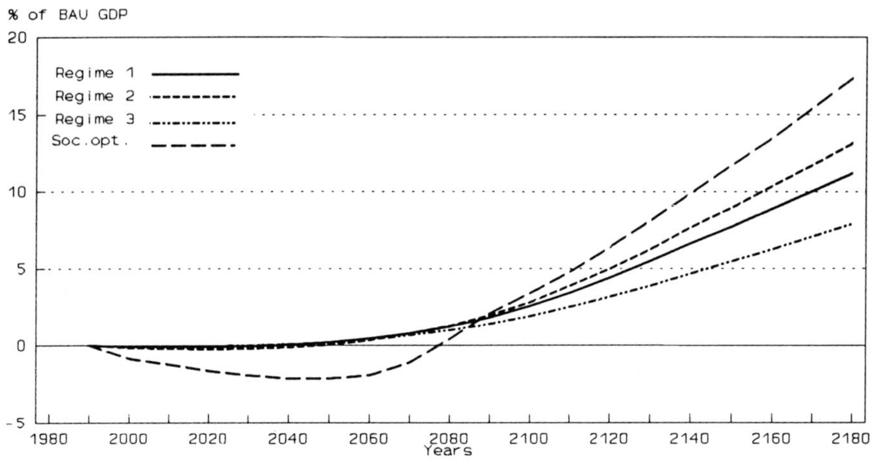

Figure 3.9 Global GDP Gains Compared to BAU, Scenario 2

socially optimal policies. The gains however are positive from the second half of the century onward, up to 31% by 2180 under the social optimum and 22% under Regime 1 (Scenario 3). The results are due to the long-term impacts of global warming, the long planning horizon and the low discount rate.[14] The problem can be compared to one of investment; the costs are immediate while the gains will come later. Here, the policy makers are willing to trade lower GDP in the near future for benefits in later periods.

The differences in the gains for the coalition and the non-cooperating regions are substantial under the social optimum as well as under limited participation. The non-cooperating countries will gain most at the end of the planning horizon, an increase in GDP up to 37% by 2180 compared to 15% in the coalition. However, the non-cooperating countries will have the highest losses in the beginning of the time period from the abatement strategies before side payments are taken into account. The high losses in the beginning are among other things due to high potential economic growth; while the high gains from cooperation in the long term are due to significant damage from climate change. This may indicate one difference between the regions inside and outside the coalition: Even if the regions outside the coalition gain substantially over the total planning period, they may be reluctant to cooperation due

[14] Although I set the rate of pure time preferences equal to zero, using a concave utility function results in an implicit discounting of future consumption (see section 3.5 below).

to high initial costs. This means that they employ a higher discount rate than the coalition.

3.4.3 Extending the Coalition

To see how the optimal policy is affected by the number of countries committed to cooperation, I compare the results from Regimes 1 and 2. The results for Regime 2 under Scenario 2 are also shown in Figures 3.4 through 3.9.

The main impression is that the optimal solutions are still somewhat far from the social optimum. However, under Scenario 3, including USA in the coalition actually means a stabilizing of total emissions just below 1990 levels in the first part of the next century (see Figure 3.5). A stabilizing of emissions at 1990 levels has been mentioned as a goal for international negotiations, for example, under the Rio conference in 1992. In the other scenarios, the emissions are higher than 1990 levels.

The differences when it comes to actual temperature increase and total GDP gains seem relatively marginal as well. However, the total side payments from the coalition to the non-cooperating countries have increased since this regime involves higher reductions, and hence higher abatement costs in the non-cooperating region. The increase in abatement can be explained from equation (3.7): Including USA in the coalition implies an increase in the damage costs taken into account by the coalition (see the left side of the equation), while it means lower gains from reduced side payments due to the damage in the non-cooperating countries (see the last part of the right side) since there are fewer countries to compensate. Thus, to attain an optimal solution, the first part of the right side has to increase, giving higher abatement in the non-cooperating countries.

3.4.4. Limited Compensation

Under Regime 3, the REST region is not compensated for emissions reductions due to transaction costs related to side payments. I assume that the emissions from REST follow the BAU path, which means that I ignore carbon leakages from actions taken by the other countries (see, e.g., Burniaux *et al.* (1992a) and Manne and Rutherford (1992)).

The emissions from the REST region accounts for about 17% in 1990 and 22% in 2230. Thus, ignoring this region under an international treaty means that still more than 75% of emissions have been accounted for. This seems like a high portion; however, as seen from the results under Regime 3 in Figures 3.4 through 3.9, the negative impacts of limited compensation are relatively more important than the positive impacts of including USA in the coalition. For instance the difference between Regimes 1 and 3 is larger than the difference between Regimes 1 and 2 with respect to temperature change: The temperature increase under Regime 3 (Scenario 2) by 2180 will be almost 5.9° C compared to 5.1° C under Regime 1. In comparison, the corresponding reduction in temperature when moving from Regime 1 to Regime 2 is only 0.5° C (see Figure 3.8).

3.5 SENSITIVITY ANALYSIS

Most of the functional forms and parameter values used in the analysis are subject for discussion, and the results will of course depend on the data employed. Sensitivity analysis is therefore essential.

The data for the *income function* are taken from the GREEN model (Burniaux *et al.* (1992a,b)). This study is rather optimistic in terms of abatement costs. Calibrating the income function using data from Manne and Richels (1992) will give lower abatement in all scenarios and regimes. For sensitivity analyses on this function, see Fankhauser and Kverndokk (1992) and Kverndokk (1993).

The studies of Peck and Teisberg (1992) and Fankhauser and Kverndokk (1992) concluded that the optimal reduction levels are very sensitive to the form of the *damage function* (the value of γ). To see the effect on the main variables in this study, I have run simulations with a linear damage function ($\gamma = 1$). Even this little reduction in the parameter value of γ has impacts on the main variables, involving for example reduced abatement and higher temperature increase. However the abatement is still high. The most sensitive variable to changes in this parameter is the side payments. These have been reduced by about 50% compared to the original levels, indicating lower gains from free riding with a linear damage function. Even if the total gains under the social optimum are higher, the higher the damage is, the total gains with limited participation seem not to be equally sensitive.

As discussed in, e.g., Cline (1992a,b) the *social discount rate* is composed of two elements: A component for pure time preference, and a component for utility-based discounting, where the latter follows from the assumption that future consumption will be higher than the current level, which means that the marginal utility from an additional unit of consumption will be lower in the future than today.

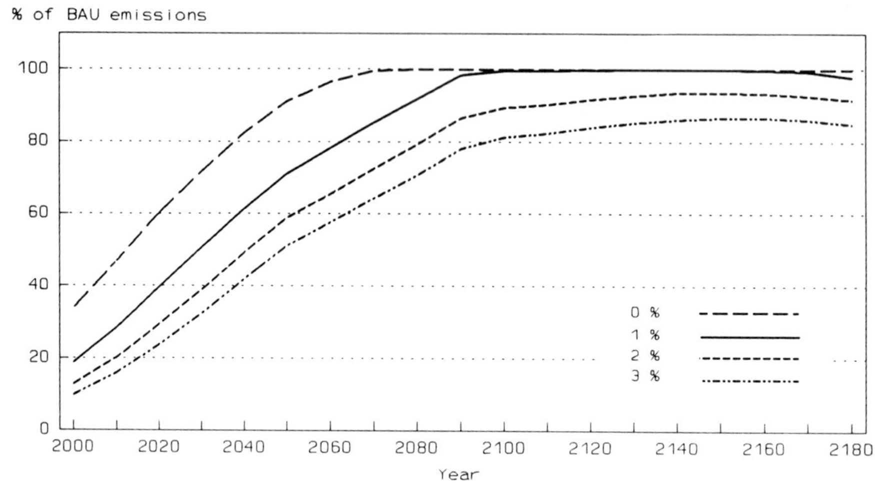

Figure 3.10 Sensitivity Analysis on the Rate of Pure Time Preferences. Social optimal Abatement relative to BAU, Scenario 2

Following the arguments of Cline (1992a,b), I have set the rate of pure time preference, ρ, equal to 0. However, this component has been the subject of long discussion (see, e.g., Lind (1982) for a survey), and is for example set equal to 3% in the studies of Nordhaus (1993a,b). Sensitivity analysis on the component of pure time preference confirms the conclusions in Cline (1992b); a high rate of pure time preference reduces the optimal abatement quite drastically in the beginning of the time horizon, but abatement rises quickly, giving high-percentage cutbacks by the end of the planning period. Increasing the rate of pure time preference to 1% reduces the social optimal abatement in 2000 under Scenario 2 by more than 40%, and delays the stage of no CO_2 emissions by two decades. A rate of 3% will correspondingly reduce the social optimal abatement by more than 70% in 2000 and give a peak of reductions at about 85% by the end of the next century. These results are shown in Figure 3.10.

Changing the optimization problem to maximize the discounted sum of utility over time, where per capita utility is a logarithmic function of per capita consumption, actually puts more weight on future consumption compared to the original optimization problem.

Thus, this is similar to a reduction in the pure rate of time preferences, giving higher abatement in the beginning of the time period.[15]

Reducing the elasticity of substitution in the utility function by setting μ equal to 1.5, which is the value applied by Cline (1992a), actually increases the utility-based discounting. Thus, this will reduce abatement in the beginning of the planning horizon, and the optimal abatement is actually within the same range as when employing a rate of pure time preferences equal to 1%.

To sum up, even if the optimal abatement paths are very sensitive to the various parameters employed, the main characteristics of limited participation under international CO_2 treaties seem robust.

3.6 CONCLUSIONS

The analysis shows that even if limited participation may have a significant influence on the climate and therefore the economy in the long run, it will not meet the recommendations from international conferences. Compared to the social optimum, there is a significant global loss of limited participation. However, the most likely alternative to coalition formation seems to be a "laissez-faire" regime, involving great damage in the long run. A treaty signed by only a group of countries may therefore be important, and the side payments to countries outside the coalition may be within politically acceptable levels, making such a policy instrument more likely. This study also emphasizes the importance of compensating as many countries as possible even in the absence of carbon leakages.

Including USA in the coalition has a relatively marginal impact on the development of the climate in the long run, even if the current emissions of USA amount to almost 25% of global emissions. This is due to the potential increase in emissions from developing countries, especially China. Therefore, the importance of future emissions abatement in the major developing countries should be emphasized.

[15] To see this, let $U(C/N) = \ln(C/N)$. We then have $\Sigma_t(1+\rho)^{-t}N\ln(C/N) = \Sigma_t(1+\rho)^{-t}(N\ln C - N\ln N)$. The latter part of this expression (NlnN) is exogenous and does not influence the optimal paths of emissions. Assuming that population, N, is increasing over time actually puts more weight on future rather than current consumption.

APPENDIX

Table 3A.1 Value of Production Parameters

	b (2015)	q^a (1990)	q^a (2090-)	$\hat{Y}^b$ (1990)	$\hat{e}^c$ (1990)
USA	2.804999	1107	299	4536.6	1338.7
EEC	2.548641	1056	276	2740.3	812.6
ROECD	3.338589	1317	331	2594.6	606.6
EX-USSR	3.080579	2069	454	1024.2	1010.1
China	5.518024	1755	443	533.2	608.2
India	5.422125	2147	482	239.2	148.9
REST	3.033441	1644	400	2913.5	1289.5

a US$ (1985) per ton of carbon
b billions US$ (1985)
c GtC

Table 3A.2 Annual Growth Rates in BAU GDP without Feedback Effects

	1990-2000	2001-2020	2021-2050	2051-2080	2081-2110	2111-2140	2141-2230
USA	2.6	2.2	1.6	1.0	1.0	0.8	0.5
EEC	2.2	1.7	1.3	1.0	1.0	0.8	0.5
ROECD	3.2	2.4	2.0	1.0	1.0	0.8	0.5
EX-USSR	2.6	2.1	1.6	1.0	1.0	0.8	0.5
China	4.6	4.4	3.4	2.5	2.0	1.5	1.0
India	4.6	4.5	3.4	2.5	2.0	1.5	1.0
REST	3.7	3.4	2.7	1.5	1.0	0.8	0.5
WORLD	3.0	2.6	2.1	1.4	1.2	1.0	0.7

Table 3A.3 Annual Growth Rates in BAU CO_2 Emissions

	1990-2000	2001-2020	2021-2050	2051-2080	2081-2110	2111-2140	2141-2230
USA	1.1	0.8	0.6	0.5	0	-1.0	-1.0
EEC	0.8	0.7	0.6	0.5	0	-1.0	-1.0
ROECD	2.0	1.5	1.3	0.5	0	-1.0	-1.0
EX-USSR	1.9	1.3	1.0	0.5	0	-1.0	-1.0
China	3.7	3.5	3.2	2.5	2.0	1.0	0.5
India	3.7	3.7	3.7	2.5	2.0	1.0	0.5
REST	2.5	2.4	2.0	1.5	1.0	0	0
WORLD	2.0	1.8	1.8	1.5	1.2	0.4	0.3

Table 3A.4 Per cent GDP Loss due to 2.5 ℃ Actual Temperature Increase (k-values)

	Total damage: 1 % of GWP	Total damage: 2 % of GWP	Total damage: 3 % of GWP
USA	0.867	1.733	2.6
ROECD	0.933	1.867	2.8
EEC	1.0	2.0	3.0
EX-USSR	0.467	0.933	1.4
China	4.067	8.133	12.2
India and REST	1.533	3.067	4.6

Table 3A.5 Value of Stock Parameters

Parameter	Value	Meaning
$X_{Exog\text{-}CO2}$	0.61[a] 0.35[b]	exogenous emissions of CO_2 (ppm)
Q^P	280	preindustrial CO_2 concentration (ppm)
Z(0)	73	additional concentration of CO_2 (ppm) from preindustrial time to 1990
$L_{CO2}(0)$	120	lifetime of CO_2 (years) in 1990

a emissions in 1990
b annual rate of increase

Table 3A.6 Value of Temperature Parameters

Parameter	Value	Meaning
A	3.607	Parameter in potential temperature function
B	20.328	Constant in potential temperature function
λ	0.75	Total warming coefficient (°C/W m^{-2})
α	0.025	Lag parameter

REFERENCES

Barrett, S. (1990) The Problem of Global Environmental Protection. *Oxford Review of Economic Policy*. 6, 68-79.

Barrett, S. (1991) *The Paradox of International Environmental Agreements*. Mimeo, London BusinessSchool.

Brooke, A., D. Kendrick and A. Meeraus (1992) *GAMS: A User's Guide - Release 2.25*. Redwood City, The Scientific Press.

Burniaux, J. M., J. P. Martin, G. Nicoletti and J. Oliveira Martins (1992a) *The Costs of Reducing CO_2 Emissions: Evidence from GREEN*. Working Paper No. 115, Economics Department, OECD, Paris.

Burniaux, J. M., J. P. Martin, G. Nicoletti and J. Oliveira Martins (1992b) *GREEN - - A Multi-sector, Multi-region Dynamic General Equilibrium Model for Quantifying* *the Costs of Curbing CO_2 Emissions: A Technical Manual*. Working Paper No. 116, Economics Department, OECD, Paris.

Carraro, C. and D. Siniscalco (1992) The International Dimension of Environmental Policy. *European Economic Review*, 36, 379-387.

Carraro, C. and D. Siniscalco (1993) Strategies for the International Protection of the Environment. *Journal of Public Economics*, 52, 309-328.

Cline, W. R. (1991) Scientific Basis for the Greenhouse Effect. *The Economic Journal*, 101, 904-919.

Cline, W. R. (1992a) *The Economics of Global Warming*. Washington DC, Institute for International Economics.

Cline, W. R. (1992b) *Socially Efficient Abatement of Carbon Emissions*. Paper presented at the Second CICERO Seminar on Climate Change, Oslo, 29 November - 2 December.

Eyckmans, J., S. Proost and E. Schokkaert (1992) Efficiency and Distribution in Greenhouse Negotiations. *Kyklos*, 46, 363-397.

Fankhauser, S. (1992) *Global Warming Damage Costs: Some Monetary Estimates*. GEC Working Paper 92-29, Centre for Social and Economic Research on the Global Environment (CSERGE), University College London and University of East Anglia.

Fankhauser, S. (1994) Estimating the Social Costs of Greenhouse Gas Emissions. Forthcoming in *The Energy Journal*.

Fankhauser, S. and S. Kverndokk (1992) *The Global Warming Game - Simulations of a CO_2 Reduction Agreement*. Memorandum No. 13, Department of Economics, University of Oslo, and GEC Working Paper 92-10, Centre for Social and Economic Research on the Global Environment (CSERGE), University College London and University of East Anglia.

Hoel, M. (1993) Efficient Climate Policy in the Presence of Free Riders. Forthcoming in *Journal of Environmental Economics and Management*.

Hoel, M. and I. Isaksen (1993) *Efficient Abatement of Different Greenhouse Gases*. Memorandum No. 5, Department of Economics, University of Oslo.

Houghton, J. T., G. J. Jenkins and J. Ephraums (eds.) (1990) *Climate Change - The IPCC Scientific Assessment*. (Report from Working Group I, The Intergovernmental Panel on Climate Change). WMO, UNEP, IPCC. Cambridge, Cambridge University Press.

Houghton, J. T., B. A. Callander and S. K. Varney (eds.) (1992) *Climate Change 1992 - The Supplementary Report to the IPCC Scientific Assessment*. (Report from Working Group I, The Intergovernmental Panel on Climate Change). WMO, UNEP, IPCC. Cambridge, Cambridge University Press.

Kverndokk, S. (1993) Global CO_2 Agreements: A Cost-effective Approach. *The Energy Journal*, 14, 91-112.

Lind, R. (ed.) (1982) *Discounting for Time and Risk in Energy Policy*. Baltimore, John Hopkins University Press.

Manne, A. S. and R. G. Richels (1992) *Buying Greenhouse Insurance - The Economic Costs of CO_2 Emission Limits*. Cambridge, MA, MIT-Press.

Manne, A. S. and T. Rutherford (1992) *International Trade in Oil, Gas and Carbon Emission Rights: An Intertemporal Equilibrium Model*. Paper presented at the International Institute for Applied Systems Analysis, Laxemburg, Austria, September.

Nordhaus, W. D. (1991a) To Slow or not to Slow: the Economics of the Greenhouse Effect. *The Economic Journal*, 101, 920-937.

Nordhaus, W. D. (1991b) A Sketch of the Economics of the Greenhouse Effect. *The American Economic Review*, 81, 146-150.

Nordhaus, W. D. (1993a) Rolling the 'DICE': An Optimal Transition Path for Controlling Greenhouse Gases. *Resources and Energy Economics*, 15, 27-50.

Nordhaus, W.D. (1993b) Optimal Greenhouse Gas Reductions and Tax Policy in the 'DICE' Model. *The American Economic Review*, 83, 313-317.

Peck, S. C. and T. J. Teisberg (1992) CETA: A Model for Carbon Emissions Trajectory Assessment. *The Energy Journal*, 13, 55-77.

Peck, S. C. and T. J. Teisberg (1993a) CO_2 Emissions Control: Comparing Policy Instruments. *Energy Policy*, 21, 222-230.

Peck, S. C. and T. J. Teisberg (1993b) Global Warming Uncertainties and the Value of Information: An Analysis Using CETA. *Resource and Energy Economics*, 15, 71-97.

Chapter 4 THE TIMING OF CO_2 CONTROL IN THE FACE OF UNCERTAINTY AND LEARNING

Charles D. Kolstad[1]

ABSTRACT

This chapter concerns the optimal regulation of greenhouse gases that lead to global climate change. In particular, the focus is on uncertainty and learning (which, over time, resolves uncertainty). An empirical stochastic model of climate-economy interactions is presented, along with results on the tension between postponing control until more is known vs. acting now before irreversible climate change takes place.

4.1 INTRODUCTION

Uncertainty is a dominant characteristic of environmental externalities, including the accumulation of greenhouse gases leading to climate change. We understand well neither the effects of climate change nor the costs of controlling greenhouse gases. This is one reason considerable sums are expended in trying to better understand this problem. An additional factor frequently comes into play having to do with the cumulative or stock effects of greenhouse gases. It is not the *emissions* of greenhouse gases that *directly* cause adverse effects; rather it is the stock of these gases that may lead to climate change and these stocks change slowly with a great deal of momentum. These two aspects of the problem--stock effects and uncertainty--lead to a tension between instituting control and delaying control.[2] Some in society will desire control of greenhouse gases before climate change is well understood. Others in society may urge delaying control until the problem is clearly delineated. If, *ex post*, the problem turns out to be less severe than expected then those urging delay will

[1] Research supported by a grant from the Research Board of the University of Illinois, by NSF Grant SES-92-09943 and by USAID cooperative agreement DHR-5555-A-00-1086. Superb research assistance from Ying Zhang is gratefully acknowledged. An earlier version of this paper was presented at the International Workshop on Costs, Impacts and Possible Benefits of CO_2 Mitigation at IIASA in Austria in September 1992 (see Kolstad, 1993). Thanks to Prof. William Nordhaus of Yale for providing an early version of his 'DICE' model.

[2] There are other examples with these basic characteristics: hazardous wastes and groundwater, acid rain, species extinction, pesticide accumulation.

have been proved correct (*ex post*). If on the other hand, the problem turns out to be more severe than expected, then delay can be very costly indeed.

This chapter concerns one of the most fundamental questions in the climate change/greenhouse gas control policy arena: when and to what extent to regulate the generation of greenhouse gases when uncertainty exists and learning is taking place. Thus this chapter seeks to determine how the *fact* that we are learning about climate change influences our actions *today* to control greenhouse gases. In our application, there is uncertainty on the damage from climate change. While there has been some work related to this question (Manne and Richels, 1992; Peck and Teisberg, 1992, 1993; Nordhaus, 1991), explicit treatment of the learning process has yet to appear in the empirical literature on climate change.[3] Our approach to the problem is to adapt a simple optimal growth economy-climate model (Nordhaus, 1992a, 1993) to include uncertainty and learning.

Two primary results emerge from our analysis. If emission control is perfectly reversible (no sunk capital), then the fact that one is learning does not appreciably affect today's emission control policies. This is because of the long lags in emissions contributing to temperature change. In the case where emission control investments, once made, become sunk costs, then rapid learning does tend to *modestly* reduce optimal current period emissions.

The next section of the chapter reviews some important contributions to the theoretical literature on learning, as well as existing empirical analyses of climate change and learning. The subsequent section presents the model of optimal regulation used in this study. The case of uncertainty in the disutility of pollution is then considered, followed by results.

4.2 BACKGROUND

4.2.1 Irreversibilities and Stock Externalities

A major literature has developed in the area of investment under uncertainty in the presence of externalities. Arrow and Fisher (1974) initiated much of the work in this area by focusing on a two period model with uncertainty about the benefits of an environmental asset that is to be exploited (eg, a canyon flooded to make electricity). With some uncertainty resolved between the two periods and the impossibility of undoing development of the environmental asset, it turns out to be optimal to bias

[3] See Kolstad (1993) and Cunha-e-sa and Kolstad (1992) for a discussion of theoretical issues surrounding learning and stock externalities.

development in favor of preservation of the environmental asset. Henry (1974) published similar results. In essence, taking an irreversible action has a cost in terms of reducing the value of information. Arrow and Fisher (1974) introduced the notion of quasi-option value, the value of the information gained by waiting before exploiting the environmental asset. Since then, there has been a considerable literature on irreversibilities and on quasi-option value (eg, see Fisher and Hanemann, 1987, 1990; Freeman, 1984; Olson, 1990; Conrad 1980; and Miller and Lad, 1984). Of course there is also a large literature on option value in the field of finance. In particular, a number of recent papers concern the optimal timing of capital investments (eg, oil field development) when learning is taking place (eg, oil field exploration); see Paddock et al. (1988).

Another related literature, primarily from the early 1970's concerns optimal growth in the presence of environmental externalities, particularly stock externalities. This was a natural extension of the optimal growth models that were popular in the 1960's and early 1970's. An important and characteristic paper in this genre is that of Keeler et al (1971). In that paper a simple optimal growth model is posited where utility is a function of consumption and a stock of pollution. Optimal paths for accumulation of capital and pollution are developed for several different types of pollution control. Other papers of this type include Plourde (1972), d'Arge and Kogiku (1973), Smith (1972), Plourde and Yeung (1989) and Forster (1973). Cropper (1976) also considers such a model of optimal growth but focuses on catastrophic environmental effects--the ultimate in irreversibilities.

4.2.2 Learning

There are three basic types of learning which are potentially applicable to global warming. One is active learning whereby observations on the state of the economy and the climate system convey information about uncertainty. Thus by perturbing emissions, one can obtain information about uncertain parameters (eg. see Balvers and Cosimano, 1990). A second type of learning is purchased learning whereby knowledge is purchased and the amount of knowledge purchased (R&D expenditures) depends on its cost and benefits (see Grossman and Shapiro, 1980). A third type of learning can be called autonomous learning where the mere passage of time reduces uncertainty. It is this third type of learning that we examine in this chapter.

4.2.2.1 Information Structures

The typical approach to including autonomous learning in models of irreversibility is to posit a two or three period model where uncertainty changes from one period to the next. Miller and Lad (1984) use a two period model with an *ex ante* probability distribution on period i benefits (b_i) of $f(b_1,b_2)$. After observing period one benefits, the *ex post* marginal distribution is obtained: $f(b_1,b_2 | b_1)$. While this is clearly learning, we need a way to parameterize the rate of learning so that the effects of the *rate* of learning can be deduced. Jones and Ostroy (1984), Olson (1990) and Marshak and Miyasawa (1968) provide such a framework through the concept of an ordering on information structures. Starting with a set of states of nature and an informative message, an information structure consists of a prior on the probabilities of receiving specific messages, along with a conditional probability on states of nature, given a specific message. Of two information structures with the same prior on states of nature, the one that has the greater variability in terms of possible posteriors is viewed as being "more informative." This is equivalent to the more informative structure yielding a higher attainable expected utility when the consumption bundle depends on the state of nature (see Jones and Ostroy, 1984, who point out that more flexibility can only be advantageous). Thus if two learning processes yield two comparable information structures, then the structure that is more informative corresponds to greater learning.

To quantify this concept of learning further, suppose there is a set of possible states of nature, indexed by $s=1,...,S$. Furthermore, suppose there is a finite set, Y, of possible "messages" containing information on the state of nature. Suppose the prior on receiving particular messages is q (dimension equal to the size of Y) and the conditional probability on states of nature (after the message $y \epsilon Y$ has been received) is $\pi(y)$. We use the term "prior" to refer to a probability distribution on state, before the message is received and posterior to refer to distributions assuming a message has been received. Let Π be a matrix with columns consisting of $\pi(y)$ with a different column for each y. Thus Π has S rows and the same number of columns as members of Y. (Π,q) is an information structure. A first goal is to develop an economically relevant ordering on information structures. A standard definition of the comparative value of information is provided by Jones and Ostroy (1984) (see also Laffont, 1989).

4.2.2.2 A Special Parameterization of Learning

We consider a special restriction on the set of comparable information structures. In particular, if there are S possible states of nature, we assume a message consists of a noisy signal as to the true state of nature and there are S possible noisy signals. Let

$\lambda\epsilon[0,1]$ reflect the level of information in the signal with 0 being no information and 1 being perfect information. Thus given a prior $\bar{\pi}$ we define the *star-shaped* information structure (Π,q) where $q\equiv\bar{\pi}$ and the sth column of Π is

$$\pi^s = (1-\lambda)\bar{\pi}+\lambda e_s \tag{4.1}$$

where e_s consists of all zeros except with a one in the sth position. Clearly $\pi_q=\Pi\bar{\pi}=\bar{\pi}$. Furthermore if $\lambda=0$, each column of Π is $\bar{\pi}$ and if $\lambda=1$, $\Pi=I$.

As an example, suppose you can receive one of three messages indicating whether the state of nature is 1, 2 or 3. We thus assume that number of possible messages equals the number of possible states-of-nature, which need not be the case. A message that conveyed the maximum amount of information would resolve all uncertainty on the state of nature. If the message is too noisy to contain any information, then the posterior on states of nature is the same as the prior. This is illustrated in Figure 4.1 where the simplex of probabilities on states of nature is shown. The prior is $\bar{\pi}$. The set of posteriors associated with a star-shaped spreading of beliefs, spread all the way out to the vertices, is shown by the three lines radiating out from $\bar{\pi}$. Perfect learning would move you to one of the three vertices following receipt of the message. Less learning would move you to one of the three points marked with circles. Even less learning would move you to one of the three points marked with x's after receiving the message.

The advantage of representing learning by this star-shaped spreading of beliefs is that the process can be parameterized by the λ in equation (4.1). The disadvantage is that we have eliminated perfectly legitimate and orderable learning processes (emanating from $\bar{\pi}$ in Figure 4.1).

4.2.3 Economy-Climate Models

Economic models have played a critical role in the formulation of environmental policy in the US over the past three decades. The main function of these models has been to simulate the economy's response to particular environmental regulations. Before putting a regulation in place, Congress or regulatory bodies desire to estimate the economic effects as well as the environmental effects of there regulations. That is precisely what an economic model can do, at least in theory.

One of the first economics papers in the global warming area is by Nordhaus (1977) and one of the earliest models is due to Nordhaus and Yohe (1983). They utilize a highly aggregated model, specifying in a single equation the relationship between world GNP and inputs of non-energy factors (such as labor), fossil fuels and nonfossil fuels. Technical change is explicitly represented. Using this highly simplified representation of the world economy, the authors focus on the effect of uncertainty in the underlying parameters on levels of CO_2 over the next 125 years. They are also able to infer which aspects of their model most affect atmospheric CO_2 levels.

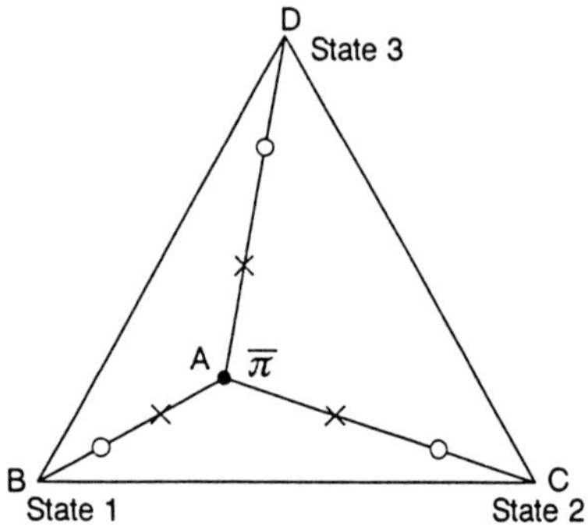

Figure 4.1 Star-shaped spreading of beliefs from $\bar{\pi}$.

William Nordhaus' work has also evolved considerably since the early 1980's. In recent papers (Nordhaus, 1992ab, 1993) he augments his economic model by incorporating equations representing the evolution of the atmosphere in response to greenhouse gas emissions. He has also conducted a useful and thorough review of the costs of control of greenhouse gas emissions (Nordhaus, 1991).

Also in the early 1980's a much more detailed model of the relationship between CO_2 and world energy demand was developed at Oak Ridge Associated Universities (Edmonds and Reilly, 1983). The model divides the world into nine regions. In each region aggregate energy demand is a function of prices and income. Supply of energy is represented in some detail, with various technological options represented separately. Using the model, they demonstrate the effect of carbon taxes, either on a world-wide basis or just for the US. Thus they are able to demonstrate the effect of regulation as well as the difficulty in controlling the problem at the sub-global level.

Edmonds and Reilly have developed their model further in recent years. In a recent paper (Darmstadter and Edmonds, 1989), they demonstrate the dramatic effect uncertainty can have on future CO_2 emissions. For example, they show that there is a 5% probability that CO_2 emissions will actually be substantially lower in 2050 than at present, at least given the probability distributions they assume for their exogenous parameters.

Recent entrants into the greenhouse gas analysis arena, though not newcomers to energy modeling, are Alan Manne and Richard Richels. They have extended Manne's ETA-Macro model (Manne, 1981) to include the generation of greenhouse gases (Manne and Richels, 1990). The model is used to determine the level of a carbon tax that would be necessary to support particular CO_2 emission goals for the US. Related to this question, they have used their model to look at the value of R&D in climate change by looking at the payoff from resolving uncertainty (Manne and Richels, 1991b, 1992).

The Manne and Richels framework has proved to be very popular in greenhouse policy circles. Peck and Teisberg (1992, 1993) have introduced a damage function into the Manne and Richels model to examine the influence of the curvature of the greenhouse gas damage function on optimal control policies. Manne and Richels (1991a) have substantially extended their model (referred to as the "Global 2100" Model) by considering several distinct regions of the world.

There are several other models of the economics of CO_2 generation that should be mentioned. Marks et al, (1989) and Dixon and Johnson (1989) have examined the effect of CO_2 emission controls on the Australian economy using the ORANI general equilibrium model. Richard Kosobud has developed a series of models, sometimes alone, sometimes in conjunction with others, to examine specific greenhouse gas issues (Kosobud, 1989, 1990).

4.3 A STOCHASTIC MODEL WITH LEARNING

4.3.1 The Stochastic Growth Model

In this section we present a general model of the dynamic evolution of an economy, corporating emission control, pollution accumulation and pollution damage. To a large extent it is a standard optimal growth model, although some aspects having to do with the climate are nonstandard. It is based on the climate-economy model of William Nordhaus (Nordhaus, 1992b). His model is deterministic however, and our model is stochastic.

The model is not regionally differentiated and involves the maximization of the net present value of expected utility. Utility is enhanced by consumption and depressed by pollution damage. Output can be channeled to consumption, emission control or investment. Uncertainty enters in that several states of the world (s) are possible and one wishes to maximize expected utility. The following represents such a model. The parameter h_t can be ignored (assumed constant) for the time being; it represents the history of learning and will be used later when we consider learning.

$$\max_{I,E} \sum_t (1+\rho)^{-t} \sum_{h_t} \mu(h_t,t) \sum_s \pi_s(h_t,t) u[c(h_t,t), d(h_t,t), s] L(t) \tag{4.2}$$

$$s.t.\ I(t,h_t), E(t,h_t) \geq 0 \tag{4.3}$$

$$Y(t,h_t) = f[K(t,h_t), L(t), E(t,h_t), t] \tag{4.4}$$

$$c(t,h_t) = [Y(t,h_t) - I(t,h_t)]/L(t) \tag{4.4a}$$

$$d(t,h_t) = g[T(t,h_t), Y(t,h_t)]/L(t) \tag{4.4b}$$

$$E(t+1,h_{t+1}) \leq e\left[E(t,h_t), Y(t,h_t),\ Y(t+1,h_{t+1})\right] \tag{4.4c}$$

$$K(t+1,h_{t+1}) = (1-\delta_K)K(t,h_t) + I(t,h_t) \tag{4.4d}$$

$$M(t+1,h_{t+1}) = (1-\delta_M)M(t,h_t) + \beta E(t,h_t) \tag{4.5}$$

$$T(t+1,h_{t+1}) = s[T(t,h_t), M(t,h_t), 0(t,h_t)] \tag{4.5a}$$

$$0(t+1,h_{t+1}) = r[T(t,h_t), 0(t,h_t)] \tag{4.5b}$$

where

I	= investment	(control)
E	= emissions of greenhouse gases	(control)
K	= capital stock	(state)
M	= stock of greenhouse gases	(state)
T	= mean atmospheric temperature	(state)
O	= mean deep ocean temperature	(state)
c	= per-capita consumption	
d	= per-capita climate damage	
Y	= gross output of goods and services	
ρ	= pure social rate of time preference	

δ_K	= capital depreciation rate
δ_M	= greenhouse gas decay rate
L	= population/labour supply
t	= time
s	= state-of-the world
β	= greenhouse gas emission factor
π_s	= probability of state s given learning history h_t: $\Sigma\pi_s=1$
μ	= probability of learning history h_t

Equations (4.2 - 4.4) constitute the basic economic model and equations (4.5) describe the evolution of the climate. Equation 4.4c is used to constrain the emission rate of greenhouse gases based on previous control decisions; i.e. investment in emission control constrains future emission choices. The links between the economic model and the climate are E and T. Emissions of CO_2 (E) increase the stock of carbon dioxide (M) which increase average global temperature (T) which causes damage per capita (d) which, finally, yields disutility. The goal is to choose the investment path and emission path that maximize expected utility.

4.3.2 Learning

Introducing learning into this model involves introducing a second set of states corresponding to different messages that might be received (see discussion of eqn. 4.1). Each message yields a different outcome of the learning process where an outcome is a new probability on states of nature, $\pi_s(t)$. Let Y_t be the set of possible single period outcomes of the learning process at time t. One can think of these as messages as was previously discussed (or see Laffont, 1989). For instance, Y_t could contain three elements, $Y_t = \{y_1, y_2, y_3\}$ where y_1 = learning which increases the likelihood that global warming is serious, y_2 = learning which indicates global warming is a modest problem. y_3 = learning which indicates that global warming is not serious. Before learning occurs we do not know whether y_1, y_2 or y_3 will be realized although we do know the probabilities of elements of Y_t occurring.

While Y_t indicates the possible outcomes of the learning process at t, to know the current state of knowledge, it is also important to know the learning that has preceded t. We call this the history of learning, $H_t = \{(y_o, \ldots, y_t) \mid y_i \in Y_i \; \forall \; 0 \leq i \leq t\}$. Notationally, H_t contains the learning that occurred up to time period t. An element of H_t is a particular history.

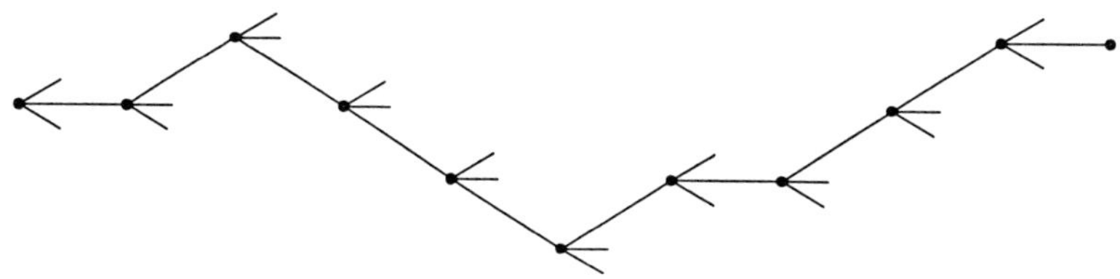

Figure 4.2 The Learning History: (0,1,-1,-1,-1,1,0,1,1,0,1,1,0).

For instance, consider a ten period world in which learning can proceed in three directions {-1, 0, 1} at any point in time. Figure 4.2 illustrates the history (0, 1, -1, -1, -1, 1, 0, 1, 1, 0). If we partition any h_t as (h_{t-1}, y_t), then we define the "predecessor" and "most recent" functions, $\varphi: H_t \rightarrow H_{t-1}$ and $\psi: H_t \rightarrow Y_t$ as $\varphi(h_t) = h_{t-1}$, $\psi(h_t) = y_t$. The function ψ indicates the most recent learning whereas φ indicates learning that occurred earlier. This allows one to functionally represent the learning path and to compute the probability vector on states of the world, $\pi(h_t,t)$. Define the transition matrix $\Pi_t(h_t)$ such that each column is a posterior probability vector corresponding to a different element of Y_{t+1}. Thus $\pi(h_t,t)$ is the column of $\Pi_{t-1}(\varphi(h_t))$ corresponding to $\psi(h_t)$. Furthermore if $q_t(h_t)$ is the probability vector associated with different elements of Y_{t+1}, then

$$\pi(h_t,t) = \Pi_t(h_t) q_t(h_t) \tag{4.6}$$

$(\Pi_t(h_t), q_t(h_t))$ is a learning structure as described earlier.

It is "easy" to modify model (4.2 - 4.5) to incorporate this learning. All of the variables in the model are already indexed by h_t we need only define how π (knowledge) and μ (probability of a history) evolve:

$$\pi(h_{t+1}, t+1) = \begin{cases} (1-\lambda)\pi(h_t, t) + \lambda e_s \\ (1-\lambda)\pi(h_t, t) \end{cases}$$

if s = ŝ
otherwise (4.7a)

$$\mu(h_{t+1}, t+1) = \mu(h_t, t)\, \pi(h_t, t) \tag{4.7b}$$

where

$$h_t = \varphi(h_{t+1}) \tag{4.7c}$$

$$\hat{s} = \psi(h_{t+1}) \tag{4.7d}$$

where e_s is a vector of 0's and 1's with a 1 in the sth position and, 0 in the rest of the positions and λ is the rate of learning. The message space is assumed to be the same as the space of possible states of the world. Thus messages are noisy indications of the true state of the world.

There are many ways uncertainty can enter a model such as this. We assume uncertainty in the damage from global warming. Specifically, we write utility as

$$u[c(t), d(t), s] = \log[c(t) - \Delta_s d(t)]. \tag{4.8}$$

We further assume there are two states of the world, B and L, corresponding to global warming being a big problem (B) vs. global warming being a little problem (L). We assume $\pi_B(t=0) = .2$ and $\pi_L(t=0) = .8$ with $\Delta_B = 5$ and $\Delta_L = 0$. Thus damage could be five times as high as assumed by Nordhaus ($\Delta_B = 5$). This is somewhat arbitrary but yields an expected value of delta of 1 and reflects the fact that damage could be serious. The variance of Δ is 4. Figure 4.3 shows how the expected variance of Δ changes with learning at various learning rates.

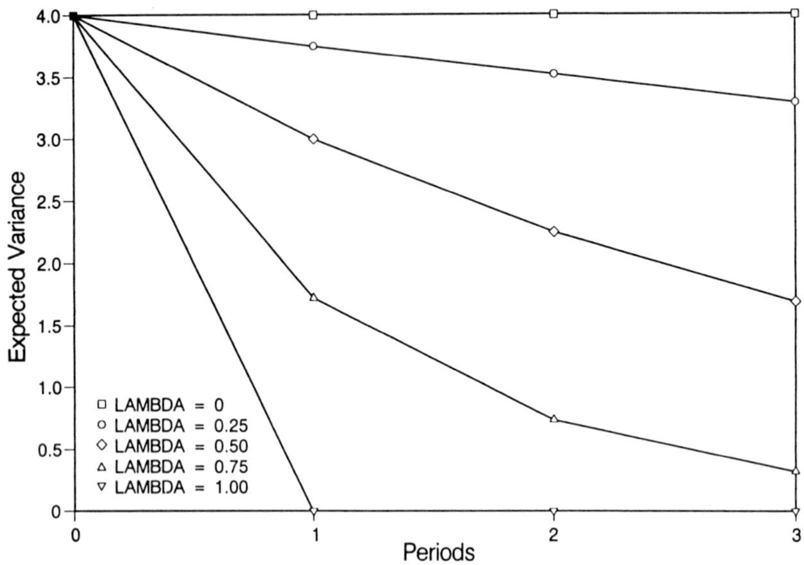

Figure 4.3 Evolution of Expected Variance of [$\pi_o(\Delta=5)=.2$; $\pi_o(\Delta=0)=.8$].

4.4 RESULTS

The model described by eqn (4.2 - 4.8) has been implemented using time points at 10-year intervals beginning in 1965.[4] See the appendix for details on the implementation. The first three points (1965-85) are used as calibration and control of emissions is fixed at zero. Optimal emission control levels are computed beginning in 1995. Learning has occurred in 2005, 2015 and 2025. No learning occurs thereafter. Learning can be of two types, B or L, corresponding to suggesting global warming is a big problem (B) vs. suggesting global warming is a little problem (L). Figure 4.4 illustrates the possible paths learning can take. If the probability of B in 1965 is $\pi(0)$, it will stay at that value through 1995. At that point the probability change depends on how learning progresses. In 2025 and thereafter there will be eight possible values

[4] The model is solved using GAMS/MINOS. I am grateful to Lars Mathiesen for suggesting a stochastic programming approach.

of this probability, depending on the learning history. All of the variables in the model must be indexed on the path learning takes.

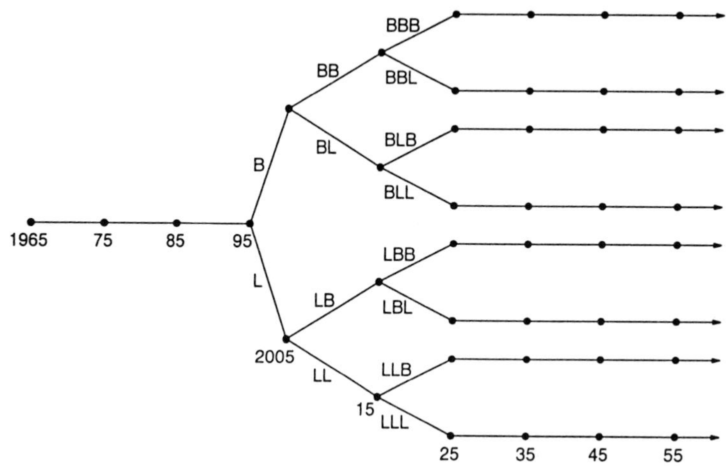

Figure 4.4 Three-period Learning.

The model described in the previous section was an infinite horizon model. Such a model takes a fair amount of computer time to solve so we have chosen to approximate this with a 20 period/200 year finite horizon model. Figure 4.5 shows, for the case of the deterministic model (4.2 - 4.5), the effect of the horizon on optimal emission control rates.[5] Clearly the control level in 1995 is largely unaffected for horizons in excess of 20 periods.

[5] The figure shows the solution to the deterministic model using eqn (8) as an objective with $\Delta_s = 1$ for S = B,L. This is the expected value of delta, and no learning occurs.

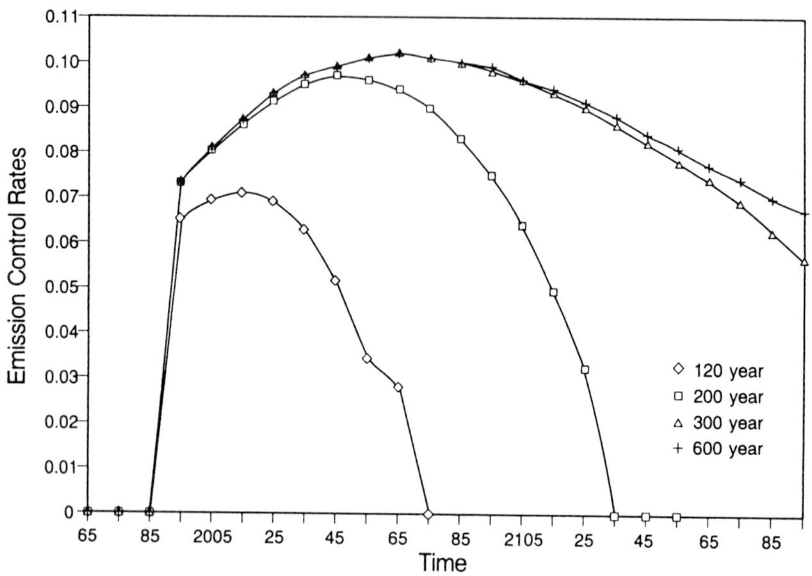

Figure 4.5 Optimal Emission Control Rates.

Model (4.2 - 4.8) was solved under two conditions. In one, emissions were allowed to take any positive value. In the other, the emission control rate, χ, was restricted to be monotonically non-decreasing, reflecting the fact that emission control investments tend to involve sunk costs: once a control level is implemented, it is unlikely to be decreased at a later date.

Focusing on the year 1995, with reversible emission control rates (i.e., eqn 4.4c is relaxed), we find that optimal control levels for greenhouse gases are virtually unaffected by the rate of learning. If one over controls today, then that error can be corrected in the future. Thus the fact that learning is taking place does not impact current decisions to control emission.

It is for this reason that we focus on emission control rates that are monotonically non-decreasing. In this case, once an emission control level is implemented it cannot later be reduced, even if climate change turns out to be less significant. Figure 4.6 shows how the level of emission control in 1995 is affected by the rate of learning, λ. Recall that $\lambda=0$ corresponds to no learning whereas $\lambda=1$ corresponds to resolution of all uncertainty in one period. Thus when learning is occurring rapidly, it pays to reduce greenhouse gas control rates from approximately 7% to as low as 3%. More modest learning rates involve more modest reductions in control rates, but reductions nevertheless.

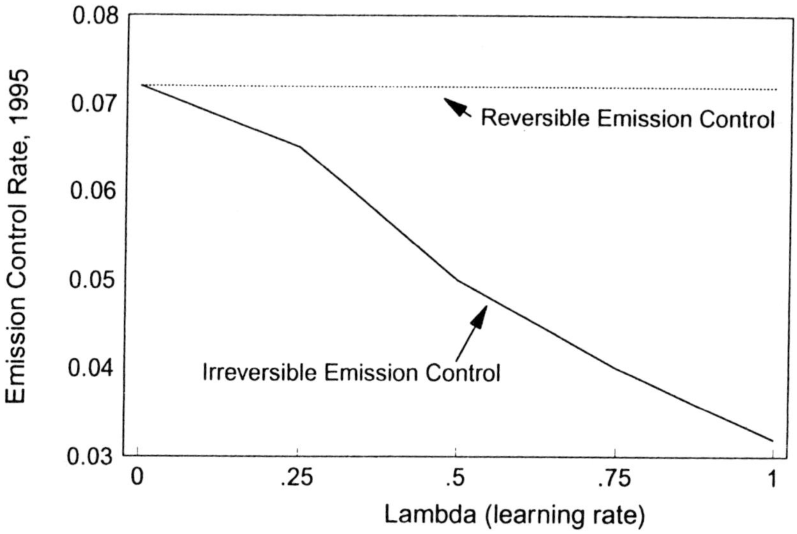

Figure 4.6 1995 GHG Control Rates vs Learning Rates.

The reason for lower emission control levels with more rapid learning is that it is better to defer irreversible control decisions until more is known. It is interesting that the irreversibility in control capital dominates the climate irreversibility. This is no doubt because of the long lags involved in climate change. Figure 4.7 shows for $\lambda=0.5$, how atmospheric temperature increases vary with learning histories.[6] Atmospheric temperature changes come long after the emissions occur. This suggests the potential payoff from control measures that are reversible.

It is interesting to compare this result to that of Manne and Richels(1992). While their model is substantially different than ours, they show that immediate resolution of uncertainty (very rapid learning) results in lower emission control rates (higher emissions) compared to the case when uncertainty is not resolved. This is qualitatively the same as our result. Figure 4.8 shows how emission control evolves with learning over time for $\lambda=0.5$. In 1995, control goes to 5%, before learning occurs. In 2005 one period of learning occurs. If that learning is B, control goes to 11%; otherwise it stays

[6] The learning histories are shown in the figure. The path BBL, for instance, corresponds to learning in the direction B in the first two periods, followed by learning in the direction L.

at 5%. After another period of learning, the range of control levels grows even further.

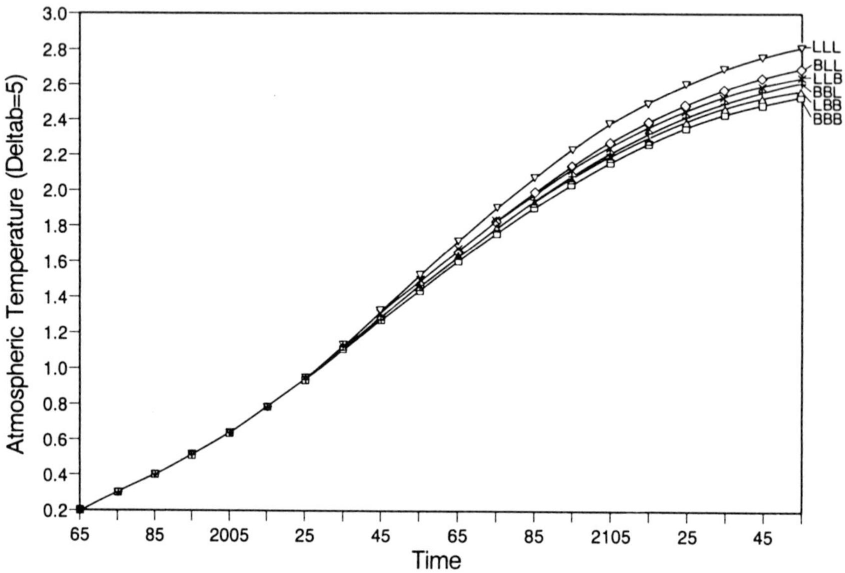

Figure 4.7 The Evolution of Atmospheric Temperature.

Figure 4.9 shows the value of information as a function of λ. The expected value of perfect information is the difference between the net present value of expected consumption less damage if emission control can be made completely contingent on the state of nature and the same figure with uncertainty and non-state-dependent controls. The expected value of perfect information (EVPI) shows what it could be worth to resolve all uncertainty instantaneously. Even modest learning rates substantially reduce the EVPI. This is consistent with Peck and Teisberg (1993) who show only modest value to resolving uncertainty.

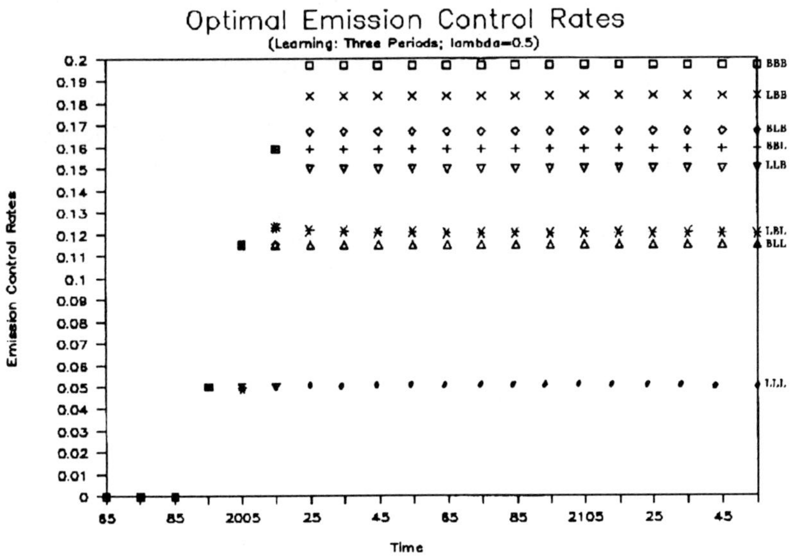

Figure 4.8 Emission Control Paths.

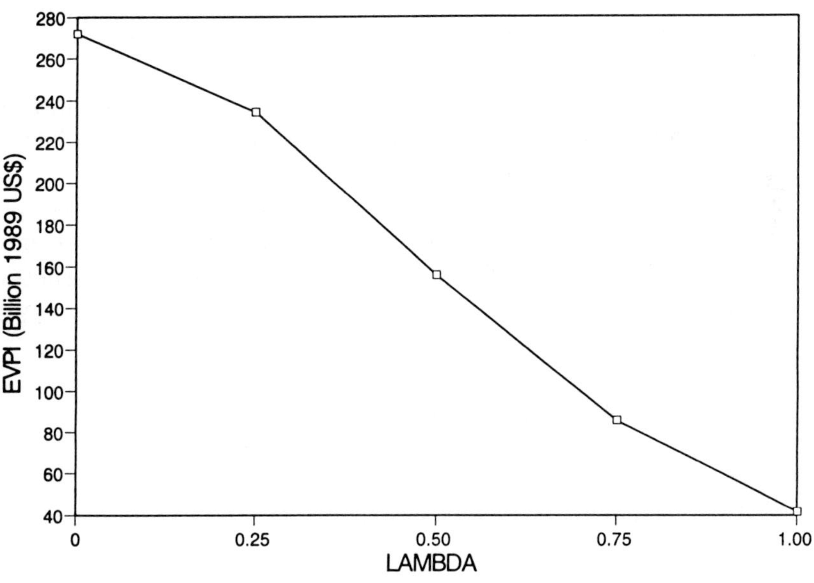

Figure 4.9 Expected Value of Perfect Information.

4.5 POLICY IMPLICATIONS

The results of this simple model do suggest that rapid learning biases current period CO_2 control levels downward. Thus these results tend to support those that argue that policy-makers should take a wait-and-see approach to CO_2 mitigation. There are two important qualifiers to this result, however. One is that learning only biases control downward and does not eliminate the desirability of some control. Secondly, the assumption of complete irreversibility of emission control investment may be extreme and certainty suggests the development of control strategies which are reversible; this is related to but not quite the same as a no-regrets strategy where actions are taken that we will never wish to reverse.

The most important caveat of all, of course, is that the model presented here is highly stylistic and is really only intended as an illustration or research tool; it is certainly not intended to be used to develop policy.

4.6 CONCLUSIONS

In this chapter we have presented an empirical model of learning with uncertainty in the context of the control of greenhouse gases. The particular focus of the model is on the conflicting irreversibilities of emission control investment and the stock effect of greenhouse gas accumulation. The issue is which of these irreversibilities dominates in determining today's emission control level.

Our conclusion is that there is no identifiable irreversibility effect associated with greenhouse gas accumulation. Furthermore, there is an effect of irreversibility in control investment. The net result is that with irreversibilities in control capital and significant learning, today's emission control levels should be reduced relative to either the no-learning case or the case of no control capital irreversibilities. Another implication is that a high priority should be placed on identifying reversible emission control.

APPENDIX: MODEL IMPLEMENTATION

In implementing the model (4.2 - 4.8), we have followed closely the deterministic DICE model (Nordhaus, 1993). Time units are ten year intervals; thus, suppressing h_t for the moment, the functions f,e,g,s and r in equations (4.4 - 4.5) are defined as f: $$Y(t) = [1 - 0.0686\chi(t)^{2.887}] A(t) K(t)^{0.25} L(t)^{0.75} \tag{4A.1}$$

where $$\chi(t) = 1 - \frac{E(t)}{\sigma(t)\, Y(t)} \quad (0 \leq \chi \leq 1) \tag{4A.2}$$

e: $$\sigma(t+1)\, Y(t+1) - E(t+1)) \geq (1 - \delta_E) \{\sigma(t)\, Y(t) - E(t)\} \tag{4A.3}$$

g: $$d(t) = \frac{0.00148\, Y(t)\, T(t)^2}{L(t)[1 + 0.00148\, T(t)^2]} \tag{4A.4}$$

s: $$T(t+1) = T(t) + 0.226\{F(t) - 1.41\, T(t) - 0.440[T(t) - 0(t)]\} \tag{4A.5}$$

where $$F(t) = 4.1 \log [M(t)/590] / \log(2) \tag{4A.6}$$

r: $$0(t+1) = 0(t) + 0.02\ [T(t) - 0(t)] \tag{4A.7}$$

where A(t) and σ(t) are exogenous technology change parameters (see Nordhaus, 1993). The damage equation (4A.4) appears to be different than used by Nordhaus but some manipulation demonstrates the form is exactly as in Nordhaus (1992b). Starting values for state variables (in 1965) are T_o = 0.2(°C), O_o = 0.1(°C), M_o = 677 (gigatons) and K_o = 18.11 (trillion US$). In addition δ_M = 0.0833 and β = 0.64 (see eqn 4.5). Furthermore, δ_K = 0.35 and δ_E = 0 or 1. Remember that δ_M, δ_E, and δ_K are per decade. The pure rate of time preference is 3% per annum.

REFERENCES

Arrow, K.J. and A.C. Fisher (1974) *Environmental Preservation, Uncertainty, and Irreversibility*. Quarterly J. of Economics, 88:312-19.

Balvers, R. and T. Cosimano (1990) *Actively Learning about Demand and the Dynamics of Price Adjustment*. Economic J., 100:882-898.

Conrad, John (1980) *Quasi-Option Value and the Expected Value of Information*. Quarterly J. Economics, 92:813-19.

Cropper, M.L. (1976) *Regulating Activities with Catastrophic Environmental Effects*. J. Env. Econ. Mgmt., 3:1-15.

Cunha-e-sa, M.A. and C.D. Kolstad (1992) *Controlling a Stock Externality in the Presence of Active Learning*. unpublished manuscript, University of Illinois.

D'Arge, R.C. and K.C. Kogiku (1973) *Economic Growth and the Environment*. Rev. Econ. Stud., 40:61-77.

Darmstadtor, J. and J. Edmonds (1989) *Human Development and Carbon Dioxide Emissions: The Current Picture and Long-term Prospects*. Chapter 3 in N.J. Rosenberg et al. (eds) Greenhouse Warming: Abatement and Adaptation (Resources for the Future, Washington, DC.

Dixon, Peter B. and David T. Johnson *Estimation of the Macroeconomic Effects on Australia of Attempting to Reduce CO_2 Emissions by 20 percent by 2005*. University of Melbourne draft manuscript.

Edmonds, J. and J. Reilly (1983) *Global Energy and CO_2 to the Year 2050*. Energy Journal, 4(3):21-47.

Fisher, A.C. and W.M. Hanemann (1987) *Quasi-Option Value: Some Misconceptions Dispelled*. J. Env. Econ. & Mgmt, 14:183-90.

Fisher, A.C. and W.M. Hanemann (1990) *Information and the Dynamics of Environmental Protection: The Concept of the Critical Period*. Scand. J. of Economics, 92:399-414.

Forster, B.A. (1973) *Optimal Consumption Planning in a Polluted Environment*. Econ. Record, 49:534-545.

Freeman, A. Myrick (1984) *The Quasi-Option Value of Irreversible Development*. J. Env. Econ. & Mgmt., 11:292-94.

Grossman, S. and C. Shapiro (1986) *Optimal Dynamic R&D Programs*. Rand J., 17:581-593.

Henry, Claude (1974) *Option Values in the Economics of Irreplaceable Assets*. Rev. Econ. Studies Symposium, pp. 89-104.

Jones, R.A. and J.M. Ostroy (1984) *Flexibility and Uncertainty*. Rev. Economic Studies, 51:13-32.

Keeler, E., M. Spence and R. Zeckhauser (1971) *The Optimal Control of Pollution*. J. Economic Theory, 4:19-34.

Kolstad, C.D. (1993) *Looking vs. Leaping: The Timing of CO_2 Control in the Face of Uncertainty and Learning*. pp. 63-82 in Y. Kaya, N. Nakicenovic, W.D. Nordhaus and F.L. Toth (eds), *Costs, Impacts and Benefits of CO_2 Mitigation*. IIASA, Austria.

Kosobud, R.F. (1989) *Economic Analysis of Greenhouse Effect Policy Issues: East and West*. University of Illinois Institute for Government and Public Affairs Working Paper #10, Urbana, IL.

Kosobud, R.F.(1990) *How Much Greenhouse Policy Intervention, and When? A Great Lakes Regional Study*. paper presented at AAAS meeting, New Orleans.

Laffont, Jean-Jacques (1989) *The Economics of Uncertainty and Information*. MIT Press, Cambridge, Mass..

Manne, A.S. and R.G. Richels (1990) *CO_2 Emission Limits: An Economic Cost Analysis for the USA*. Energy Journal, 11:51-74.

Manne, A.S. and R.G. Richels (1991a) *Global CO_2 Emission Reductions--The Impacts of Rising Energy Costs*. Energy Journal, 12:87-107.

Manne, A.S. and R.G. Richels (1991b) *Buying Greenhouse Insurance*. Energy Policy, 19:543-552.

Manne, A.S. and R.G. Richels (1992) *Buying Greenhouse Insurance: The Economic Costs of CO_2 Emission Limits*. MIT Press, Cambridge, MA.

Manne, A.S. (1981) *ETA-MACRO: A User's Guide*. Electric Power Research Institute Report EA-1724, Palo Alto, California.

Marks, R.E., D.L. Swan, P.B. Dixon, D.T. Johnson, P. McLennan and R. Schodde (1989) *The Feasibility and Implications for Australia of the Adoption of the Toronto Proposal for CArbon Dioxide Emissions*. Report to CRA Ltd.

Marshak, J. and K. Miyasawa (1968) *Economic Comparability of Information Systems*. Int'l Econ. Rev., 9:137-74.

Miller, J.R. and F. Lad (1984) *Flexibility, Learning and Irreversibility in Environmental Decisions: A Bayesian Approach*. J. Env. Econ. & Mgmt., 11:161-72.

Nordhaus, W.D., (1977) *Economic Growth and Climate: The Case of Carbon Dioxide*. Amer. Econ. Rev.

Nordhaus, W.D. and G.W. Yohe (1983) *Future Paths of Energy and Carbon Dioxide Emissions*. in National Research Council Carbon Dioxide Assessment Committee, Changing Climate (National Academy Press, Washington, D.C..

Nordhaus, W.D. (1991) *To Slow or Not to Slow: The Economics of the Greenhouse Effect*. Economic Journal.

Nordhaus, W.D. (1992a) *An Optimal Transition Path for Controlling Greenhouse Gases*. Science, 258:1315-1319.

Nordhaus, W.D. (1992b) *The 'DICE' Model: Background and Structure of a Dynamic Integrated Climate-Economy Model of the Economics of Global Warming*. Cowles Foundation Discussion Paper 1009, New Haven, CT.

Nordhaus, W.D. (1993) *Rolling the 'DICE': An Optimal Transition Path for Controlling Greenhouse Gases*. Resource and Energy Economics, 15:27-50.

Olson, Lars J. (1990) *Environmental Preservation with Production*. J. Env. Econ. & Mgmt., 18:88-96.

Paddock, J.L., D.R. Siegel and J.L. Smith (1988) *Option Valuation of Claims on Real Assets: The Case of Offshore Petroleum Leases*. Quart. J. Econ., 103:497-508.

Peck, S.C. and T.J. Teisberg (1992) *CETA: A Model for Carbon Emissions Trajectory Assessment*. The Energy Journal, 13:55-77.

Peck, S.C. and T.J. Teisberg (1993) *Global Warming Uncertainties and the Value of Information: An Analysis Using CETA*. Resource and Energy Economics, 1571-97.

Plourde, C.G. (1972) *A Model of Waste Accumulation and Disposal*. Canad. J. Econ., 5:119-125.

Plourde, C. and D. Yeung (1989) *A Model of Industrial Pollution in a Stochastic Environment*. J. Env. Econ. & Mgt., 16:97-105.

Smith, V.L. (1972) *Dynamics of Waste Accumulation: Disposal Versus Recycling.* Quart. J. Econ., 86:600-616.

Chapter 5 A GLOBAL MARKETABLE PERMITS APPROACH TO CO_2 MITIGATION: IMPLICATIONS FOR U.S. ENERGY DEMAND

Adam Rose, Brandt Stevens, Ping-Cheng Li[1]

ABSTRACT

This chapter presents a three-part model to analyze the economic impacts of a marketable permits approach to CO_2 reduction. The three components include: 1) a global political economy module to determine the efficient levels of CO_2 reduction for all countries and the equilibrium permit price, 2) an abatement strategy module to determine the direct costs and permit transactions by each sector of the U.S. economy, and 3) a general equilibrium module to determine the broader effects on the U.S. economy and feedback effects on energy industries. A preliminary version of the model is applied to analyzing the short-run impacts on U.S. energy industries of the country's optimal mitigation strategy. Simulations are performed for two mitigation tactics: "no-regrets" conservation and changes in the product mix.

5.1 INTRODUCTION

The controversy continues regarding the merits of limiting the emissions of greenhouse gases. Still momentum is building toward a global agreement. Several policy instruments have been proposed to implement an agreement, the carbon tax being the most widely supported and most thoroughly analyzed (see, e.g., Manne and Richels, 1991; Whalley and Wigle, 1991; Nordhaus, 1993).

[1] An earlier version of this paper was presented at the Annual Meeting of the Mineral Economics and Management Society, Reno, Nevada, February 1993; and at the Annual Meeting of the European Association of Environmental and Resource Economists, Fountainebleau, France, July 1993. The authors wish to thank Charles Kolstad for his helpful comments as a discussant at the latter conference. The authors wish to thank Oscar Frias and Michelle Leitzinger for their assistance. The views expressed in this paper are those of the authors and do not represent any official position.

In the U.S., however, strong support exists for a marketable permits approach (see, e.g., Task Force, 1991).[2] Among the several reasons for favoring permits are: claims of its relative superiority as a policy instrument (see, e.g., Barrett et al., 1992), consistency with the emission trading theme of various Clean Air Act Amendments, and the lack of experience in the U.S. with energy taxes. To date, however, there have been no comprehensive studies of the composition of a global agreement based on a marketable permits system and its implications for various U.S. industries.

This chapter summarizes a three-part model for analyzing the impacts on U.S. energy demand of a marketable permits treaty for CO_2 mitigation. It is composed of:

1. A global political economy module to determine the efficient or cost-effective level of CO_2 reduction. It also evaluates the efficiency and equity implications of alternative permit allocations.
2. An abatement strategy module to determine the direct costs and permit transactions by each sector of the U.S. economy.
3. A general equilibrium module to determine the broader effects on the U.S. economy and feedback effects on the energy industries.

This chapter addresses important modeling issues, yields some unique policy insights, and illustrates the capabilities of the modeling framework. The majority of the numerical results in the literature on global warming are subject to large errors, and this is also true of our results from the application of module 5.3.

Still, the chapter offers some general principles that should prove useful to other analysts and to policy-makers. First, we demonstrate that the assignment of property rights for CO_2 emissions implies a unique post-trading abatement strategy for the U.S. in the short run, no matter how a fixed quantity of permits is assigned globally. Also, we show why modeling the abatement response under the permits instrument is more complicated than modeling carbon taxes in the long run. Finally, we find that a sizable difference is likely to exist between the partial equilibrium and general equilibrium mitigation response for the economy as a whole. At the same time, we find that even a *no regrets* strategy will have a strongly negative impact on U.S. energy industries.

[2] Note that this approach has gained favor with international agencies as well. For example, this paper is an outgrowth of work by Adam Rose for the United Nations Conference on Trade and Development on a marketable permits approach (see Barrett et al., 1992). Also, at the Earth Summit, the head of the United Nations Environment Programme spoke in favor of this instrument. Moreover, if a global treaty on CO_2 mitigation only specifies reduction quotas, it may be left to individual nations to choose how to meet these targets internally. Moreover, the emphasis of the recent Earth Summit on "joint implementation," which calls for bilateral or multilateral negotiations between countries to reduce CO_2 emissions, may result in an international "offset" program analogous to the permits approach (see Barrett et al., 1992).

5.2 OVERVIEW OF THE MODEL

A schematic depiction of the model is presented in Figure 5.1. The initial inputs are a set of cost functions and benefit functions relating to CO_2 abatement for each major emitter or receptor nation. We realize such data are very tenuous, but note that the structure of the model allows for continuous refinement and updating.

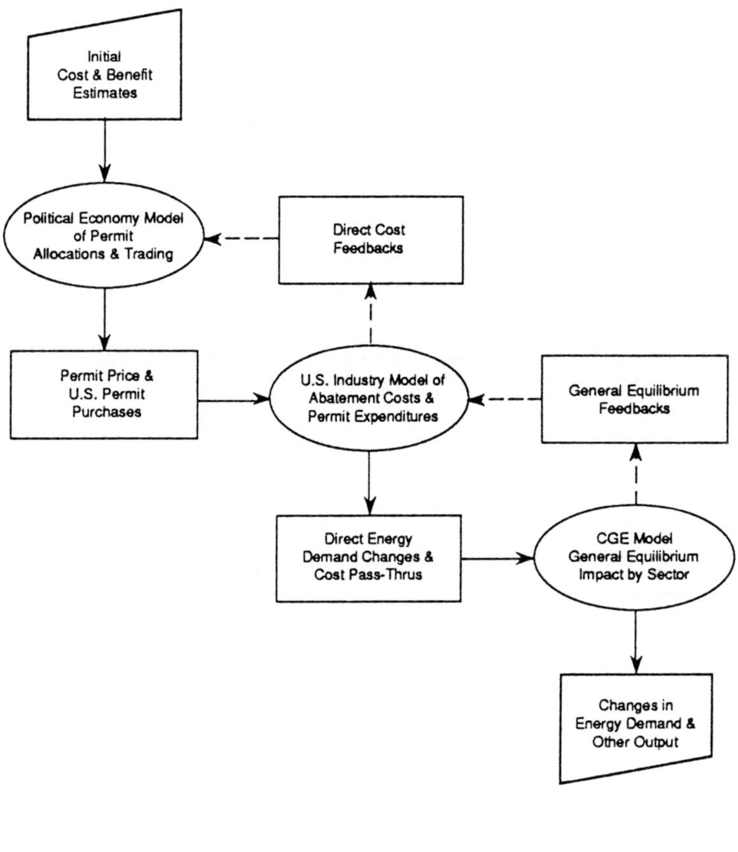

Figure 5.1 Flow diagram of the model

The first module addresses the political economy of a global CO_2 agreement. It consists of a non-linear programming model based on the benefit and cost functions fed into it and an objective function to maximize global welfare. It takes as its starting point specifications of permit assignments corresponding to alternative equity criteria for their allocation. The module is able to determine the globally efficient level of CO_2 reduction and its value at the margin, i.e., the equilibrium permit price. It also yields the least-cost response of the U.S. to any allocation, including specification of the total number of permits bought or sold. Interestingly, as will be shown below, there is a unique post-trading abatement requirement for the U.S (and all other countries as well); hence, mitigation strategy scenarios merely differ by the number of permits to be transacted.

The second module is an abatement strategy model for the U.S. Because of a lack of consensus on such strategies, it is flexible enough to utilize a range of cost functions, reflecting differences in assumptions about substitution elasticities, conservation potential, and end-of-pipe abatement (see, e.g., Nordhaus, 1991; Cline, 1992). This module calculates a set of partial equilibrium responses to the outcomes of the first module. Specifically, it identifies the number of permits that will be purchased by each sector, how abatement costs and permit cost increases translate into fuel price increases, and how conservation results in a direct reduction in energy demand. Note from Figure 5.1 that alternative abatement cost functions can be fed back into module 1 to determine the sensitivity of global abatement requirements and the equilibrium permit price to conditions in the U.S.

The third module is a general equilibrium model that analyzes the inter-market linkages of global warming policy to determine the broader effects on the U.S. economy and feedback effects on energy sectors themselves. Examples of the latter include increased energy demand directly and indirectly from the manufacture of energy-saving equipment, or, say, new nuclear power plants, as well as further multiplier effects of projected decreases in coal and oil production.[3] Ultimately, we hope to adapt the computable general equilibrium (CGE) model of Lin (1991) for this purpose. This chapter performs a more limited analysis with a U.S. input-output model that can incorporate limited changes in prices and technology.[4] Note the feedback loop here as well. It involves adjusting industry abatement responses to general equilibrium effects, such as changes in demand and prices. These feedbacks can be transmitted all the way to module 5.1, to take general equilibrium costs (and benefits) into account.

[3] The multiplier effects of these cutbacks might be quite disparate between regions of the country (e.g., the "oil-patch" and Appalachia). However, the model cannot capture these regional differential effects.

[4] I-O analysis is more restrictive than its CGE counterpart, but does have one advantage at an initial stage of analysis. Rather than a simultaneous solution, which facilely yields a bottom-line, the recursive nature of our I-O algorithm requires us to specify and check the various linkages in the chain of causation of economic impacts every step of the way.

5.3 GLOBAL POLITICAL ECONOMY MODULE

The rationale for the marketable permits approach stems from the work of Coase (1960). What has become known as the Coase Theorem states: externalities can be eliminated effectively through market transactions if property rights can be assigned, regardless of who receives the rights (assuming no significant transactions costs or income effects). This approach has been applied to pollution problems in the United States since the 1970s, and has also been implemented in a number of other countries (see Tietenberg, 1992).

A marketable permits approach has several appealing qualities in the context of the global agreement on CO_2 (see, e.g., Epstein and Gupta, 1991; Barrett et al., 1992). The global problem requires a global solution, but since there is no supranational authority to impose an agreement, policy instruments that promote voluntary action are especially attractive. In this regard, a marketable permits approach is able to ensure that any CO_2 reduction is achieved at lowest global cost. Countries are more likely to engage in an agreement that they perceive as fair, and since the Coase Theorem guarantees that pursuit of equity (through alternative permit assignments) will not undercut efficiency, this issue can be addressed directly. Finally, by internalizing the externality and utilizing market institutions, this approach is self-enforcing.

This module is able to determine the costs and benefits to individual nations of CO_2 agreements based on alternative equity criteria (see Table 5.1).[5] It incorporates data on individual country costs and benefits (adapted from the work of Nordhaus, 1993, and others) into a non-linear programming module to calculate the efficient CO_2 reduction, equilibrium permit price, as well as post-trading benefits, mitigation costs, and permit transactions for all countries (see Rose and Stevens, 1993a).

We briefly illustrate the workings of the model, beginning with the application of a static version of the Sovereignty criterion, which calls for a proportional cutback in current account emissions for each country equal to 20% of a year 2000 baseline (considered necessary to cap CO_2 concentrations at twice their current levels).

Our analysis is undertaken for eight countries or country groups, listed in column 1 of Table 5.2, representative of various levels of development as denoted by their Gross Domestic Product (GDP) in column 2. Columns 3 and 4 of Table 5.2 list estimates of benefits (environmental damage avoided) from mitigation of 20% of fossil fuel CO_2 emissions in the year 2000 and the associated direct costs, respectively.

[5] For an in-depth discussion of these criteria (based primarily on altruistic motives), the reader is referred to the chapter by Rose in Barrett et al. (1992); but cf. the chapters by Barrett in that volume, which examines agreements based on alternative concepts of self-interest.

At this stage of scientific inquiry, the numbers are crude, but are indicative of relative policy impacts.

Table 5.1 Alternative equity criteria for global warming policy

Criterion	General Operational Rule	Operational Rule for CO_2 Permits
Horizontal	Equalize net welfare change across nations (net cost of abatement as proportion of GDP equal for each nation)[a]	Allocate permits to equalize net welfare change (net cost of abatement as proportion of GDP equal for each nation)[a]
Vertical	Progressively share net welfare change (net cost proportions inversely correlated with per capita GDP)[a]	Progressively distribute permits (net cost proportions inversely correlated with per capita GDP)[a]
Ability to Pay	Equalize abatement costs across nations (gross cost of abatement as proportion of GDP equal for each nation)[b]	Allocate permits to equalize abatement costs (gross cost of abatement as proportion of GDP equal for each nation)[b]
Sovereignty	Cut back emissions proportionally across all nations	Allocate permits in proportion to emissions
Egalitarian	Cut back emissions in proportion to population	Allocate permits in proportion to population
Market Justice	Make greater use of markets	Auction permits to highest bidder
Consensus	Seek a political solution promoting stability	Distribute permits so (power weighted) majority of nations are satisfied
Compen-sation	Compensate net losing nations	Distribute permits so no nation suffers a net loss of welfare
Rawls' Maximin	Maximize the net benefit to the poorest nations	Distribute large proportion of permits to poorest nations
Environ-mental	Cut back emissions to maximize environmental values	Limit for vulnerable ecosystems (e.g., forests)

a. Net cost equal to the sum of mitigation benefits - abatement costs + permit sales revenues - permit purchase costs.

b. Gross cost refers to abatement cost only.

The net change in national welfare, before permit trading, from an application of the Sovereignty criterion is presented in column 5 of Table 5.2. All eight countries/regions would stand to reap positive net benefits except the C.I.S. The sum of net benefits is $15.31 billion. Permit trading will take place as long as some countries have marginal abatement costs greater than or equal to the world market price of permits and some countries have abatement costs less than or equal to this price. It is usually assumed, and generally borne out by cost studies, that abatement costs are relatively higher in developed countries than developing ones. Trading will thus lower the total costs to the former and provide a transfer to the latter, which in effect also lowers the net abatement costs of developing countries.

Table 5.2 Changes in national welfare due to static sovereignty permit assignments (cost-effectiveness case: 20% abatement)

Country (Area)	Per Capita GDP[b]	All Cases	(Without Trading)		With trading		
		Benefit	Cost	Net Change	Cost	Trading	Net Change
	(1)	(2)	(3)	(4)=(2)-(3)	(5)	(6)	(7)=(2)-(5)-(6)
United States	$20,725	$16.98	$9.13	$7.85	$3.64	$4.19	$9.15
Canada	18,648	1.07	.89	.18	.30	.43	.34
Western Europe[c]	14,322	10.27	6.67	3.60	2.08	3.24	4.95
C.I.S.	3,491[d]	2.28	2.90	-.62	7.72	-6.14	.70
Brazil	2,167	1.52	.10	1.42	.53	-.63	1.62
Central Africa[e]	475	.14	.02	.12	.02	-.01	.12
Indonesia	527	.47	.06	.41	.30	-.36	.53
China	375	5.38	3.03	2.35	3.72	-.73	2.39
Total	--	$38.11	$22.80	$15.31	$18.31	$0.00	$19.80

a. All figures in billions of 1990 dollars except Per Capita GDP.
b. Data for 1989 from World Bank (1991).
c. Western Europe includes the European Community countries: Belgium, Denmark, France, Germany, Greece, Ireland, Italy, Luxembourg, Netherlands, Portugal, Spain, and the United Kingdom.
d. Data for 1988 for Soviet Union from Euromonitor (1991).
e. Central Africa includes: Burundi, Cameroon, Central African Republic, Congo, Equatorial Guinea, Gabon, Rwanda, and Zaire.

The effect of trading on the welfare of the eight world regions is presented in the last three columns of Table 5.2. The results emanate from our model's determination of optimal individual abatement levels by equalizing marginal costs across countries. They indicate that the U.S., Canada, and Western Europe would be permit buyers and the other five countries/regions would be permit sellers. Nearly 205 million permits would be traded with a total value of $7.86 billion at the equilibrium price of

$38.35 per ton of CO_2. The optimal abatement level for the U.S. is approximately 12.8%, thus necessitating purchase of 109.3 million permits.

Permit trading has several positive implications. First, the equalization of marginal costs of abatement saves this global group $4.49 billion, or almost 20% of the $22.80 billion of abatement costs associated with the uniform emission reduction requirement. The net cost savings (including expenditures on permits) to the three Western industrialized countries/regions is $2.80 billion, or about 17% of their costs at uniform abatement levels. In addition, permit trading is estimated to be a source of revenue for developing and transitional economies. In fact, for Brazil and Indonesia, the permit revenues more than offset their total abatement costs. Finally, all of the countries examined receive positive net benefits from a 20% CO_2 emission reduction policy, though note the U.S. and Western Europe receive the highest net change overall and the greatest gains from trading.

Table 5.3 Permit trades for the U.S. under alternative allocation criteria

Allocation criterion	Baseline CO_2 emissions (thousands)	CO_2 emission reductions (thousands)	Initial permits bought or allocated (thousands)	Number of permits bought or sold (thousands)	Value of permits (billions)
Sovereignty	1,516,668	194,067	1,213,334	109,267	$4.19
Egalitarian	1,516,668	194,067	414,995	907,610	$34.80
Horizontal	1,516,668	194,067	1,337,056	-14,455	-$0.55
Market justice	1,516,668	194,067	1,289,167	33,433	$1.28
Rawlsian	1,516,668	194,067	1,130,111	192,490	$5.96

Table 5.3 shows the effects of five different equity criteria on the U.S. demand for permits. In the Market Justice case, we assume the U.S. and Western Europe can exercise market power, though this effect is modest. Even under this case, the U.S. remains a net purchaser of permits. The Horizontal equity case presented here assumes that permits are distributed according to relative proportions of global GDP (as opposed to the total net welfare change, which is a preferred but computationally more complicated measure), and, in this case, the U.S. receives sufficient permits initially to be able to reduce its overall abatement level to 12.8% and still have permits to sell (see also Rose and Stevens, 1993b, for an analysis of the other equity criteria).

A definitive finding of our previous analyses is that trading substantially erodes the differences in real world implications of various equity criteria. Note from Table 5.3, there is a striking uniformity in the post-trading abatement levels for the U.S. (and for each of the other countries entering a CO_2 treaty), which translates into a strong similarity in net welfare changes (not shown). This is due to the uniqueness of the equilibrium outcome of the Coase Theorem. That is, no matter how permits are assigned there is a single, least-cost mix of abatement levels between countries. Put another way, this means that post-trading abatement costs will always be the same for any given cost-effectiveness target no matter what the initial allocation of permits. Since these abatement costs will be a major proportion of the net costs (including permit transactions), this feature will significantly reduce the variance between nations. Only those countries engaging in large permit transactions will have noticeably varying net welfare changes across the various equity criteria. All of this has very important policy implications. It, in effect, narrows the distinct number of equity criteria that may need to be examined, and hence could speed the negotiation process. Two countries may differ in their first choice of equity principles to apply to permit assignments, but if the two principles have the same welfare implications, a deadlock can be avoided.

5.4 ABATEMENT STRATEGY MODULE

The results of the previous section narrow the range of considerations relating to the CO_2 abatement level required of the U.S. as a whole. Specifically, the U.S. will always find it most efficient to abate 12.8% of its baseline CO_2 emissions and to purchase permits (at \$38.35 per ton) for any deficit between this level and the number of permits it receives (no matter what the allocation). The number of permit transactions can vary over a sizable range, and the variation increases as we consider the several possible institutional arrangements for implementing the permit options within this country. In the short run, all that matters is the permit price and its affects on the firm's marginal cost. Permit expenditures or revenues pertain only to average costs, and thus their effects, if any, are only relevant to long-run pricing or entry/exit considerations.

It might be helpful to examine the workings of a carbon tax, and how it is modeled, as a point of reference. The simplest version is a tax per unit of carbon, which would be effectively applied to carbon-emitting goods (for the purpose of simplification, let us confine ourselves for now to fossil fuels--coal, oil, and gas). This raises the price of these inputs, and firms and consumers adjust accordingly by means of various options: conservation, fossil fuel substitution, substitution of other fuels for fossil fuels, etc. From a modeling standpoint, the analysis is quite simple in terms of feeding direct changes in fuel prices into a model that contains mitigation cost functions by sector (see below).

However, the tax instrument does not map into a unique marketable permits counterpart, but instead several alternatives are possible (cf., Pezzey, 1992):

1. Auctioning the entirety of permits needed to achieve the U.S. abatement target.
2. Distributing permits assigned to the U.S. to firms and households free ofcharge, and letting them buy and sell freely in the world market.
3. Distributing permits to achieve U.S. abatement targets to firms and households and allowing for trading within the U.S. only.
   a. The sub-case where the government distributes the entirety of the permits freely and, hence, subsidizes CO_2 abatement.
   b. The sub-case where the government auctions a subset of permits to obtain the revenue it needs to purchase permits in the world market.

Depending on the case, the firm payouts and government revenues differ between the carbon tax and marketable permits approaches. Under the tax option, emitters face costs over the entire range of emissions. There are costs of mitigation up to a point and then a tax penalty for all carbon emitted (in terms of purchases of basic fuels). In the case of permits, when they are assigned free of charge, some emissions do not incur any costs. Hence, total expenditures differ between the two policy instruments, as does the ability to affect the cost of production and, hence, prices and other variables in the long run.

The differences can be illustrated by the use of some realistic numerical examples, based on the following assumptions:

- U.S. carbon emissions equal to 1.5 billion tons
- World abatement target equal to 20%
- Allocations based on the sovereignty criterion, i.e., permits equal to 80% of emissions for all countries
- U. S. efficient abatement target equal to 12.8%
- Permit price equal to \$40 per ton
- Linear abatement cost function, C, with the properties

Case 1 is essentially the same as a carbon tax in the short run and long-run. Firms and households will purchase permits so that an average of 88% of baseline CO_2 can continue to be emitted. This yields \$52.8 billion in revenue. Given our linear cost function, total abatement costs would be \$3.6 billion = $.5 \times 180 \times 10^6 \times \$40$.

Under Case 2, 1.2 billion permits would be distributed free of charge, and U.S. firms and households would purchase another 120 million permits (worth \$4.8 billion) on the world market. As in the previous case, and in the remainder of the cases, total abatement costs would be \$3.6 billion.

Under Case 3a, the government would need \$4.8 billion to go into the world market and purchase the additional permits. Spreading this revenue requirement over all permits would reduce the permit price to \$3.64; however, the value of the permits would still be \$40 per unit. A more reasonable approach would simply be to auction 120 million permits to the highest bidder (at \$40 per unit). For Case 3b, the government would allocate 1.32 billion permits free of charge and absorb the \$4.8 billion cost of buying permits in the world market. This, in effect, would be a large subsidy to carbon emitters in the U.S. (even larger than the total mitigation costs for the entire country).

Modeling Case 1 would be the same as that of a carbon tax--adding the auction price to the price of primary energy and then running these price changes through the model. However, under Cases 2 and 3, the effect on long-run average price is lower because emitters are only penalized for a portion of CO_2 released to the atmosphere.

The empirical estimation of abatement strategies for each sector of the economy can be complicated. We propose a pragmatic approach, which utilizes conservation levels and elasticity measures that capture the essence of most tactics in an overall mitigation strategy and which capitalizes on the availability of parameters in the literature.

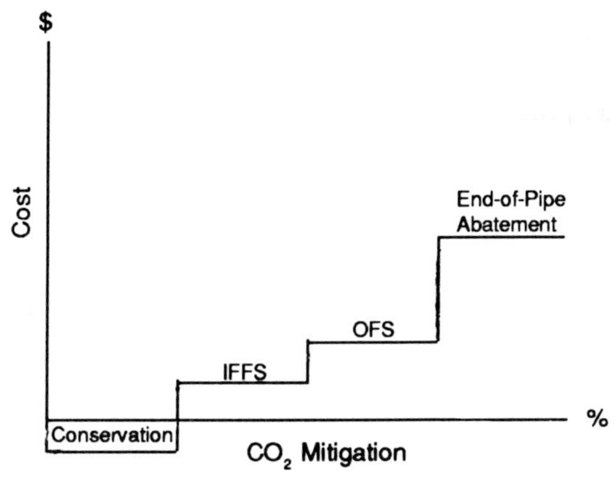

Figure 5.2 CO_2 cost function

The two most prominent CO_2 mitigation tactics are conservation and input substitution. We depict these responses in Figure 5.2 by a step function because of lack of sectorally detailed data on some mitigation tactics and the use of constant elasticities (which in some production functions imply constant input value shares). For our conceptual analysis at this point, the step function poses no problem, but it implies there will be some error in its empirical counterpart when compared with a (more realistic) continuous cost function. Note also that we are utilizing the notion of the KLEM production or cost function, where the acronym letters stand for Capital, Labor, Energy, and Materials, respectively. Also, note the function is doubly nested, meaning it allows for substitution across and within these aggregates with the energy subaggregate, including across and within the categories of fossil fuels and alternative energy sources.

The first step of the cost function refers to *no regrets* (costless or even cost-saving) conservation. This would be either a technological innovation or a move toward the efficiency frontier as a result of correcting a misallocation (e.g., eliminating energy-wasting practices).[6] For our empirical analysis, we could use existing parameters such as Manne and Richels' (1991) concept of autonomous energy efficiency improvements (AEEI). Their estimates have been rather small--on the order of 0.5% to 1.0% per year--though these could be modified by more optimistic estimates of costless conservation in the literature (e.g., OTA, 1990; NAS, 1990; and Lovins and Lovins, 1991), which are in the range of 20% (or more) overall that could be implemented by the year 2000.[7]

Another major form of conservation is price-induced, i.e., a response to changes in relative prices, as would be stimulated by our initial change in fuel prices as a result of a carbon tax or various types of permit regimes. All incur some costs unless the elasticity of substitution is infinite. Basically, we categorize these options under the headings below following Cline (1992):

- IFFS--Inter-fossil fuel substitution
- OFS--Other factor substitution
- PMS--Product mix substitution
- NFFS--Non-fossil fuel substitution

The final category of mitigation tactics, though limited in the near-term, refers to "end-of-pipe" options. These usually involve additional capital equipment and hence represent a type of negative technological change.[8]

Finally, we note some limitations of this approach, with these shortcomings also being applicable to most CGE and macro-forecasting analyses, since they utilize constant elasticity measures. Implicitly, each tactic would be used in sequence up to the permit price. This means that many tactics would be used up to their limit. For

[6] Such practices would include turning off unnecessary heating and lighting.

[7] Still another form of costless conservation would be the substitution of inputs under conditions of infinite elasticity of substitution. This may be possible, say, for multiple-fuel capable boilers and would result in a complete switch from one fuel to another in this case, and complete switches for other types of input substitutions, though such possibilities are very limited.

[8] This refers to the subset of new equipment purchases that result only in reductions in emissions and no cost savings. Of course, there are a large number of examples of process integrated technological changes that do lead to cost savings, but they are included in the various conservation tactics mentioned above.

example, there could be complete substitution away from coal in many sectors and substitutions away from fossil fuels in general (assuming this is the less costly of the various other tactics). Of course, there are physical limits to such substitution, i.e., there are no alternatives for gasoline and electricity in many processes. Moreover, elasticity measures are typically based on the historical range of experience. Our cost function goes beyond that, and, without some explicit constraints, is likely to be overly flexible. Some realism could be incorporated through external estimates of no-regrets conservation (e.g., engineering estimates of conservation practices relating to each fuel by sector of use).

Still, these limitations aside, one can readily foresee the implications for energy use in general and for those fuels with the highest unit emissions (coal). Their cutbacks are likely to dominate mitigation strategies in most sectors.

5.5 GENERAL EQUILIBRIUM IMPACTS

Ideally our analysis of sectoral and economy-wide impacts would be undertaken with a computable general equilibrium model. The model developed by Lin (1991) is being adapted for this purpose. In the meantime, we present some experiments with the use of an input-output (I-O) model, which provide insights into key issues, though in a relatively more restrictive and less elegant manner. We utilize the 1990 IMPLAN U.S. I-O Table (U.S. Forest Service, 1993), which is a non-survey based update of the official U.S. Department of Commerce 1982 "benchmark" table. The table is in an "industry by industry" format, and we have aggregated it to the 2-digit Bureau of Economic Analysis Industry Classification levels (82 sectors).

There are, however, some relative advantages to the following I-O simulations. Since they allow for limited price responses and substitution possibilities, they might serve as a useful estimate of the short-run response, or otherwise as upper bound impacts.[9] Second, since the analysis is performed in a recursive manner, and thus separates the responses to various tactics embodied in the direct mitigation cost function.

First, let us note two basic considerations. We assume permits are assigned according to cases discussed in the previous section in a way that enables us to view the effect of permit prices, in the short run, as equivalent to a carbon tax imposed on individual fuels. The direct cost increases, based on our $38.35 estimate of the

[9] The short run may be relatively long in terms of clock time given the gestation period in building large power plants and given long-term contracts prevalent for many fuels, especially coal.

equilibrium permit price and data adapted from Montgomery (1992) on fuel price increases, are as follows:

Fuel	Cost per MM BTU[10]	Cost per physical unit	Cost as % of producer price
Natural Gas	$0.58	$0.61/TCF[11]	34.5%
Oil	$0.77	$4.99/barrel	24.9%
Coal	$0.96	$23.01/ton	88.2%

Also, we assume permits are assigned as a fixed proportion of baseline emissions across all sectors.[12] Recall from the Coase Theorem that the assignment of permits does not really matter to the efficient mitigation strategy intranationally as well. That is, within the U.S., there will be a unique pattern of abatement levels across sectors following trading, no matter how the permits are initially distributed.

5.5.1 Basic Results

We first examine the conservation tactic. This is often labeled the *no-regrets* strategy, since many analysts believe an extensive array of costless conservation measures exist. We specify an extreme case in which the 12.8% cutback in CO_2 emissions (corresponding to the optimal abatement response of the U.S. to a global 20% target) is met entirely by conservation in the purchase of all fuels in all intermediate (goods used to produce other goods) sectors and final demand sectors (personal consumption, investment, and government spending).

The simple multiplier approximations of general equilibrium effects are shown in Table 5.4, including total gross output and employment effects on individual fuels and on the economy as a whole.[13] Obviously the impacts are uneven across sectors and

[10] Million British thermal units.

[11] Thousand Cubic Feet

[12] This corresponds to the Sovereignty equity criterion, though the underlying principles differ in the intranational context. This type of allocation is typical in environmental policy, and is often (and mistakenly) justified by reference to Horizontal equity (see Rose, 1992).

[13] These experiments were undertaken using the latest benchmark U.S. Input-Output Table updated to 1990 in the IMPLAN System and further updated by us to the year 2000 (see U.S. Forest Service, 1993). We remind the reader of the inherent limitations of I-O tables. Note that our analyses are performed using a closed version of the I-O table, and therefore we are able to calculate direct, indirect, and induced (Type II multiplier) effects. For the sake of brevity, we will refer to both types of higher-order effects as *indirect*.

are most heavily borne by the energy industries. Output and employment reductions in the fossil energy sectors range from a high of 30.6% in oil and gas extraction to a low of 16.0% in refined petroleum. The economy-wide impact is a reduction of only 2.3%. This may seem small at first, but note that in absolute terms it would mean the loss of 2,500,000 jobs. The results indicate that second-order impacts will achieve some additional mitigation; so that there is actually a need for direct conservation cutbacks of less than 12.8% because of the approximately 10% energy use reductions achieved through direct and indirect output declines. The direct reductions necessary may only be about two-thirds of those originally stipulated and dramatize the potential importance of general equilibrium feedbacks into module 5.1. The only good news here is that the energy sectors can achieve emission reductions without having to undertake any specific mitigation other than their own output reduction (if baseline emissions are defined in a standard way).[14]

Table 5.4 Economic impacts of the basic conservation response, year 2000

	Gross Output (million $1990)		Employment (person years)	
Sector	Level	Percent	Level	Percent
Coal	-$5,540	-26.4	-54,443	-26.4
Oil/gas	-81,979	-30.6	-227,750	-30.6
Refined petroleum	-35,825	-16.0	-24,916	-16.0
Electric utilities	-36,580	-16.7	-117,671	-16.7
Gas utilities	-21,871	-20.9	-47,273	-20.9

The 12.8% CO_2 reduction target could be implemented by mandating conservation as opposed to a market incentive approach using permits or taxes. The command and control approach could leave prices unchanged (assuming again that conservation up to the target level is costless). The incentive approach, on the other hand, would raise the price of basic fuels and essentially inspire what amounts to price-induced conservation. The analysis of our first tactic corresponds more closely to the command and control case, or to an incentive approach where each sector's optimal response is a 12.8% direct reduction in energy intensity. Actually, the latter is highly unlikely for two reasons. First, the optimal response will most assuredly be differential

[14] Of course, in the long run, some offset to the output and hence energy cutbacks will be achieved through the direct and indirect inputs to the production of energy-saving equipment.

levels of conservation across sectors. Second, it will also most definitely call for differential cutbacks in the use of fuels given their different CO_2 emission rates. Because of the difficulty of modeling such non-linear and price-related responses with an I-O model, we do not perform the analysis of this case in this chapter. We can, however, draw some conclusions about the results. They would yield lower overall fuel reductions than the uniform conservation case but with a much higher percentage of the reduction borne by the coal industry.

Table 5.5 Economic impacts of the product mix response, year 2000

Sector	Price change (percent)	Gross output (million $ 1990)	Employment (person years)
Coal	92.6	-$ 727	-7,144
Oil/gas	31.6	-6,664	-18,531
Refined petroleum	16.4	-5,621	-3,909
Electric utilities	8.0	-3,380	-10,873
Gas utilities	15.2	-1,815	-3,923
Total	2.9[a]	-$191,836	-1,476,767

[a] Weighted Average

Price effects can be analyzed, to some extent, in the context of our I-O model. We can calculate a set of energy price multipliers of the impacts of the permit policy on energy and all other sectors. These are presented for individual fuels and the economy as a whole in the first column of Table 5.5. We then utilize these results to simulate changing product mix as a mitigation tactic. We multiply the price changes by a set of final demand elasticities to determine the direct and indirect impacts in various sectors.

The results are presented in Table 5.5 in terms of direct and indirect reductions in output and employment for energy by fossil fuel type and for the economy as a whole. They show this tactic can have a significant impact across many sectors (not just energy). The economy-wide price increase of 2.9% results in a modest reduction in energy industry output, in comparison to Table 5.4, but a sizable economy-wide output reduction of 1.4%. As with our first tactic, it translates into "mitigation" of greenhouse gas emissions via cutbacks in output, and thus reduces the number or intensity of other tactics that need to be utilized. In essence, if one is modeling a market incentive approach, the results presented in Tables 5.4 and 5 should be combined, yielding an economy-wide gross output reduction of 3.7% and energy industry output reductions as high as 33% (oil and gas extraction). Again, we note that the emission reductions stemming from the general equilibrium responses are much larger than those initially required so that the combined results would only be relevant if a very myopic policy target were to be implemented.[15]

Overall, these simulations indicate some general equilibrium feedback effects that have a bearing on the estimate of the cost of mitigation presented in the previous section. These effects, whether measured in the manner of our analysis or as consumer/producer surplus (dead-weight) losses, cause the abatement cost curve initially used in module 5.2 to shift upward significantly. Ideally, these results would also be fed back to module 5.1. This is an ambitious undertaking in which work has only begun.

5.5.2 Sensitivity Analysis

An issue that arises in the previous analysis is the manner in which conservation is costless. If the fuel input per unit of output decreases by 12.8%, might this set into motion production or price responses that could offset in part, in full, or more than fully, the direct and indirect reductions stemming from energy conservation itself.

There are several possibilities. First, if the conservation is totally due to inefficiencies, then there is a cost reduction equal to 12.8% of fuel input costs, which are in the range of 1.0% to 5.0% in most sectors. This could be passed on entirely as a price reduction, though this is unlikely. Any reduction in the direct price of each good, however, would potentially reduce the price of all goods in which it is a direct and indirect input, thereby generating price multiplier effects analogous to those discussed in the previous subsection. This would increase the purchasing power of consumers and lead to direct and indirect increases in output and employment.

Second, higher profits that would also be associated with the cost reduction could be paid out as higher returns to capital. A similar outcome would result if conservation stems from substitution of capital for energy. Either of these would result in a direct and indirect stimulus to the economy.

Third, the cost decrease would raise the productivity of labor and warrant higher returns to this primary factor. A similar positive stimulus would ensue from the substitution of labor for energy.

Fourth, if the conservation is brought about through the substitution of energy-saving equipment, then there is an expansionary stimulus due to the manufacturing

[15] Note that this analysis still only analyzes the response to price rises in final goods and services. Adjustments to increases in input prices would also set off a set of responses in the production of intermediate goods likely further accentuating these results.

of that equipment. There may also be expansions due to increases in the demand for other direct inputs (the general class of materials) substituted for energy.

Table 5.6 Output impacts of conservation cost-savings respending alternatives, year 2000[a]

Sector	Capital, 50%	Labor, 75%	Labor, 100%	Combustion equipment	Price reduction
Coal	$ 124	$ 188,000	$ 250,000	$ 298,000	$ 152,000
Oil/Gas	1,870	2,814	3,763	3,026	2,647
Refined Petroleum	1,895	2,851	3,813	2,711	3,342
Electric Utilities	2,013	3,029	4,051	3,894	2,307
Gas Utilities	975,000	1,466	1,961	1,658	1,204
Economy-wide	$110,011	$165,518	$221,364	$261,275	$165,250

[a]In millions of 1990 dollars.

Table 5.7 Output impacts of the conservation response net of respending alternatives, year 2000[a]

Sector	Capital, 50%	Labor, 75%	Labor, 100%	Combustion equipment	Price reduction
Coal	-$ 5,416	-$ 5,352	-$ 5,290	-$ 5,242	-$ 5,388
Oil/Gas	-80,109	-79,165	-78,216	-78,953	-79,332
Refined Petroleum	-33,930	-32,974	-32,012	-33,114	-32,483
Electric Utilities	-34,567	-33,551	-32,529	-32,686	-34,273
Gas Utilities	-20,896	-20,405	-19,910	-20,213	-20,667
Economy-wide	-$210,165	-$154,658	-$98,812	-$58,901	-$154,926

[a]In millions of 1990 dollars.

To examine these possibilities, we ran a set of sensitivity tests. The simulations were run by assigning some proportion of the cost savings to each of the four alternatives. For example, since some portion of profits is subject to corporate income taxes, another portion is retained earnings, and because a portion of what is paid out as dividends is saved, we only used 50% of the cost-savings as the direct stimulus in each sector attributable to the direct capital offset. In the case of labor, we examined an upper-bound estimate of 100% of the cost savings and a 75% level, the latter reflecting individual income taxes and savings leakages. For the equipment case, we assigned the entire cost savings level to the combustion sector as a proxy for all possible energy-saving equipment (and material substitution). In the price case, we simply reduce sectoral prices by 12.8% of total direct fossil fuel costs.

The output impacts of conservation cost-savings respending alternatives is presented in Table 5.6. They range from $110 billion for the capital spending alternative to $261 billion for energy-saving combustion equipment. Still, all five alternatives fall short of offsetting the negative effects on output for the economy in general and the energy sectors in particular (see Table 5.7). The variation in the energy sector offsets is less than 5% between alternatives.

Table 5.8 Output impacts of the conservation response net of an equal mix of respending alternatives, year 2000[a]

Sector	Conservation gross response (1)	Respending mix effect (2)	Conservation net response (3) = (1) + (2)
Coal	-$ 5,540	$ 191,000	-$ 5,349
Oil/Gas	-81,979	2,589	-79,390
Refined Petroleum	-35,825	2,700	-33,125
Electric Utilities	-36,580	2,811	-33,769
Gas Utilities	-21,871	1,326	-20,545
Economy-wide	$320,176	$175,514	-$144,662

[a]In millions of 1990 dollars.

Table 5.9 Output impacts of the conservation response net of a predicted mix[a] of respending alternatives, year 2000[b]

Sector	Conservation gross response (1)	Respending mix effect (2)	Conservation net response (3) = (1) + 2
Coal	-$ 5,540	$ 209,000	-$ 5,331
Oil/Gas	-81,979	2,599	-79,380
Refined Petroleum	-35,825	2,557	-33,268
Electric Utilities	-36,580	2,998	-33,582
Gas Utilities	-21,871	1,369	-20,502
Economy-wide	$320,176	$187,142	-$133,034

[a]Weights: capital (.3), labor (.2), combustion equipment (.4), and prices (.1).
[b]In millions of 1990 dollars.

There are two reasons the offsets are smaller than the $320 billion direct conservation output reduction. First, three of the offset alternatives suffer from initial leakages for a variety of reasons (retained earnings, taxes, and savings). Second, the multiplier impacts of energy sector production would appear to be greater than the offsets. Direct comparisons of these multiplier impacts can be made with the labor (100%) and combustion equipment alternatives.[16]

The overall outcome of our experiment depends on which respending alternatives are operative. Of course, the most likely response will be a mix of alternatives. The results of two such simulations are presented in Tables 5.8 and 5.9. The first gives each of the four responses equal weight (we utilized only one of the labor alternatives the 75% respending version). The result is a $145 billion *decrease* in total gross output over the whole economy in the year 2000, still a 1.04% decrease over baseline. Table 5.9 presents the results of an assignment of differential weights based on the judgment of the authors. For each of the energy sectors, the results are hardly different from Table 5.8, and, for the economy as a whole, the net effect is reduced slightly to a $133 billion reduction in total gross output, a decrease of 0.96%. This result indicates that there is likely to be an overall economic loss to the economy from pursuit of a 12.8% energy conservation reduction. Even if we cut this target in half to account for conservation accompanying output reductions, the output impacts are likely to be negative, given our use of a linear model and because none of the offset effects is as large in absolute value as the conservation stimulus.

5.6 CONCLUSION

We have analyzed three crucial aspects of the implementation of the marketable permits approach to global warming policy. First, we demonstrated that, beginning at the global level, there is a unique equilibrium permit price that can bring about the most cost-effective response among nations, irrespective of how permits are initially assigned. Second, we showed how to translate this global permit price into a national policy variable in a manner similar to modeling the imposition of the carbon tax. Third, we simulated the partial equilibrium and general equilibrium impacts of two of the possible industry response strategies. Our results indicate that general equilibrium effects may be helpful in fine-tuning global warming policy to minimize

[16] Note that we converted the combustion equipment expenditures into annualized equivalents, and thus our simulation results represent an average of impacts over the life of the equipment. To the extent that the equipment will have to be manufactured and put into place in the early stages of the policy time horizon, this offset alternative is actually likely to follow a cyclical path, significantly larger in the first few years and somewhat smaller in the remaining years.

the negative effect on the U.S. economy. Still, it appears that global warming mitigation strategies will have a slightly negative impact on the U.S. economy as a whole and a very strongly negative impact on the energy sectors even under the so-called *no regrets* strategy.

We should note that the above conclusion did not explicitly take the benefits of global warming policies into account. Our analysis indicates that the projected total benefits of CO_2 mitigation will exceed the costs in the U.S. under most permit allocation rules. There is considerable uncertainty over the gross benefits, but it is still unfortunate that national accounts omit such environmental factors. Still, these additional considerations are unlikely to offset the negative impacts on the energy industries.

REFERENCES

Barrett, S., M. Grubb, K. Roland, A. Rose, R. Sandor, and T. Tietenberg (1992) *Combating Global Warming: A Global System of Tradeable Carbon Emission Entitlements*. UNCTAD, Geneva, Switzerland.

Cline, W. (1992) *The Economics of Global Warming*. Institute for International Economics, Washington DC.

Coase, R. (1960) The Problem of Social Cost. *Journal of Law and Economics*. 3: 1-44.

Energy Information Administration (1988) *International Energy Annual*. Washington DC.

Epstein, J. and R. Gupta (1991) *Controlling the Greenhouse Effect: Five Global Regimes Compared*. The Brookings Institution, Washington DC.

Intergovernmental Panel on Climate Change (1991) *Climate Change: The IPCC Scientific Assessment (revised)*. Cambridge University Press, New York.

Lin, S. (1991) *A Computable General Equilibrium Model for U.S. Mineral Policy Analysis*. unpublished Ph.D. dissertation, Department of Mineral Economics, West Virginia University, Morgantown.

Manne, A. and R. Richels (1991) Global CO_2 Emission Reductions: The Impacts of Rising Energy Costs. *Energy Journal*. 12: 87-107.

Montgomery, W.D. (1992) *The Carbon Tax, The Environment, and Economic Growth*. paper presented at the Symposium on Enhancing Environmental Quality Through Economic Growth, Washington DC.

Nordhaus, W. (1991) The Cost of Slowing Climate Change: A Survey. *Energy Journal*. 12: 35-67.

Nordhaus, W. (1993) Rolling the DICE: The Optimal Transition Path for Controlling Greenhouse Gases. *Resource and Energy Economics*. 15:27-50

Pezzey, J. (1992) The Symmetry between Controlling Pollution by Price and by Quantity. *Canadian Journal of Economics*. 25: 983-91.

Rose, A. (1992) Equity Considerations of Tradeable Carbon Entitlements. in S. Barrett et al., *Combating Global Warming*.

Rose, A. and B. Stevens (1993a) The Efficiency and Equity of Marketable Permits for CO_2 Emissions. *Resource and Energy Economics*. 15: 117-46.

Rose, A. and B. Stevens (1993b) *Equity Aspects of the Marketable Permits Approach to Global Warming Policy*. paper presented in an Association of Environmental and Resource Economists Session at the Annual Meetings of the Southern Economic Association, New Orleans.

Task Force on the Comprehensive Approach to Climate Change (1991) *A Comprehensive Approach to Addressing Potential Climate Change*. U.S. Department of Justice, Environment and Natural Resources Division, Washington DC.

Tietenberg, T. (1992) Relevant Experience with Tradeable Entitlements in S. Barrett et al *Combating Global Warming*.

U.S. Forest Service (1993) *IMPLAN Modeling System*. Rocky Mountain Experiment Station, Ft. Collins.

Whalley, J. and R. Wiggle (1991) Cutting CO_2 Emissions: The Effects of Alternative Policy Approaches. *Energy Journal*. 12: 109-24.

World Bank (1991) *World Development Report 1991*, Oxford University Press New York.

Chapter 6 IMPLEMENTATION ISSUES FOR GLOBALLY TRADEABLE CARBON ENTITLEMENTS

Tom Tietenberg[1]

ABSTRACT

Tradeable permit approaches to pollution control have been successful implemented in the United States since 1977 and are now being used in several other countries. One proposed new application of this approach involves its use in controlling global warming. While tradeable permits have enjoyed some considerable success in the various domestic contexts, by no means does that guarantee their success in the international context. In this chapter the implementation issues associated with crafting a workable international tradeable entitlement system for greenhouse gas emissions are explored using the UNCED Convention as a point of departure. Drawing upon the experience gained from both the existing applications of tradeable permits and the monitoring and enforcement of other international agreements, this analysis focuses on the certification, monitoring, and enforcement functions associated with a transferable permit system. Special attention is given to assuring that the system serves current purposes while facilitating further development to a more mature, comprehensive system.

6.1 INTRODUCTION

The United Nations Framework Convention on Climate Change, opened for signature in Rio at the Earth Summit during June, 1992, calls for reducing greenhouse gas emissions to 1990 levels by the end of the decade.[2]

How can those reductions be achieved? One clue is provided in the Convention itself. Principle 3 under Article 3 states "measures to deal with climate change should be cost-effective so as to ensure global benefits at the lowest possible cost." One

[1] This work was originally commissioned by and prepared for the United Nations Conference on Trade and Development Secretariat. The material is copyrighted by the United Nations. Permission has been granted to the author for its use in this volume.

[2] The Convention will enter into force three months after the 50th nation has ratified it. As of September, 1993, 36 nations had deposited instruments of ratification.

approach which is, in principle, consistent with the cost-effectiveness criterion involves tradeable entitlements.

In principle the case for a tradeable entitlements system is strong.[3] In the short run it offers the possibility of reaching the environmental goals at a lower cost than would be possible if each country were limited to reduction options within its own borders. Because it allows the separation of the financing of control from the implementation of control, it facilitates transboundary cost sharing (an item of particular importance to both the developing countries and the transition economies of Eastern Europe). Tradeable entitlements also facilitate the mobilization of private capital for controlling global warming; private capital is likely to be a critically important component of any effective global warming strategy as long as public capital remains insufficient. Finally, and perhaps most importantly, tradeable entitlements facilitate the development and implementation of new methods of controlling global warming. By offering greater flexibility in how targets are achieved (as well as by providing economic incentives for the adoption and use of unconventional approaches), tradeable entitlements significantly lower the long run cost. Lowering long run cost, in turn promotes a greater international acceptance of the idea of limits on greenhouse gases and reduces the difficulties associated with assuring compliance.

How an international tradeable entitlements system might be fashioned from the current situation, however, is not at all obvious. Facilitating immediate progress without jeopardizing a smooth evolution to a more mature, comprehensive system requires careful attention to the implementation details. Beginning with a simple system with the understanding that it could evolve to a more complete system presumes some knowledge about the more complete system. Otherwise the process could well proceed down a path which might ultimately lead in an incompatible, and difficult to change, direction.

The initial steps laid out by the convention were necessarily hesitant. Since the Convention is comprehensive neither in terms of administrative details nor participants, achievement of its objectives depends on an institutional structure which is still being crafted. Appropriately defined institutions and administrative procedures facilitate a natural evolution of this simple system by opening the agreement to more signatories, by facilitating the expansion of coverage to all greenhouse gases, and by providing the foundation for tighter certification, monitoring and enforcement procedures. Recognizing that an international system for controlling greenhouse gases must evolve slowly over time, this report evaluates a particular evolutionary path.

[3] We only summarize here a few major points. For a complete discussion of the advantages of this approach see United Nations Conference on Trade and Development (1992).

In this chapter the various implementation issues associated with establishing a tradable entitlement system for greenhouse gases are examined using the UNCED Convention as a point of departure. Using this particular point of departure establishes certain parameters of our inquiry. In accordance with the convention we shall focus, at least initially, on carbon dioxide rather than on the totality of all greenhouse gases and we recognize that initial limitations on carbon will affect only the industrialized nations with the developing countries expected to join subsequently.[4]

6.2 DEFINING THE EVOLUTIONARY STAGES

The initial steps laid out by the convention were necessarily hesitant. Since the Convention is comprehensive neither in terms of pollutants nor participants, achievement of its objectives depends on an institutional structure which is still being crafted. Appropriately defined institutions and administrative procedures facilitate a natural evolution of this simple system by opening the agreement to more signatories, by facilitating the expansion of coverage to all greenhouse gases, and by providing the foundation for tighter certification, monitoring and enforcement procedures. Recognizing that an international system for controlling greenhouse gases must evolve slowly over time, this chapter envisions a particular evolutionary path.

6.2.1 The Joint Implementation Stage

The first stage has already been established by the Framework Convention. It involves a set of nations (listed in Annex I of the convention) which have accepted a goal of stabilizing greenhouse gas emissions at 1990 levels by 2000. Since the *laissez-faire* outcome would normally involve increasing emissions, emission reductions will probably be required to meet the goal. Other nations have subscribed to the general principles of the convention, but not the stabilization goal.

Two aspects of this agreement are sufficiently important to the design of a greenhouse gas entitlement system that they deserve emphasis. First, stabilization is stated as a goal, not a requirement. Second, it applies only to the listed nations; other nations are under no equivalent constraint.

[4] Principle 1 under Article 3 states "the developed country parties should take the lead in combatting climate change and the adverse effects thereof." This is implemented in practice by delineating, under Article 4, certain special obligations to the parties listed in Annex 1 to the convention. The parties listed in Annex 1 are developed countries and several Eastern European nations.

As the Annex I nations seek to reduce emissions sufficiently to meet the goal, they have two choices--they can either find the reductions within their own borders or they can seek them in other nations. The possibility for a transborder approach to emission reductions is clearly anticipated in the convention by section 2(a) of Article 4 which reads "these Parties may implement such policies and measures jointly with other parties and may assist other Parties in contributing to the achievement of the objective of the convention and, in particular, that of this subparagraph."

The most natural vehicle for a transborder approach to emission reduction is an offset version of a tradeable entitlement policy. Under this policy a nation could satisfy its emission reduction requirements with a combination of domestic emissions reductions and international offsets. Offsets are emission reduction credits which, once certified, can be transferred across borders. They could be acquired either from other Annex I nations which voluntarily reduce emissions more than required by the stabilization goal or from other nations. Although the joint implementation stage bears little resemblance to an actual greenhouse gas entitlement market[5], it serves the very important purpose of launching the process and providing opportunities for the various supporting administrative institutions to "learn by doing" as they mature in their assigned roles. During this stage, for example, *ex ante* estimates or emission reductions could be validated by *ex post* monitoring, using this experience to produce superior *ex ante* estimates in the future.

6.2.2 The Initial Entitlement Stage

In the second phase of this process the stabilization goal would be replaced by specific emission requirements for each of the signatory nations. The resulting levels of authorized emissions would then provide a suitable basis for the assignment of actual emission entitlements for carbon dioxide among the signatory nations.[6]

[5] This was reinforced by the August, 1993 meeting in Geneva at which an agreement was reached that, until the year 2000, joint implementation projects should be considered additional to and separate from the commitments already agreed to by the developed nations in the Convention.

[6] Although trades in other greenhouse gases would be possible in this stage subject to the certification procedure, we shall focus in this stage on a tradable carbon entitlement, the most likely candidate for the first formal system. Most analysts have noted that sources and gases other than fossil fuel CO_2 are not sufficiently able to be monitored for inclusion in an entitlements system at present. For a review see D.G. Victor (1992). Not counting the Clinton administration plan announced in October, 22 of the remaining 23 OECD nations have adopted targets (Turkey has not). All of these except Canada's and Australia's are in terms of carbon emissions only. (The Netherlands has adopted two targets, one for carbon and the other for all greenhouse gases). For a review of all but the U.S. target see International Energy Agency (1992).

Although the normal presumption for Annex I nations would be that they would receive entitlements equal to their 1990 emission levels, other possibilities exist. If, for example, scientific evidence suggests the need for more drastic reductions (as was the case with the ozone depletion gases controlled by the Montreal Protocol), fewer entitlements could be granted. If the problem of global warming is found to be less harmful than originally thought, more entitlements could be granted.

Because certified entitlements would be freely transferable among signatory nations, the informal and largely experimental trading system of the joint implementation phase would be replaced in the initial entitlement stage with a more formal (and, therefore, a more efficient) version. Organized exchanges would be established to lower transactions costs while a reporting network would provide a public means of accountability. Nonsignatory nations could participate in the process either by signing the agreement (and thereby immediately receiving an authorized level of tradeable entitlements) or by creating offsetting reductions which were certified by an agency designated by the Conference of Parties.

This second stage would offer significant opportunities and incentives for developing nations to join the agreement. In return for their acceptance of ultimately binding limits on greenhouse gas emissions as negotiated with the Conference of Parties,[7] nations could become full participants in the agreement with complete access to the tradeable entitlement market. Full participation in that market would provide access to large amounts of private capital as well as to an efficient means of technology transfer.[8]

The alternative, remaining a nonsignatory nation and participating only through the occasion sale of specially created offsets, would offer less flexibility and less access to private capital.[9]

[7] The assumption in most discussions of this subject is that developing countries would receive an allocation of entitlements well in excess of current needs. (See, for example, A. Rose and T. Tietenberg (1993). The resulting surplus could be leased to industrialized nations until such time as the local development program created a need for them. The private capital generated by these leases could be used to pursue high priority development goals (as defined by the developing country, not by any international lending agency).

[8] Why, some might ask, is private capital necessary when the Global Environmental Facility and an environmental fund (created, but yet to be implemented) within the climate change agreement are prepared to dispense public funds? Private sources of capital seem important for two reasons: (1) public funds are not likely to be sufficient to fill the need and (2) the private capital raised by leasing excess entitlements comes with fewer strings attached.

[9] The complete absence of any tradable entitlement system would be the worst possible outcome for developing countries because they would only have access to the public funds provided by international agencies.

The second stage does not presume any particular domestic strategy for achieving reductions. The reductions could be achieved by traditional command and control regulations, transferable emission permits for domestic sources, or various kinds of "green fees". Combinations of these policies are also possible as long as the resulting emissions do not exceed the amounts authorized by the allocated entitlements.[10]

6.2.3 The Culmination Stage

The second stage primarily involves trades in carbon entitlements between sources within signatory nations. Developing countries would be allowed to "opt in" by means of the offset certification procedure or they could become a signatory. In any event the market would be limited in terms of both the number of participants and the degree of participation.

The final stage involves an entitlements market which is completely inclusive in terms of both participation of all nations and coverage of all greenhouse gases. Our objective in defining this stage is to highlight the conditions which would be necessary to achieve this final stage as well as to discuss the evolution of both procedures and institutions which would be necessary to achieve them.

6.3 DESIGNING A GREENHOUSE GAS TRADING SYSTEM

Key Characteristics of the System

The List of Controlled Substances. Creating a workable transferable permit system must begin with agreement on the substances that will be subject to the entitlement system. What gases would be involved?

Though a comprehensive approach incorporating all greenhouse gases allows the global warming objective to be achieved with the widest selection of options, reducing the cost of achieving any given target[11], in the current environment a comprehensive approach is not feasible. In the harsh realities associated with implementing a workable global system of control, practical solutions dominate less practical solutions,

[10] It could also reserve some proportion, known typically as the growth increment, to accommodate expected future growth. This growth increment could be used to encourage the development of new employment generating activities by reserving entitlements for this purpose.

[11] For a strong articulation of this point of view see the discussion in Stewart and Wiener (1991) and U. S. Department of Justice (1991).

even those which may appear superior in theory.[12] Carbon dioxide is much easier to monitor and to document than other gases such as methane.[13] Since an effective transferable entitlement system presumes a capability to calculate the effects of various control strategies on emissions, incorporating all sources of all greenhouse gases from the beginning would run the significant risk that fraudulent trades could result--trades that appear to involve equivalent reductions in greenhouse gases, but in fact do not.

Fortunately, starting with a system based on carbon dioxide does not preclude its subsequent expansion to include other greenhouse gases. The decision to incorporate other gases could be made on a case-by-case basis. Reductions in other greenhouse gases would be tradeable, contingent on a demonstration that the effect of the proposed reductions on global warming would be equivalent, even considering all of the associated monitoring and enforcement problems. Some sources of other greenhouse gases can already meet this test; others can not. While those meeting the test could be incorporated immediately into the system, incorporation of the others would await a demonstration of this capability.

The Entitlement. Though as a practical matter entitlements can be imposed at any stage in the life cycle of a greenhouse gas--extraction or production, consumption or emission--, all other things being equal entitlements defined in terms of emissions would be preferred; this would produce the most cost-effective outcome. With an emissions target the tendency of the market to seek the least cost means of control would be focused on reducing emissions, which, of course, is the objective.[14] But all other things are never equal, so intervening in other stages of the life cycle can become a viable, if less desirable, alternative.[15] Different entitlement systems could be established for the case of carbon dioxide emissions from fossil fuels. Requiring entitlements for fossil fuels as they are extracted, as they move through the wholesale

[12] Richard Elliot Benedick, one of the negotiators of the Montreal Protocol, also suggests that an early insistence on a comprehensive approach could be quite counterproductive by "bogging down the negotiators and unnecessarily prolonging the entire process." See Benedick (1991,11).

[13] A comprehensive treatment of this point of view is presented in Swart (1991). Victor (1991b) also suggests that a comprehensive approach is impractical, given the uncertainty of monitoring gases other than CO_2.

[14] If the ratio of emissions to produced substance is controllable, then focusing on production provides no incentives to reduce this ratio, even when that may be the preferred method of emission reduction. This problem has arisen with the Montreal Protocol, since it controls production and consumption rather than emissions. This structure provided inefficient incentives for signatory countries to test suggested technologies or to develop new technologies for the recovery and destruction of the controlled substances. See Bohm (1991).

[15] In the Montreal Protocol, for example, the allowances of controlled substances are defined in terms of production and consumption rather than emissions.

distribution channels, as they are sold to final consumers or as the emissions are injected into the atmosphere by the combustion source are all possible points of control.

To facilitate a market in entitlement transactions the entitlement should be defined in terms of a homogeneous unit such as a ton of carbon dioxide emissions. To the extent that the entitlement involves another gas or an activity other than emission, the results must be translated back into a "ton of carbon dioxide equivalent".

Entitlement Duration. Entitlements would be dated in annual terms--1991, 1992, etc. and could be used during the specified year or after, but not before. Each entitlement would allow a specific amount of emissions. Once the amount authorized by the entitlement had been emitted, the entitlement would become worthless.

Domestic Complementary Policies. How individual signatory governments would meet their authorized emission target would be left up to them. Some might choose to use a traditional regulatory system, while others might choose to use emission charges. The effective functioning of a tradeable carbon dioxide entitlement system does not presume or require any particular choice.

The Private Trading Variant. Transferable entitlement systems could be designed where all trades took place exclusively between governments, exclusively between private parties, or with some mix of the two. Exclusive government trading could be initiated whenever actual greenhouse gas emissions were lower than allowed by the treaty in some countries and higher than allowed in others. Tradeable entitlements could be created by a generous treaty definition of "allowable" emissions, by certified reductions in either controlled or uncontrolled greenhouse gases, or by certified increases in greenhouse gas absorption capabilities.

Governments could choose to facilitate the participation of trades by private organizations by allocating their entitlements among the various competing sources within its borders.[16] In theory entitlements could either be auctioned off, with the sources purchasing them from their respective governments at the market-clearing price, or distributed to each source on the basis of some allocation rule (typically historical use). Theory suggest that either approach will ultimately result in a cost-effective allocation of the control responsibility among the various polluters as long as they are all price-takers, transactions costs are low, and the entitlements are fully transferable. *Any* free distribution of entitlements among existing sources is compatible with cost-effectiveness because the after-market in which firms can buy or sell entitlements eliminates any excess costs associated with the initial allocation.

[16] It could also reserve some proportion, known typically as the growth increment, to accommodate expected future growth. This growth increment could be used to encourage the development of new employment generating activities by providing entitlements at below-market prices.

This highly significant result implies that under the right conditions governments can use this initial allocation of entitlements to pursue within-country distributional goals without interfering with cost-effectiveness.

Both free distribution and the auction market have significant disadvantages. A main disadvantage of the free distribution approach is that it does not generate any revenue for the government. A main disadvantage of the auction approach is that it raises the financial burden of the polluting firms, a significant deterrent in an increasingly competitive global market.[17]

The middle ground between these two domestic distribution approaches can be exploited to produce a workable alternative. The government could, for example, allocate the entitlements on a grandfathered basis, but charge an annual fee to those receiving them.[18] Because this fee would be lower than the auction price, it would put less of a financial burden on the polluters. Yet the ability to collect an annual fee would make it possible for governments to derive some revenue from the system, revenue that could be used to bolster monitoring and enforcement efforts. The ability to collect this revenue could well serve even to increase the attractiveness of becoming a signatory nation.

Once completed, the initial within-country allocation of entitlements would serve as a baseline for trades of controlled substances. The entitlements would define allowable emissions of greenhouse gases; any certified reductions from that baseline would entitle the bearer to sell the excess entitlements to another party seeking to justify a larger allowable emission level, than the one to which they would otherwise be authorized.

Entitlement Supply. Once the initial allocation has been completed, it is necessary to decide the conditions under which the supply of entitlements could be expanded. The general rule is that supply increases should not undermine the goal of global warming.

One possible source of permit supply is provided by participants who undertake activities which absorb or destroy greenhouse gases. Currently, for example, it is possible to destroy CFC's so that they do not contribute to global warming. By allowing the certified destruction of greenhouse gases to create entitlements, the development of destruction techniques and capability would be encouraged.

Another source of entitlement expansion could come about from the development of new greenhouse gas absorption capabilities. The most obvious example is the

[17] See the support for this statement in Tietenberg (1990).

[18] An alternative approach, suggested in Kägeson (1991), envisions allocating 80% of the permits by a grandfathering approach and the remaining 20% by an auction approach. His study focused on the limited objective of implementing a greenhouse gas permit system only among the EC countries.

development of new biomass plantations to sequester carbon through photosynthesis or, perhaps, saving forests scheduled for burning (thereby preventing the liberation of the sequestered carbon). Allowing these offsetting activities to create new entitlements provides a powerful inducement for them to be undertaken and, therefore, is a desirable objective. However, this path is not without its problems.

Generally, only offsetting activities which go beyond some prespecified baseline qualify for entitlements. The baseline establishes the "normal" or "expected" amount of planting or harvesting activity. Without the baseline it would be possible for a country, for example, to threaten to burn all its forest for the sake of acquiring valuable global warming entitlements once they could be convinced not to carry out their threat. In essence without a properly defined baseline, incentives for counterproductive strategic behavior are introduced.

While the within-country initial allocation of entitlements would automatically define a trading baseline for controlled substances in signatory nations, it does not define a baseline for uncontrolled greenhouse gases, for emission reductions from nonsignatory nations, or from activities which could sequester greenhouse gases. Although this baseline would have to be separately defined, it could be determined on a case-by-case basis. Trades involving these activities would only be approved upon completion of the certification process (described below in the section on monitoring and enforcement). Determining the appropriate baseline would be a normal component of the certification process.

Trading Rules. Once the entitlement system is defined and the initial allocation of entitlements has been completed, the rules governing entitlement trades must be determined. How are the trading ratios determined?

For global gases same-pollutant trades should normally take place at a one-to-one ratio. For example, a one-ton emission entitlement originally held by a source in country x should, when sold, entitle the acquiring trader to a one ton emission in any country.[19]

Interpollutant trades would not generally be conducted on a one-for-one basis because the different greenhouse gases do not have the same affect on global warming. The trading ratios for interpollutant trades should reflect the relative contributions of the gases. The objective should be to assure that global warming would not be adversely affected by the trade.

Ideally, it would be possible to develop a table of equivalences, which would be published and available to all traders in advance. Unfortunately any such table would necessarily represent a simplification of the atmospheric chemistry involved in global

[19] The problem is complemented somewhat when the greenhouse gases also have local affects. In this case the local effects can be handled by local regulations. No trade violating a local regulation would be permitted.

warming. In particular since the reactions take place at markedly different rates, the contribution of each gas to global warming is a function of time. Reductions in some gases would have a more immediate impact than others.[20] The practical implication of these chemical reactions for policy is that the relative contribution of any two gases to global warming is not a constant. And the construction of a table of equivalences presumes that the relative contribution can be expressed as a constant.

Some proposed trades would necessarily involve some uncertainty. Suppose, for example, that the contribution of some particular greenhouse gas to global warming was not known with precision or that the capability to monitor emissions would be less than perfect (introducing the possibility that post-trade emissions would in fact be higher than expected). As an alternative to banning all such trades, one way to confront and to deal realistically with this uncertainty is to account for it in the trading ratios.

When the proposed trade involves some uncertainty about the effects, "worst case" trading ratios can be used. Suppose, for example, that the ability to monitor emissions from a new source is accurate only within 50% of the actual. This trade could be permitted as long as two times as much reduction of this gas was required compared to a case in which monitoring was perfect. Adjusting trading ratios as a response to uncertainty offers regulators a considerable amount of discretionary flexibility.

The Clearinghouse and Banking Function. The global greenhouse trading system should involve a provision for a clearinghouse to provide a centralized location for entitlement transactions. With a clearinghouse potential purchasers do not have to spend excessive amounts of time and resources searching out potential trades; they merely have to check the clearinghouse.

Entitlement banks should provide a safe haven for certified, but unused credits. In terms of the security they offer the depositor, banked entitlements should be considered the equivalent entitlements in use. One problem that has arisen in the American system of emissions trading is that some of the entitlements held in banks (which in that program are controlled by the states) were confiscated as a means of quickly improving air quality. Confiscation, of course, had the unintended, but nonetheless very real, effect of destroying any incentive for firms to create banked credits. Regulatory authorities should not be allowed to confiscate banked credits if the proper incentives are to be maintained. Further air quality improvements can be secured while maintaining the integrity of banked entitlements, a topic discussed subsequently in this chapter.

[20] For example, a 1 ppm concentration of methane at a point in time contributes 26 times more to global warming than an equivalent amount of CO_2, but the atmospheric lifetime of CO_2, is 10 to 20 times longer.

Leasing. Leasing recognizes and responds to the temporal variability in the need for entitlements, a point that has special validity in the greenhouse gas context.

As an inducement to the participation of developing countries in a comprehensive global warming agreement, the initial international allocation of entitlements is likely to grant a surplus to the developing countries. This surplus would provide an opportunity to earn revenue (from trading those surplus entitlements to the industrialized nations), and an opportunity to create an entitlement reserve to facilitate and accommodate future economic growth. Unfortunately, outright sale of the entitlements can exploit only one of these opportunities; it would provide revenue only by selling entitlements from the reserve.

Leasing provides an opportunity to fulfill simultaneously both objectives. Leased entitlements are transferred to a new owner for a fixed, prespecified period of time. (In other words a nation could lease some portion of its 1991-1994 entitlements, but retain the 1995 and subsequent entitlements). This arrangement would work well for both the industrialized and developing countries. Because restructuring the industrial system takes time, the industrialized nations could use leased entitlements to buy time while the restructuring takes place. Meanwhile the developing countries would be receiving much needed revenue from the leased entitlements. As the developing countries' needs for entitlements to support economic growth grew, they could retain the later entitlements as a guaranteed source. Leasing would facilitate the transition to a new sustainable regime of energy use in both the developing and industrialized world.

6.3.1 Special Design Issues

The special circumstances associated with establishing a global version of a tradeable entitlement system for greenhouse gases give rise to some potential pitfalls. By anticipating these problems it is possible to design the system to avoid them or to mitigate their importance.

Market Power. Market power can exist in entitlement markets on either the buyer or seller side. When traders attempt to control the price (rather than react to the market-determined price), the cost effectiveness of the system can be affected.[21]

[21] Only if the control authority initially allocates the permits so a cost-effective allocation would be achieved even in the absence of any trading will the possibility of market power be completely eliminated. In this special case cost-effectiveness would be achieved even in the presence of market power on either the buyer or seller side because no trading would take place, eliminating the possibility of exploiting any market power.

The degree to which market power can be exercised either internationally or, if the private variant is chosen, domestically is a function of how permits are initially allocated. The larger is the share of the initial allocation of permits to colluding sources, the larger is the potential deviation of ultimate control costs from the least-cost allocation. When the colluding sources are initially allocated an excess of entitlements, higher control costs can be inflicted on others by withholding some permits from the market. However, when colluding sources are initially allocated a deficiency of entitlements, to reduce prices they must necessarily bear higher control cost as the means of reducing demand for the permits.

The evidence suggests that market power has rarely been a problem in existing uses of tradeable entitlements and is not expected to be much of a problem in a global entitlement market either. Successful cartels are difficult to establish and maintain for any commodity, and greenhouse gas entitlements would not be an exception. Successful cartelization on the seller side depends upon an ability to capture a substantial proportion of the available supply of entitlements available for sale. All other things being equal, the larger the supply which falls outside the control of the cartel, the lower the cartel's ability to wield its influence.

Because this would be a global market, the number of potential buyers and sellers is very large indeed. The market would not be thin. Since market power can only arise whenever either the buying activity or the selling activity is concentrated in the hands of a few nations, as long as the baseline (initial entitlement allocation) is defined in such a way as to create a large number of buyers and sellers, cartelization is unlikely.

Even if the initial allocation does not precipitate a large number of buyers and sellers, the number can be increased by other means. For example, allowing new entitlements to be created by the certified destruction of greenhouse gases and by certified increases in greenhouse gas absorption capabilities creates alternative supplies of entitlements. Allowing both international trades and the participation of private sources in these trades would also increase the number of traders.

Finally, as Grubb and Sebenius (1991) point out, the ultimate protection against the accumulation of market power is the capability all nations have to withdraw from the agreement. Any nation or group of nations which has been damaged by a cartelization of global warming entitlements can simply withdraw from the agreement and be unconstrained in their emissions. The right of withdrawal provides a significant check on the benefits of cartelization.

Changing Control Needs. It is necessary to build into a global greenhouse entitlements system a process for change. With as much uncertainty as currently exists with respect to global warming, change is inevitable. Yet change, if not done carefully, has the potential to undermine the system. Tradeable entitlements systems depend on entitlement bearers having secure ownership rights to the entitlements. Insecure property rights undermine the incentive to invest in the activities which free up the entitlements for sale. Can the desire to constantly incorporate change be made compatible with the desire to preserve sufficient security for the entitlements?

Fortunately, compatibility is possible. To the extent the change can be anticipated it should be announced in advance. If initial entitlement allocations turn out, in retrospect, to be too high over the long run and further reductions are necessary, the schedule of further reductions should be published in advance. Uncertainty is reduced when the changes are known in advance.

Second, even when the ultimate changes are not known in advance, it is possible to have a stable process for incorporating those changes. Reviews of the the aggregate emission target should be conducted on a periodic, announced schedule. The process and criteria for deciding whether a change is needed and, if so, how large a change is needed should be made transparent to interested parties.

Finally, the process for incorporating changes should be known in advance as well. Suppose the latest scientific evidence suggests that further emission reductions are necessary. In general all entitlements, whether banked or in use to justify emissions, should be reduced proportionately or the changes should be taken out of the growth increment. Government should *not* confiscate banked credits just because they are handy and not currently in use.[22] Confiscation destroys the incentive to invest in emissions reduction for the purpose of creating excess entitlements.

Hoarding. In the emissions trading program developed in the United States fewer entitlement trades have taken place than expected. In part this is due to the fact that some sources have hoarded created credits; instead of selling them, they have retained them for possible future use. The effect has been thinner markets and fewer trades than might otherwise have been possible. Would hoarding plague a global market in greenhouse entitlements? If so, what can be done to counteract it?

The key to assessing the likelihood of hoarding in a global market is to understand the motivations which give rise to hoarding behavior. Hoarding is a response to risk and it intensifies the very problem to which it is a response. Faced with the risk that future entitlements might not be available for sale, firms creating credits retain them to preserve their future options should they decide to expand operations.

Price risk was another incentive for hoarding. Prices were individually negotiated and were subject to considerable variability depending on the available supply of credits. Because emissions trading transactions have taken place in local markets with few buyers and sellers, demand and supply conditions could change rather quickly; the

[22] While this proposition may appear to be self-evident, in fact confiscation of banked credits has taken place in the United States.

large number of participants which are the precondition for a stable and smoothly functioning market were not present.[23]

Is this likely to be a problem for a global market? Probably not. Its global scope would provide a large number of participants on both the buyer and seller side of the market. Thin markets would be much less of a problem, even for within country trades.

Certain design features of the global system also mitigate against the occurrence of hoarding. Allowing entitlement supply to be increased by the development of greenhouse gas absorption or destruction activities provides an alternative source of entitlements. Facilitating the transboundary flow of entitlements would also contribute to the type of vigorous market which destroys the preconditions for hoarding.

Recent developments in the markets for the sulfur allowances (the entitlements in the U. S. acid rain reduction program) also provide some hope for reducing the risk associated with both the availability and price of future entitlements. A spot market for these allowances already exists. The first auction of allowances took place at the Chicago Board of Trade during 1993. Large brokerage firms have shown some interest in participating.

Organized markets for futures contracts, which have existed for well over one hundred years, bring together those who want to minimize their risk with those who are willing to take the risk for a price.[24] An active futures market affords an opportunity for purchasers to lock in the future price and availability of entitlements, thus eliminating the need to hoard entitlements as a hedge against this risk.

The evolution of these financial markets is an important complement to the evolution of the cost-effective control of greenhouse gases. They facilitate trades by revealing prices and by reducing the risk associated with investments to reduce greenhouse gases.

Other financial devices are also being developed to facilitate the evolution of this system by reducing risk. In the United States, for example, insurance pools are currently being set up to allow the utilities to pool their risks. Since penalties associated with violating the sulfur standards are rather severe, utilities selling entitlements run the risk that some future unanticipated event (such as an equipment malfunction) could make them vulnerable to a penalty assessment. The traditional strategy to deal with this contingency is to hold more entitlements than necessary. With an insurance pool utilities would band together to share the costs of penalties levied for truly unanticipated events, providing greater security for entitlement sellers. Greater security translates into a greater willingness to sell.

[23] The effect of this type of sequential trading is modeled and analyzed in Atkinson and Tietenberg (1991).

[24] For an examination of futures markets in general see Carlton (1984).

Designers of the global market should, however, guard against the use of artificial (and ultimately counterproductive) strategies to strip hoarders of their entitlements. For example, one such counterproductive strategy which has been introduced by some states in the United States is to require banked credits to be sold or used within three years or be lost. This "use it or lose it" provision undermines the incentive to undertake emission reduction strategies which would free up entitlements to be sold. Assuring vigorous markets and the introduction of futures trading provide an effective policy response to any risk that might be associated with these markets; artificial strategies to undermine hoarding simply make the problem worse by reducing the flow of entitlements available for sale.

6.4 IMPLEMENTING A GLOBAL ENTITLEMENT SYSTEM

While implementing a global entitlement system to control greenhouse gases is in the narrow sense an unprecedented event, in the broader sense a large number of international precedents exist.[25] While this is not the appropriate place to review, or even summarize, these precedents, it is important to realize that most of the individual steps involved have been undertaken before in the context of one or more international agreements.

6.4.1 Initial Steps

An international system for controlling greenhouse gases must evolve slowly over time. The initial steps will necessarily be small as the parties will be hesitant to undertake large commitments.[26]

As long as the initial steps are consistent with the evolution of a cost-effective and equitable system, a slow moving process can be an advantage rather than a disadvantage. Starting small gives both the institutions and the parties a chance to

[25] Levi (1991), for example, notes the striking parallels between the requirements for a greenhouse gas permit system and the arrangements which grew out of the Bretton Woods Agreement of 1944. Some of the specific aspects of global warming believed to have direct parallels in the situation facing the Bretton Woods negotiators are the need: (1) to structure an agreement such that both developed and developing countries would find it in their interest to participate, (2) to balance emissions reductions with efforts to increase absorption capacity, (3) to provide technical assistance to countries too poor or too small to know the most efficient way to contribute atmospheric improvement, (4) to monitor aggregate and individual country achievements and punish offenders, and (5) to have the means to follow a well-defined path of net reduction in greenhouse gas emissions so as to achieve climatic stability.

[26] For examples of proposals for initial steps see Dudek and LeBlanc (1990) and Roland (1991).

adjust and to become familiar with the system. Since most initial efforts will be precedent setting, it will take a while to work them out. Once the precedents have been established, however, the process will become smoother, quicker and better able to handle a larger number of participants, gases and entitlement trades.

Institutions will evolve along with attitudes. The clearinghouse and bank will be able to develop effective procedures for processing entitlement transactions. Signatory nations will become better at tracking and reporting emissions. Nonsignatory nations will learn how to participate in the process from the outside and become familiar with the advantages of become a signatory. Maturation of the associated financial markets will lay the groundwork for reducing the transactions costs associated with trading entitlements, a process which has already been triggered by the U. S. sulfur allowance program.

The first step in implementing a transferable carbon entitlement system would require establishing carbon dioxide emission entitlements with deadlines for achievement for all signatory nations. Non-signatory nations would be treated as a group; their projected behavior would be incorporated in the forecasted trends which form the basis for the signatory nations' allocation. Signatory nations would be free to trade these entitlements with other signatory nations subject to a certification process (described below).

Should negotiators deem it desirable to facilitate the transfer of technology to the less developed countries, charging an annual fee for the carbon entitlements allocated to the signatory nations would supply a funding source.

A designated international organization would have the responsibility for receiving monitoring reports from the signatory nations, developing monitoring estimates for the nonsignatory nations, putting together an annual report summarizing progress toward the treaty-specified targets (including for signatory nations a country-by-country assessment of compliance) and issuing this report to the public. This agency would also have the responsibility for (1) providing technical assistance to all nations on available means of combatting global warming (assistance to signatory nations would be subsidized), (2) maintaining an up-to-date accounting system of national obligations and achievements which incorporates the changes occasioned by trades and (3) establishing and running a greenhouse gas entitlement bank to store certified entitlements available for international trade.

Signatory nations would be required to report to the monitoring authority annually on the details of their compliance with treaty obligations, including a detailed estimate of the sources of emissions within its boundaries.

Though nonsignatory nations would not be assigned entitlements, they could participate in the system by serving as an alternative source of carbon emission reductions. A set of rules governing the process for acquiring greenhouse gas entitlements from nonsignatory nations would be established by negotiation. These rules should allow nonsignatories to create entitlements for subsequent sale to signatory nations subject to a certification process. The rules should allow interpollutant trades as long as the traders can demonstrate that the potential for global warming is reduced (or at least not increased) by the trade. Leasing, as well as sales, of flow entitlements should be approved.

Approval for entitlement trades among *signatory* nations would hinge on a certification process run by the selling country. Certification could in principle be extended to: (1) emission reductions which are quantifiable, surplus, and enforceable, (2) entitlements created by enhanced greenhouse gas absorption or by greenhouse gas destruction. Approval for entitlement trades involving *nonsignatory* nations would be similar with the additional requirement that the certification must secure the approval of the designated international authority as well as the approval of the selling country.

Specific procedures for modifying the key aspects of the agreement should be adopted. These procedures would recognize and incorporate better scientific understanding of atmospheric chemistry, improvements in monitoring and other changing circumstances. Specific aspects of the document subject to revision would include the aggregate emission targets and compliance deadlines, and the list of controlled substances. The document should also specify procedures for admitting new signatories.

6.4.2 Evolving To A More Complete System

Beginning with a simple system with the understanding that it could evolve to a more complete system presumes some knowledge about the more complete system. Otherwise the process might well start down a path which could ultimately lead in a rather different direction. What would a more complete system look like?

The rules governing a more complicated system represent a natural evolution of the simple system by opening the agreement to more signatories, by facilitating the expansion of coverage to all greenhouse gases, and by providing the framework for tighter certification, monitoring and enforcement procedures. Principles for the evolution of this more complete system include:

1. Expand the number of participants in the transferable carbon dioxide entitlements program by establishing procedures for admission of new signatory nations.
2. Expand the list of "controlled substances" in response to deepening scientific understanding and improving monitoring technologies. The list should be subject to periodic review with specific procedures for modification announced in advance. To the extent that direct emissions monitoring is not practical, acceptable proxies should be developed.

3. The transferable entitlements program could be expanded to include private participants. Signatory nations would be encouraged to allocate their allotment of carbon dioxide entitlements among domestic users. Because it is not practical to involve small users (such as individual households), these entitlements should be allocated to the point in the distribution chain offering the maximum opportunities for emission reduction.
4. While initially the limited system would necessarily be based primarily on trust, as the system evolved more complete monitoring and enforcement procedures could be developed. The primary responsibility for enforcement of entitlement obligations imposed on individual sources would fall upon the host nation. Signatory nations should be required to adopt domestic procedures for receiving and processing complaints of violation, including complaints from public or private organizations in other signatory nations. The designated monitoring authority should provide an appeals opportunity for all complaints not resolved at the national level.
5. The designated monitoring and enforcement authority would have two main sanctions at its disposal for dealing with those nations failing to comply with their treaty obligations: (1) adverse publicity and (2) veto power over international greenhouse gas trades. Adverse publicity, a traditional international sanction, can be a powerful stimulus for compliance.[27] The veto power would be limited to specific grounds. The designated authority would merely have to determine whether or not the evidence for certifying the entitlements for sale was sufficient. A positive determination would authorize the transboundary trade. A negative determination would be accompanied, where possible, by a listing of remediable steps which, if taken, could lead to trade approval. Entitlements could be stored in the bank only after securing certification approval.

6.5 MONITORING AND ENFORCEMENT

Although it would be inappropriate to attempt to catalogue all of the monitoring and enforcement activities that would be associated with each of the unique sources of greenhouse gases, it is essential that general monitoring and enforcement strategies be identified as well as the institutional requirements to implement them. In this section we undertake that task, paying particular attention to the relative roles for private organizations, nations and international agencies.

[27] See the examples provided in Chayes and Chayes (1991).

6.5.1 Monitoring

Nations which contain the individual pollution sources would have the major responsibility for monitoring and enforcing the activities of those sources. Ample precedent for this approach has been established by previous international agreements.[28]

Self-reporting of their emission levels by the sources to national enforcement agencies would be the key to the system supplemented by cross-checks provided by other sources of data. A standard practice in environmental regulation, self-reporting has the virtue that it is relatively inexpensive, and when backed by the appropriate sanctions for misrepresentation, it provides remarkably accurate information.[29]

The reports should be simple, yet revealing, with respect to the compliance status of the reports. One model of this approach is provided by the discharge monitoring reports required by the U.S. Clean Water Act. Under that act the National Pollutant Discharge Elimination System (NPDES) requires polluters to file with EPA discharge monitoring reports (DMRs) which list discharge levels and method of discharge.

This combination of suitably defined effluent standards coupled with monitoring reports that directly reveal the compliance status of the source has proved in the United States to be an effective enforcement vehicle, not only for EPA, but for private enforcers as well. Citizens have access to the discharge monitoring reports, and the courts have generally accepted the information contained in them as proof of violations in enforcement actions brought by private parties.[30] By allowing citizens to use these reports as the basis for their claims, monitoring reports have significantly reduced the cost of of both public and private enforcement activities.[31] The discharge monitoring report approach has proven so successful for the control of water

[28] Since 1973, for example, the Washington Convention on International Trade in Endangered Species of Wild Fauna and Flora has established worldwide trade controls based on a permit system administered completely by the individual nations. For other examples see Sand (1991, 259-261).

[29] In the United States virtually all environmental statutes now make the falsification of these environmental compliance reports a criminal offense with the responsible individual subject to personal fines and imprisonment. For further details see Segerson and Tietenberg (1992).

[30] See *Student Public Interest Research Group v. Fritzsch, Dodge & Olcott, Inc.* 579 F. Supp at 1538.

[31] For an analysis of the determinants and effects of supplementing public enforcement with private enforcement as revealed by the U. S. experience see Naysnerski and Tietenberg (1992)

pollution in the United States that it was incorporated into air pollution control legislation in 1991.[32]

Private monitoring has also become an important check of the self-reporting system. Many bodies of the water in the United States are now protected and monitored by private associations of citizens concerned about that particular body of water. One increasingly typical activity of these voluntary, privately financed associations is hiring a full time "riverkeeper" or "harborkeeper" to continuously monitor polluting activities on that body of water and to monitor the water quality itself. Successful legal actions brought against municipalities and other polluters along the Hudson River in New York State have not only been a powerful force for cleaning it up, but have established some institutional and legal precedents which are now being following in other parts of the country.

Computer systems and software have now advanced to the point where it has become relatively easy to develop automated systems to aggregate and to integrate all of these reports, to compare them to alternative sources of data for purposes of validation, and to identify and prioritize suspected cases of noncompliance for action.[33] Both the requisite hardware and software could be easily be transferred to nations who sought them.

The direct measurement of emissions is expensive, but fortunately it is also unnecessary. Because of the close link between the carbon content of the fuel and the amount of carbon dioxide emissions, monitoring the flow of fuels is not a bad proxy for monitoring emissions.[34]

This is an important point because monitoring the flow of fuels is much easier than monitoring emissions. While record keeping systems are already in place to keep track of the flow of carbon-based fuels across international borders, no counterpart exists for emissions.

Individual governments would also have a responsibility for assuring that all greenhouse gas entitlement trades are integrated into their emissions inventory. To assure that their national entitlement allocations square with their emissions all

[32] Section 504 of the 1991 Amendments to the Clean Air Act require facilities to certify that they are in compliance with all applicable regulations and permit conditions. These certifications are similar to the discharge monitoring reports required under the Clean Water Act, and they are also subject to sanctions for false or improper certification. See Unterberger (1991, 1636).

[33] The nature of the system used to support the U. S. lead trading system is described in Nussbaum (1991). Kete (1991) describes in some detail the system currently being developed to support the sulfur oxide trading system that is an integral part of the U. S. approach to reducing acid rain.

[34] It is, however, not a perfect proxy. Since combustion is the source of the carbon dioxide emissions, oil used for lubrication does not pose the same problem as oil used as a fuel. As a practical matter this can be handled by giving production facilities a credit for all oil sold for lubrication.

governments would at least have to be notified of trades. Following a trade from country x to country y of z tons, country y's total allocation would be increased by z and country x's decreased by z. While the total would be unaffected, the allocation of that total among the countries would have been affected. When the entitlements are leased, rather than sold, the adjustments must be made when the lease begins and when it ends. To fulfill these responsibilities each national government would have to be notified of all trades and leases affecting sources within its boundaries. Depending on the wishes of the treaty negotiators, national approval of transboundary trades may or may not be required.

Prior to trading any entitlements the selling source would have to have the entitlements certified. Only certified entitlements can be traded. The national government responsible for monitoring and enforcing the activities of the selling source would have the responsibility for the certification process. Certification criteria should be included in any global warming agreement but the U. S. certification process provides one possible model.

Three conditions would normally be necessary for certification--the emission reductions represented by the entitlements offered for trade must be enforceable, surplus,[35] and quantifiable. This certification process would be required either for entitlements freed up by emission reductions or for entitlements created by greenhouse gas absorption or destruction capabilities. The burden of proof is on the seller; he/she must convince the certifying authority that these conditions have been met. The certification process is necessary to prevent fraudulent trades (those which appear to result in no degradation or an improvement in the global climate, but which in fact intensify the problem). The certification process could also be used to eliminate substitutions which are not desired outcomes of the trading process.[36]

How would these monitoring and enforcement activities be financed? Private entitles could pay a fee to the government for each entitlement received. This revenue could be used to finance the record-keeping and surveillance necessary to run the system. Even fees which are too low to have any significant incentive can still have the potential to generate considerable revenue.

A smaller, but important monitoring role would be reserved for a newly designated international monitoring authority. One department in this organization

[35] Surplus generally means reductions not otherwise required by law. For example, reductions in CFC's in fulfillment of the Montreal Protocol process would not be considered surplus, but reductions over and above those required would.

[36] Some have expressed concern, for example, that a trading system would encourage the substitution of nuclear or hydro power for fossil fuels. Should this concern receive general support, the certification process would be the appropriate institutional arrangement to assure that these substitutions could not be the basis for trades.

would be set aside to receive the individual country reports, to integrate them with estimates of emission levels for the nonsignatory countries, to validate these country estimates as far as possible by cross-checking it with independent sources, and to match reported achievements with treaty obligations on a country-by-country basis. This organization would then issue an annual report containing the country-by-country compliance information, a summary of progress toward the overall global warming goals, and recommendations for the future.

While this type of *ex post* monitoring would clearly not be appropriate for all environmental problems (particularly those posing immediate health risk), it does seem appropriate for controlling greenhouse gases. Given the rather long lags which characterize the gap between the emission of greenhouse gases and the perceived effects of the higher concentration, the delays associated with *ex post* monitoring seem inconsequential.[37]

6.5.2 Enforcement

Menu of Transboundary Enforcement Options. What works at the national level would not work or even be necessary in the international context. Oversight of the national enforcers requires a rather different approach. As Professor Louis Henkin puts it in his book *How Nations Behave* despite some conspicuous departures,"*almost all nations observe almost all principles of international law and almost all of their obligations almost all of the time*" (emphasis in the original).[38]

Compliance with the norms of international law by the nation states subject to those laws is driven less by the economic motivations that tend to drive profit-making firms than by the desire to be a part of maintaining a stable, dynamic equilibrium in international relations. This basic tendency can certainly be reinforced by an appropriately designed global warming agreement; rules universally perceived as fair are more likely to be obeyed than rules perceived as unfair.

It can also be reinforced by instituting what Chayes and Chayes (1991, 290-304) call the principles of "accountability" and "transparency". According to these principles governments should be held accountable for their behavior by rendering their performance transparent to scrutiny by the international community. Requiring the submission of, and making public, the annual reports to the international monitoring authority would provide the foundation for applying these principles.

[37] This point is developed in Haites (1991).

[38] Cited in Chayes and Chayes (1991, 289).

Skeptics are likely to suggest that public awareness of noncompliance is not a sufficient condition for compliance. That may be true, but it is also true that the power of public awareness should not be underestimated. In the United States, for example, many toxic air pollutants had not been regulated prior to the passage of the 1990 Clean Air Act Amendments. The Emergency Planning and Community Right-to-Know Act of 1986, however, required some 20,000 plants to report annual emissions of 320 potentially carcinogenic chemicals. These reports were then made available to the public. Publicized by newspapers and environmental groups, they became a rallying point for action. Even before the discussions about increasing the regulation of these substances became serious, businesses were beginning to make significant reductions in toxic emissions to counter the adverse publicity. A 39% reduction in toxic emissions has already been achieved by Monsanto. Dow Chemical plans a 50% reduction by 1995; and DuPont plans a 63 % reduction by 1993.[39]

An alternative to taking legal action against a noncomplying party is provided by defining a complaint process to make all noncompliance transparent and to hold responsible parties accountable. One model of this approach, for example, is provided by the 1957 Rome Treaty establishing the European Economic Community.[40] Article 155 of that treaty designated the EEC Commission as responsible for implementing the treaty's provisions and Article 169 empowers it to initiate proceedings against any noncomplying party.

The proceedings involve four stages. In the first step "letters of formal notice" are sent by the commission to member states believed to be in violation of an EEC environmental directive. In the second step the member state is given the opportunity to respond to the notice, filing any information with the commission that it believes supports its position. In the third step the commission issues a "reasoned opinion" confirming the infringement or dismissing the case. Finally, should continuing noncompliance continue, the EEC has the authority to initiate legal proceedings before the European Court of Justice.

Private actions are an important component of this process. As Sand (1991, 272) puts it: What may be the most significant feature of this procedure, however, is mentioned nowhere in the treaty and evolved only gradually in the course of its implementation; more than half of the infringement proceedings initiated against member states were based not upon the commission's own monitoring of compliance but on citizen complaints- from private individuals, associations (such as Greenpeace and Friends of the Earth), or entire municipalities.

[39] Howe (1991, 35).

[40] Sand (1991) describes this custodial procedure as "one of the most important means of enforcing EEC environmental standards."

This mechanism has apparently been successful not only in encouraging complaints, but in encouraging compliance as well.[41] The increasing role for public interest, nongovernmental organizations in the process of monitoring and enforcement in international law offers a wide range of new opportunities for enhancing the degree of compliance with international agreements.[42]

Three aspects of the Rome Treaty experience are particularly worth noting. First, the process of filing and publicizing complaints is by itself a powerful force for compliance. Second, to afford standing to public interest nongovernmental organizations considerably strengthens a monitoring and enforcement regime based upon transparency and accountability. Finally, the process of putting in place an international regime for enforcement is likely to be an evolutionary process. The first steps will be hesitant and will no doubt prove frustrating for those who seek to transfer the traditional national model of sanctions into an international setting. Nonetheless, experience shows that transparency and accountability can be powerful enforcement principles which can provide a foundation for action. Should further action become necessary, these initial steps will pave the way for a further strengthening of the procedures.[43]

Domestic Enforcement. As with monitoring the primary responsibility for enforcing the entitlements held by individual sources would fall on the countries containing the regulated source. Countries would naturally be able to chose their own set of enforcement strategies but the menu of possibilities is growing rapidly. Many conventional and some unconventional strategies exist for encouraging these sources to comply.[44]

[41] The number of complaints rose from 10 in 1982 to 460 in 1989. One complaint by a citizen of the United Kingdom against his own government ultimately led to an investigation that resulted in infringement proceedings against seven member states. Sand (1991, 272).

[42] For an examination of the successful roles public interest nongovernmental organizations have already played in shaping, monitoring and enforcing international environmental law as well as a vision suggesting a greatly enhanced role in the future see Sands (1989).

[43] Many observers believe that harmonizing the competing claims of protecting the global environment and protecting national sovereignty will require the development of an enforcement process in stages. For one view of how this might be done in two phases see Holley (1991, 82-95).

[44] For a description of some of these innovative strategies and how they work in practice see the enforcement essays in Tietenberg (1992).

6.6 CONCLUSIONS

1. Starting with a transferable entitlement system based solely on carbon dioxide does not preclude its subsequent expansion to include other greenhouse gases.

   The decision to incorporate other gases could be made on a case-by-case basis. Reductions in other greenhouse gases would be tradeable contingent on a demonstration that the the effect of the proposed reductions on global warming would be equivalent.
2. To facilitate a market in entitlement transactions the entitlement should be defined in terms of a homogeneous unit such as a ton of carbon dioxide emissions.

   To the extent that the entitlement involves another gas or an activity other than emission, the results must be translated back into a "ton of carbon dioxide equivalent.
3. Entitlements would be dated in terms of years 1994, 1995, etc. The entitlements could be used during the specified year or after, but not before.

   Each entitlement would allow a specific amount of emissions. Once the amount authorized by the entitlement had been emitted, the entitlement would become worthless.
4. The effective functioning of a tradeable carbon dioxide entitlement system does not presume or require any particular domestic control policy.

   How individual signatory governments would meet their authorized emission target would be left up to them. Some might choose to use a traditional regulatory system, while others might choose to use emissions charges or transferable entitlements.
5. Transferable entitlement systems could be designed where all trades took place exclusively between governments, exclusively between private parties, or with some mix of the two.

   Exclusive government trading could be initiated immediately following the acceptance of specific emission targets by some group of signatory nations. Governments could subsequently choose, if it so wishes, to facilitate trades by private organizations by allocating their entitlements among the various competing sources within its borders.
6. The supply of entitlements could be expanded beyond treaty limits as long as the expansion does not undermine the goal of global warming.

   One possible source of expanded permit supply is provided by participants who undertake activities which destroy greenhouse gases. Another source of entitlement expansion could come about from the development and commercialization of new greenhouse gas absorption capabilities.

7. For global gases same-pollutant trades should normally take place at a one-to-one ratio. Interpollutant trades would not generally be conducted on a one-for-one basis because the different greenhouse gases do not have the same effect on global warming.

   The trading ratios for interpollutant trades should reflect the relative contributions of the gases. The objective should be to assure that global warming would not be adversely affected by the trade.
8. Adjusting trading ratios offers regulators a considerable amount of discretionary flexibility that can be used to include even reductions characterized by a considerable amount of uncertainty.

   Suppose, for example, that the contribution of some particular greenhouse gas to global warming was not known with precision or that the capability to monitor emissions would be less than perfect (introducing the possibility that post-trade emissions would in fact be higher than expected). As an alternative to banning all such trades, one way to confront and to deal realistically with this uncertainty is to account for it in the trading ratios.

   When the proposed trade involves some uncertainty about the effects, "worst case" trading ratios can be used. Suppose, for example, that the ability to monitor emissions from a new source is accurate only within 50% of the actual. This trade could be permitted as long as two times as much reduction of this gas was required compared to a case in which monitoring was perfect.
9. The global greenhouse trading system should involve a provision for a clearinghouse to provide a centralized location for entitlement transactions.

   With a clearinghouse potential purchasers do not have to spend excessive amounts of time and resources searching out potential trades, they merely have to check the the clearinghouse.
10. Market power has rarely been a problem in existing uses of tradeable entitlements and is not expected to be much of a problem in a global entitlement market either.

    Successful cartels are difficult to establish and maintain for any commodity, and carbon dioxide entitlements would not be an exception. Successful cartelization on the seller side depends upon an ability to capture a substantial proportion of the available supply of entitlements available for sale. All other things being equal the larger the supply which falls outside the control of the cartel, the lower the cartel's ability to wield its influence.
11. A process for change should be built into a global greenhouse entitlements system.

    With as much uncertainty as currently exists with respect to global warming, change is inevitable. Yet change, if not done carefully, has the potential to undermine the system. To the extent the change can be anticipated it should be announced in advance. So if initial entitlement allocations are too high over the long run and the ultimate desired reductions are known, the schedule of reductions should be published in advance. Uncertainty is reduced when the changes are known in advance. Even when the ultimate changes are not known in advance, it is possible to have a stable process for incorporating those changes.

12. The evolution of associated financial markets for trading entitlements is an important complement to the evolution of the cost-effective control of greenhouse gases.

    Mature financial markets for greenhouse gas entitlements facilitate trades by revealing prices and by reducing the risk associated with investments to reduce greenhouse gases.

13. An international system for controlling greenhouse gases must evolve slowly over time. The initial steps will necessarily be small as the parties will be hesitant to undertake large commitments.

    As long as the initial steps are consistent with the evolution of a cost-effective and equitable system, a slow moving process can be an advantage rather than a disadvantage. Starting small gives both the institutions and the parties a chance to adjust and to become familiar with the system. Since most initial efforts will be precedent setting, it will take awhile to work them out. Once the precedents have been established, however, the process will become smoother, quicker and better able to handle larger number of participants, gases and entitlement trades.

    Institutions will evolve along with attitudes. The clearinghouse and bank will be able to develop effective procedures for processing entitlement transactions. Signatory nations will become better at tracking and reporting emissions. Nonsignatory nations will learn how to participate in the process from the outside and become familiar with the advantages of becoming a signatory. Financial exchanges will lay the groundwork for markets which reduce the transactions costs associated with trading entitlements.

14. An existing or new international organization would have the responsibility for receiving monitoring reports from the signatory nations, developing monitoring estimates for the nonsignatory nations, putting together an annual report summarizing progress toward the treaty-specified targets (including for signatory nations a country-by-country assessment of compliance) and issuing this report to the public.

    Signatory nations would be required to report to this organization annually on the details of their compliance with treaty obligations, including a detailed estimate of the sources of emissions within its boundaries.

15. Approval for entitlement trades among signatory nations would hinge on a certification process run by the selling country. Approval for entitlement trades

involving nonsignatory nations would require the assent of the international monitoring organization.

Certification involving signatory and nonsignatory nations would be handled differently. While the agreement provides a natural trading benchmark for signatory nations (defined by their allocation of entitlements), it provides no trading benchmark for the nonsignatory nations. The certification procedure for nonsignatories must establish that benchmark as well as fulfill the other certification requirements.

Certification could in principle be extended to: (1) emission reductions which are permanent, quantifiable, surplus, and enforceable, (2) entitlements created by enhanced greenhouse gas absorption or by greenhouse gas destruction. Approval for entitlement trades involving nonsignatory nations would be similar with the additional requirement that the certification must secure the approval of the appropriate officials at the Global Warming Authority as well as the approval of the selling country. In this way signatories could be assured that the trading benchmark was acceptable.

16. The monitoring organization would have two main sanctions at its disposal for dealing with those nations failing to comply with their treaty obligations: (1) adverse publicity and (2) veto power over international greenhouse gas trades.

Adverse publicity, a traditional international sanction, can be a powerful stimulus for compliance. The veto power would be limited to specific grounds. The global warming authority would merely have to determine whether or not the evidence for certifying the entitlements for sale was sufficient. A positive determination would authorize the transboundary trade. A negative determination would be accompanied, where possible, by a listing of remediable steps which, if taken could lead to trade approval. Entitlements could be stored in the Global Warming Authority Bank only after securing certification approval.

REFERENCES

Atkinson, S. E. and T. H. Tietenberg (1991) Market Failure in Incentive-Based Regulation: The Case of Emissions Trading. *Journal of Environmental Economics and Management* 21(1): 17-31.

Benedick, R. E. (1991) Lessons from the 'Ozone Hole'.in: Benedick, et. al *Greenhouse Warming: Negotiating a Global Regime*. Washington, World Resources Institute. 9-12.

Bohm, P. (1991) Incomplete International Competition to Reduce CO_2 Emissions: Alternative Polices. *Journal of Environmental Economics and Management* 24 (May):258-271.

Bohm, R. (1992) Distributional Implications of Allowing International Trade in CO_2 Emission Quotas. *The World Economy*. 15 (January): 107-122.

Carlton, Dennis W. (1984) Futures Markets: Their Purpose, Their History, Their Growth, Their Successes and Failures. *The Journal of Futures Markets* 4(3):237-271.

Chayes, A. and A. H. Chayes (1991) Adjustment and Compliance Processes in International Regulatory Regimes. *Preserving the Global Environment: The Challenge of Shared Leadership*. New York, W. W. Norton & Co. 280-308.

Dudek, D. J. and A. LeBlanc (1990) Offsetting New CO_2 Emissions: A Rational First Step. *Contemporary Policy Issues* 8(3): 29-42.

Grubb, M. and J. K. Sebenius (1991) *Participation, Allocation and Adaptability in International Tradeable Emission Permit Systems for Greenhouse Gas Control.* Paris, Organization for Economic Co-operation and Development Workshop on Tradeable Permits to Reduce Greenhouse Gas Emissions. (June 26-27)

Haites, E. (1991) Tradeable Allowances and Carbon Taxes: Cost-Effective Policy Responses to Global Warming. *Energy Studies Review* 3(1): 1-19.

Holley, S. E. (1991) Global Warming: Construction and Enforcement of An International Accord. *Stanford Environmental Law Journal* 10: 44-96.

Howe, C. W. (1991) An Evaluation of U. S. Air and Water Policies. *Environment* 33 (September): 10-15, 34-36.

International Energy Agency (1992) *Climate Change Policy Initiatives* (Paris: OECD).Kägeson, Per. (1991). *Economic Instruments for Reducing Western European Carbon Dioxide Emission* Stockholm: Swedish Environmental Advisory Council, Ministry of the Environment (June 7).

Kete, N. (1991) *The Acid Rain Control Allowance Trading System*. Paris, Organization for Economic Co-operation and Development Workshop on Tradeable Permits to Reduce Greenhouse Gas Emissions. (June 26-27)

Levi, Maurice D. (1991) *Borrowing from International Financial Organizations: Learning from Bretton Woods for Dealing with Global Warming*. University of British Columbia Working Paper (January).

Naysnerski, Wendy and T. H. Tietenberg (1992) Private Enforcement of Federal Environmental Law. *Land Economics*. 68 (February): 28-48.

Nussbaum, B. (1991) Phasing Down Lead in Gasoline in the U. S.: Mandates, Incentives, Trading and Banking. Paris, Organization for Economic Co-operation and Development Workshop on Tradeable Permits to Reduce Greenhouse Gas Emissions. (June 26-27)

Priznar, F. J. (1990) Trends in Environmental Auditing. *Environmental Law Reporter: News and Analysis* 20: 10179.

Roland, K. (1991) An International Clearinghouse for Greenhouse Gas Entitlements: Some Issues for Norwegian Negotiators. a draft memo from ECON Energi, Oslo, Norway.

Rose, A. and T. Tietenberg (1993) An International System of Tradable CO2 Entitlements: Implications for Economic Development. *Journal of Environment and Development* 2(1): 1-36.

Sand, P. H. (1991) International Cooperation: The Environmental Experience. In *Preserving the Global Environment: The Challenge of Shared Leadership* ed. Jessica Tuchman Mathews. New York: W. W. Norton & Co.

Sands, P. J. (1989) The Environment, Community and International Law. *Harvard International Law Journal* 30(Spring):393-420.

Segerson, Kathleen and T. H. Tietenberg (1992) Defining Efficient Penalties, in: T. H. Tietenberg, ed. *Innovation in Environmental Policy: Economic and Legal Aspects of Recent Developments in Liability and Enforcement* (Cheltingham, England: Edward Elgar Publishing)

Stewart, R. B. and J. B. Wiener (1991) A Comprehensive Approach to Climate Change. *The American Enterprise* : 75-80.

Swart, R. (1991) *Greenhouse Gas Emissions Trading: Defining the Commodity*. Paris, Organization for Economic Co-operation and Development Workshop on Tradeable Permits to Reduce Greenhouse Gas Emissions. (June 26-27)

Tietenberg, T.H. (1990) Economic Instruments for Environmental Regulation. *Oxford Review of Economic Policy* 6:17-33.

Tietenberg, T. H. (1992) *Innovation in Environmental Policy: Economic and Legal Aspects of Recent Developments in Liability and Enforcement* (Cheltingham, England: Edward Elgar Publishing).

U. S. Department of Justice (1991) A Comprehensive Approach to Addressing Potential Climate Change. Washington, U. S. Department of Justice.

United Nations Conference on Trade and Development (1992) *Combating Global Warming: Study on a Global System of Tradable Carbon Emission Entitlements* (New York: United Nations).

Unterberger, Glenn L. (1991) Citizen Enforcement Suits: Putting Gwaltney to Rest and Setting Sights on the Clean Air Act. *Environmental Reporter: Analysis and Perspective* (4 January): 1631-1636.

Victor, D. G. (1991) Limits of Market-Based Strategies for Slowing Global Warming: The Case of Tradeable Permits. *Policy Sciences*, 24 (2): 199-222.

Victor, D. G. (1992) Practical Aspects of Implementing Greenhouse Taxes: Issues for OECD Countries, in: OECD, *Climate Change: Designing a Practical Tax System* (Paris: OECD).

PART II

INTERNATIONAL TRADE AND ENVIRONMENT

Chapter 7 TRADE AND ENVIRONMENT: A BRIEF SURVEY OF CURRENT THEORETICAL ISSUES

Ineke de Boer[1]

ABSTRACT

The recently intensified concern about international environmental problems has generated theoretical and practical questions regarding international trade and the environment, which can be seen as a special case of international environmental problems. This chapter gives a brief survey of theoretical issues on the impacts of trade, trade policy and environmental policy on trade patterns and the environment. From the literature it becomes clear that these impacts will depend mainly on the character of the externality of the traded goods and the assumed market structure but also that the impacts can be very diverse so that emprical studies are needed to analyze the actual impacts.

7.1 INTRODUCTION

Although the attention for the relation between international trade and the environment is relatively new, the importance of this relation is nowadays commonly recognized. WCED (1988), OECD (1991), Anderson and Blackhurst (1992), Low (1992), amongst others, indicate that international trade has impacts on the environment through altering the volume and international location of production and consumption. They also point to the fact that trade and the environment are heavily tied up with policy. For instance, the Brundtland Commission (WCED, 1988) and the OECD (1991) consider the exchange of natural resources for foreign currency in the developing countries as one of the most important links between international trade and the environment, in particular since trade policy in the developed countries affects the quality and the use of natural resources in the developing countries. The existence of high import tariffs on processed products, set by the developed countries, results

[1] I wish to thank my superviser professor Henk Folmer (General Department of Economics, Wageningen) for his stimulating support and useful comments. Also I am grateful to professor Arie Kuyvenhoven, Nico Heerink (both of the Department of Development Economics, Wageningen) and Alison Burrell (Department of Agricultural Economics, Wageningen) for their useful comments and suggestions.

in greater dependency of developing countries on the export of raw materials. This in turn, together with international debt problems and a fast growing population, leads to environmental damage and excess extraction of natural resources.

The social concern about international environmental problems has generated many questions regarding the interaction between international trade and the environment. Most of these questions consider either the impact of trade policy on the international environment or the impact of environmental regulation on international trade patterns. Since the environmental component is a relatively new element in trade theory, these questions raise some interesting theoretical and analytical points which have been discussed in the literature mainly over the last five years. This chapter surveys the recent literature on the major questions in both these areas, including

- What is the impact of trade on national and international welfare, including the environment?
- What impact might (regional) trade liberalization have on the welfare of countries both within and outside the region in question if the environmental impacts are considered?
- What is the impact of domestic unilateral environmental politics on international competitiveness and on patterns of production, consumption and trade?
- What is the impact of (full or partial) agreements on international trade or environmental policy on welfare?
- Is it possible to use trade policies as an incentive to enhance full cooperation in international environmental policy?

Our special interest is to discuss the theoretical issues of the impacts of certain policies on trade patterns and the environment. Issues of more practical nature, like the role of the GATT, harmonisation issues or choice of policy instruments, are not treated explicitely but only indirectly taken into acccount.

The chapter contains six sections. Section 7.2 discusses the impacts of trade on national and international welfare. It illustrates that the direction of these impacts (positive or negative) depends to a large extent on the character of the externalities of the traded good, whereas the size depends on the assumed market structure. Next is a discussion of how trade or environmental policy is able to influence these impacts. section 7.3 analyzes the impacts of trade policy, especially trade liberalization, on the environment in the absence of adequate environmental policy. Section 7.4 looks at the impacts of environmental policies on international competitiveness. The instruments for influencing international cooperation are discussed in section 7.5. The question to be asked is whether and how trade instruments can be used to enhance the multilateral cooperation that is needed to deal with multilateral or global environmental problems. Finally, section 7.6 contains the conclusions and recommendations for further research.

7.2 IMPACTS OF TRADE ON NATIONAL AND INTERNATIONAL WELFARE

In the traditional literature, gains from international trade are based on differences between production factors of various countries. These differences ensure that the marginal costs for certain products will vary amongst countries. The country with the lowest marginal production cost enjoy a comparative advantage and will produce the good for domestic and foreign consumption. Every country will improve his welfare by increasing either producer or consumer surplus. This also implies an improvement of international welfare, as welfare is defined as "consumer surplus + producer surplus + net revenue of the government".[2] This definition of welfare is often called "economic welfare". Only recently, in the 1980's, the environment has been taken into account in traditional trade theory, where the environment is mostly seen as the "third" factor of production and/or a consumption good. A country is thought to have an environmental abundance if it has a relatively large assimilative capacity or a relatively greater ability to tolerate pollutants. Social welfare (including the environmental element) will now be defined as "consumer surplus + producer surplus + net revenue government + quality of the environment". This implies that the environment is also treated as a factor that directly influences the well-being of individuals. This way of dealing with the environment makes it possible to trade-off economic welfare, against so called environmental welfare. It is essential to separate social welfare into an economic and in an environmental component because the consensus over the measurement of these two components will differ; there exists more international consensus over the measurement of the economic welfare than over environmental welfare. This is due to differences in preferences about environmental welfare, which depend in general on the level of income, education and culture. Furthermore these preferences can shift over time. This means that optimal environmental welfare is not clearly defined. Runge (1993) distinguishes four aspects of the impacts of trade on welfare:

1. Allocation of economic activities amongst countries.

[2] To consider the net revenue of the government makes only sence if it concerns national welfare, since the net revenue is just a transfer if it comes to international welfare.

Trade causes international specialization and contributes to economic efficiency. This is, in general, considered to increase international economic welfare through higher consumer or producer surplusses. The idea of increasing international environmental welfare through trade is also accepted because specialization leads to the use of relatively abundant resources and the resulting efficiency causes an economical use of that abundant resource. The effects for the environmental welfare for a specific country, however, can be positive or negative. The final result is influenced by the fact that international specialization contributes to a division of the international production chain, from extraction till consumption, over more countries. In this sense Anderson (1992) states that in a relatively small country the national environmental welfare can improve if specialization leads to a greater import of goods with production externalities or a greater export of goods with consumer externalities.

2. Scale effects.
   Trade causes an expansion of production, consumption and transportation, contributing to higher levels of GNP. In general, this results in increasing economic welfare due to higher consumer and producer surplusses. Opinions about the effects on environmental welfare will differ, however. In the view of some analysts, an expansion of production and consumption will cause the environment to deteriorate, because a growing economy is associated with more pollution. Others consider a growing GNP as a vehicle for improving the environment through raising incomes so that there is more concern for environmental protection and more money to spend on it. So, we need empirical analysis to be sure about the sign and the size of the impacts of scale effects for environmental welfare.
3. Composition of output.
   A growing GNP can lead to a relative reduction of heavy manufacturing industries and a shift to higher levels of services or technical knowledge in a country. This results, in general, in growing economic welfare. The effects for national environmental welfare though, are unsure, because it depends after all on the absolute levels of pollution. The effects for international welfare are also uncertain, because a changing composition of output in one country can lead to the creation of so called "pollution havens" in other countries. These pollution havens occur if different environmental preferences exist among countries.
4. Technologies employed in various countries.
   Trade can induce more rapid technological innovations that lead to more economic welfare. As increased value is given to environmental quality with increases in income, markets for technological innovations that improve the

environment may develop and grow. In that case it is likely that national and international environmental welfare will improve.

Runge has described these categories of impacts of trade on welfare. The accompanied most likely outcome for economic and environmental welfare are based on the "traditional trade theory". This illustrates that in all cases the impacts of trade on economic welfare are positive. However, the sign of these impacts of trade on environmental welfare can be both, positive are negative. This is due to the fact that the character of the externalities of the traded goods can be different. Therefore the impact on social welfare is not certain in all cases.

7.2.1 Externalities of the traded goods

The impacts of trade on welfare will depend on the externalities of the traded goods. In this respect the scale (national or international) and the character (unidirectional or reciprocal) of the externalities should be taken into account. An externality is *unidirectional* if a country is either a *source or a victim* of the environmental problem. In this case the economic analysis is different depending on whether there is one or more than one victim countries. An externality is said to be *reciprocal* if a country is both the *source and the victim* of the environmental problem.

Trade in whales or in tropical wood can be seen as examples of trade flows with a unidirectional character, considering a few source countries and many victim countries. The countries that extract the resources (for example Norway in the case of whales and Brazil in the case of tropical wood) are seen as the source countries and the rest of the world as the victim countries. Here, the revenue from extracting the resources for the traded goods, which can be seen as the producer surplus, is benefitting the source countries. This means an improvement in economic welfare in the source countries. The victim countries experience an improvement in economic welfare through an increase in consumer surplus (the availability of the traded goods). Individuals in source and victim countries face a drop in social welfare through their concern about the whales and the tropical forest. There exists a problem of over-exploitation if the source country does not take the effects of its action for the victim countries into account. The central and challenging question for the victim countries is how to convince the source country to protect the resource. In order to do so, the victim countries need to cooperate in order to put pressure on the source country. The difficulty is, however, the increasing incentive for "free riding" when there are more victim countries involved. For example trade in beef from cattle treated with harmful hormones or fruit polluted with pesticides are trade flows with a unidirectional character, considering one or more source countries and one victim country, where the countries of production can be seen as source countries and the consumer country as the victim. The central question here is whether these goods from different sources can be subjected to product standards in the victim market. An indirect effect of these product standards is the potential for protection of the internal market using the environment as a pretext. As an example might be the European Union's decision to ban the import of beef from cattle treated with hormones (Runge, 1993). In this setting Barbier and Rauscher (1993) investigated the (international) potential trade impacts of product standards on the production and extraction of timber.

Trade in goods with externalities in the production process, for example generating greenhouse gases, are mostly examples of international environmental problems with a *reciprocal* character. Here a country is both victim and source of the problem because it is assumed that each country benefits from using the environment. At the same time, each country is damaged by environmental deterioration from domestic use of the environment as well as from foreign use of the environment. In this case each country will face an incentive to take unilateral action because each could be potentially better off if they cooperate to protect these resources. An efficient and effective solution depends mainly on the action of the other countries or on the degree of cooperation. If no cooperation is possible each country will maximize its own benefits. The solution will most probably lead to over-utilization or to the "tragedy of the commons" (Hardin, 1968). If a sub-group of countries decides to cooperate to prevent environmental deterioration, the colluding countries will try to maximize their collective benefits (see Barrett, 1993). If the solution is stable, it has the drawback that it is not cost efficient and does not effectively protect the environment. If one assumes increasing marginal costs of abatement it is not cost efficient if only a small group of countries reduces emissions by a certain amount. Unilateral actions create incentives for "free riding" by non-participating countries who can benefit from the reduction in pollution without bearing any abatement costs. Another reason for not reaching an effective solution is that reduction of emissions by one group of countries can be offset by the non-participating group of countries through two channels. First, the possibility exists of an increase in the production of basic materials in unconstrained regions, which results in extra emissions. Second, less demand due to emission reductions of the participating countries can cause a fall in world prices of basic materials, which can then lead to an increase in polluting manufactures in the unconstrained regions. In the case of full cooperation the problem is to maximize total global benefits. This solution offers scope for reaching good management of the environment, but is very difficult to implement in practice.

7.2.2 Market structure

In the literature, supporters of "the new trade theory" state that the size of the impacts depends basically on the assumed market structure, and that the impacts of trade on economic and environmental welfare will be greater in a market with imperfect competition than in a competitive market. This is due to the existence of rivalry among interdependent firms in imperfect competitive markets where a firm must take its rivals' actions into account in setting out an optimal strategy. The problem with the new trade theory is that a generally accepted theory is still to be developed. It depends on the institutions or assumptions what the nature of competition will be (Helpman and Krugman, 1991). Within the new trade theory, the most used market structure, when the environment is taken into consideration, is the Cournot oligopoly. Competition in this market is characterized by internal economies of scale, which assumes "large" firms with imperfect competition. Furthermore, a Cournot oligopoly assumes a few sellers and many buyers, which restricts the influence on the market to the producers. The producers compete in quantities (Cournot competition) and behave non-cooperatively. Finally there are non-zero profits assumed which means that there is a restriction on the entry of new firms.

The choice of a Cournot oligopoly implies that all the interesting aspects of strategic behavior, on the side of the producers, arise in these models. That is, if non-zero profits are assumed, there are strategic gains to be obtained from trying to manipulate markets. The strategic gains exist of extracting foreign monopoly rents when domestic industry expands its market share. This can be initiated on the side of the producers . The government, however, is also able to extract foreign monopoly rents by using policy instruments for altering the rivalry between the industries in different countries to the benefit of their domestic firms and consumers. This is the main incentive for the government to intervene. The consequence is that optimal trade policy for an oligopolistic industry is not necessarily laissez faire (which is the case with traditional trade theory). The literature on the incentives for the government to extract foreign monopoly rents are mainly based on the game theoretic models of Brander and Spencer (1983, 1985) and Dixit (1984). Although the literature state that the intervention of the government can be optimal there are practical complications for implementation; such as the need for full information in order to describe the nature of competition or the change of retaliation by other governments which can result in a lower level of international welfare.

After this theoretical analysis of the influence of market structure on the size of trade impacts on welfare, we are able to compare empirical work based on perfect and imperfect competition and show that these differences in market structure lead to differences in the magnitude of the impacts. Empirical work based on traditional trade theory is, for instance, done by Lucas et al. (1992) and Low and Yeats (1992).

Their case studies on the issue of the relocation of polluting industries due to domestic environmental policy conclude that there is no direct relation between investment decisions and environmental regulations. Both papers suggest that the cost of emission control is in general just a tiny fraction of operating costs. This implies that, in this case, we can conclude that the impact on the economic and environmental welfare due to environmental policy is relatively small. Empirical work based on the new trade theory in the same field, namely on the investment decisions of plants, is done by Ulph (1993a). He assumes a Cournot oligopoly with a few fertiliser plants and concludes that the loss of welfare due to environmental policy is significant. This results from the fact that there are relatively few plants which are earning significant rents, so the loss of some plants has implications for the total national welfare. These results, from both empirical papers, are consistent with the theoretical analysis, that the magnitude of the impacts of trade on welfare depends on the assumed market structure.

7.3 INFLUENCE OF TRADE POLICY (ESPECIALLY TRADE LIBERALIZATION) ON WELFARE

Over the whole world, many goods are subject to trade policies of various nature. Politicians have an incentive to use trade policies if the net gains of protection for national welfare are positive. Trade instruments are implemented to increase economic welfare, but which at the same time have consequences (direct or indirect) for environmental welfare. Examples of such trade instruments are tariff and non-tariff barriers, production and export subsidies and international trade agreements such as voluntary export restraints, preferential and commodity agreements (Arntzen et al., 1992). These instruments affect the environment because they give rise to national or international disruptions to an effective allocation of economic activities and can cause domestic overproduction.

In general, the net gain in national welfare of domestic protection is coupled with a loss in welfare of other countries such that overall welfare declines. In that case there exists a situation in which national interests conflict with the common interest. These conflicts can refer to the economic as well as the environmental welfare. An important question is whether the impacts of (national) trade policies will conflict with optimizing international environmental welfare.

Although it is commonly recognized that there are impacts of trade policies on international environmental welfare, there exists also a broad consensus in the literature that trade policies are not seen as the best instrument to be used in solving environmental problems. This conclusion follows from the fact that the optimal instrument to use is that instrument, among the list of feasible instruments, whose base is most closely related to the source of the market failure. These instruments will in most cases be related to an appropriate international environmental policy. The use of an appropriate international environmental policy can cause an improvement in environmental welfare. With the use of trade policy this improvement is not guaranteed because trade instruments cause a disruption in the effective allocation of economic activities. For instance, an import tariff set by a country can induce economic inefficiency and extra production in that country, while it reduces the efficient production in foreign countries. Trade policy can be used, however, to enhance international cooperation as will be seen in section 7.5.

A second remark relates to the institutional framework in which the question should be answered. The principles of the GATT under the agreement in the Uruguay Round of December 1993, ensures a further liberalization of world trade. So it is more realistic to reformulate the central issue now as concerning what impact (regional) trade liberalization might have on the environmental welfare of countries both within and outside the region in question. However no clear answer to this question can be provided. It depends on the signs (negative or positive) of the four earlier described impacts of trade and trade policies. To analyze the magnitude of the impacts of trade liberalization on international welfare, empirical studies are needed.

In Anderson and Blackhurst (1992) and Low (1992) some empirical studies on the effects of liberalizing world trade can be found, all based on the traditional "comparative advantage" theory. The overall conclusion is that liberalization of trade will enhance economic welfare. The consequences for the environment though are unsure. Where appropriate environmental policies are in place, where growth is associated with environmentally friendly technological change, or where trade liberalization reduces environmentally destructive economic distortions or increases productive efficiency, the effects on the environment are likely to be positive. Practical implications to create circumstances which improve both, economic and environmental welfare are reviewed by Heerink et al. (1994).

Empirical studies with the assumption of strategic behavior are urgently needed. Richardson (1989, 1990) and Thursby and Thursby (1990) survey empirical research on how trade liberalization affects national economic welfare under imperfect competition, but the environmental welfare is not taken into account in this research.

Very much related to the above question is the question of what impact (regional) trade liberalization might have on the quality of the environment in growing economies. In the literature we find two schools of thought on the issue of the environment in growing economies. One school uses case studies to analyze the impact of opening up markets on the growth of an economy and on the quality of the environment. Lucas et al. (1992) and Birdsall and Wheeler (1992), for instance, both find that growing but closed economies tend to become more pollution intensive, whereas the opposite is true for more open economies. Both authors try to explain this development from the fact that economies with open trade regimes attract more foreign investment than closed ones. This results in a faster adoption and diffusion of clean technology for a country with an open economy. Braga (1992) comes up with another argument based on his empirical work on resource management in the forestry sector. He states that an open economy needs to be more efficient to be competitive. This implies that sawmills must work more efficiently which means a less environmentally degrading use of tropical wood.

The second school of thought is mainly theoretical. Van der Ploeg et al. (1993) and van der Ploeg and Lighthart (1993) for instance, extend the new growth theory by incorporating respectively pollution as a by-product of production and renewable resources as a factor of production. The new growth theory refers to a situation where the steady-state growth rate is no longer given but the theory endogenises this rate to allow for the influence of government policies, environmental degradation and other economic variables such as preferences and knowledge spill-overs arising from R & D. The main concern of van de Ploeg et al. (1993) is to find an optimal rate of growth together with an optimal environmental policy in order to achieve what they call a "win-win" situation. This is a situation where the care of the national environment requires economic growth, while economic growth in turn cannot take place without proper care for the environment. They conclude that a "win-win" situation is possible under two conditions: the environment must be treated as a stock and the intertemporal elasticity of substitution must exceed unity. The main concern of Van der Ploeg and Lighthart (1993) is to analyze whether international cooperation will lead to less resource use and to an improvement in environmental quality. They find that this correlation is in most cases true. They find also that international policy coordination can harm environmental quality and boost the rate of economic growth if there are, apart form international environmental externalities, other external effects on the production side present which influence the benefits of other countries. In this sence they have considered international knowledge spill-overs arising from R & D and externalities resulting from productive government spending.

7.4 INFLUENCE OF ENVIRONMENTAL POLICY ON INTERNATIONAL COMPETITIVENESS

The concern about environmental problems has generated national environmental policy and international agreements to protect the environment. An interesting point is to analyze the policy instruments which are proposed to increase environmental welfare, but which have consequences for international competitiveness. Examples of environmental instruments that affect trade patterns are (Arntzen et al., 1992):

- direct regulatory instruments, such as standards or norms. These are the primary means of achieving environmental goals at national level. We distinguish product and process standards. Product standards are related to externalities in consumption and may be used as a non-tariff barrier to trade. Process standards are related to externalities in production and may create pollution havens in the case where standards differ between countries;
- economic instruments such as subsidies, charges and tax incentives. Subsidies may cause allocation and trade distortions. The use of the other economic instruments, like emission charges,[3] product charges,[4] administrative charges, deposit refund systems, fiscal incentives and marketable permits, are at present not being used very often. Therefore most of the empirical effects of these instruments on trade is still unknown. Chapter 14 of this book explores the problems of implementing marketable permits to deal with the acid rain problem in Europe;
- international environmental agreements, such as the Montreal Protocol (1987), the International Tropical Timber Agreement (1985) or the United National Convention on International Trade in Endangered Species of Wild Flora and Fauna (1973), contain almost always one or more trade restrictions.

The central question is what the impacts of environmental policy are on international competitiveness. To answer this, the scale on which the environmental problem occurs and the structure of the market must be taken into account. We will therefore distinguish national and international environmental policy in a market of perfect and imperfect competition. This viewpoint, by the way, looks at the problem from a different angle than in the previous viewpoint in section 7.3. Because environmental policy is seen as an optimal instrument while trade policy is not considered to be optimal to improve environmental welfare.

7.4.1 National environmental policy

Domestic environmental policy is appropriate if the environmental problem occurs on a local scale. In that case a country decides to reduce and/or abate a certain amount of the domestic emissions and/or waste by setting taxes, subsidies, standards or other

[3] When it concerns externalities in production

[4] When it concerns externalities in consumption

regulations. The central question will be how these unilateral actions alter patterns of production, consumption and trade through their impact on international competitiveness. The main challenge to the international community when the environmental externalities are exclusively domestic is to agree on a set of principles which can serve as a basis for national environmental policy. In this prospect, the Polluter Pays Principle is often called as a guiding principle in national environmental policies. Harmonization of principles is highly desired in order to avoid disruptions in international competitiveness through national environmental policy (Heerink et al., 1993).

The traditional "comparative advantage" theory will study the change of specialization patterns in several cases: (i) the case where one country imposes environmental regulations or (ii) the case where all countries decide to impose environmental regulations. Studies that focus on the theoretical impacts of environmental policy on the comparative advantage of countries are for instance found in Pethig (1976), Blackhurst (1977), Siebert (1977, 1985), Asako (1979), McGuire (1982) and Baumol and Oats (1988).

The new trade theory assumes monopolistic or oligopolistic competition. Theoretical work in this area is for instance done by Markusen et al. (1993). In their approach firms make discrete decisions such as whether or not to shut down firms and/or transfer production to plants in another region (in contrast with marginal decisions over continuous variables such as prices and output as is done in the traditional theory). These authors look specifically at environmental quality and government competition in the sense that governments can compete in terms of their environmental policies when plant locations are endogenous.

7.4.2 International environmental policy

International environmental policy in the form of voluntary multilateral agreements is needed when it concerns multilateral or global environmental problems. Here the central question will be what the impact of (full or partial) agreements is on economic welfare as well as on environmental welfare of the colluding as well as of the non-colluding countries. The environmental element of welfare makes cooperation much more difficult, because countries have to agree on regional or global norms and standards for shared resources, and on the distribution between countries of the costs of environmental protection. The character of the externalities of the traded goods results in different approaches in the literature to the issue of optimal environmental policy:

1. Compensation and penalties for unidirectional environmental problems
In the literature, unidirectional environmental problems are usually connected with possibilities for the victim countries to influence the source countries' environmental policies. Most of the cases deal with various kinds of financial incentives on the side of the developed countries to influence the developing countries' resource extraction policies. An important assumption in this literature is that only the developed countries are concerned about the global environment. This implies that all the initiatives to achieve environmental welfare improvement must come from high-income countries. They can offer incentives or impose penalties in order to reach agreement with the developing countries. The nature of the incentives to be used to enhance full cooperation is important. Examples of available compensatory mechanisms are cash transfers, debt for nature swaps, technology transfers and various penalties for destruction of nature. The effect of various financial operations by the developed countries on the resource extraction policies of developing countries has been analyzed by Strand (1990, 1991) and Raucher (1989). Chichilnisky (1993) analyses the impact of different property right policies, such as land reallocation, on the resource extraction.
2. Unilateral actions with reciprocal environmental problems
Each country has an incentive to act unilaterally if the externality is reciprocal. The literature comes up with many studies which analyze the impact of a unilateral action, such as a production tax or an abatement equipment standard in one country, on the level of transnational pollution, the terms of trade and rewards for the factor of production. In traditional (neo-classical) setting without strategic behavior, Krutilla (1991) and Merrifield (1988) have done some theoretical work on the impact of unilateral environmental policy on the level of transnational pollution, the terms of trade and factor rewards. The main concern of both authors is the question of whether a unilateral action can be effective or not. Merrifield argues that unilateral actions can succeed in reducing the level of emissions, but that free trade in goods and in capital could cause foreign emissions to rise sufficiently to induce an increase in the net level of emissions. Krutilla comes to more or less the same conclusion when he considers the impacts of an environmental tax of a large country (which affects the world prices of traded commodities) in an open economy. He states that environmental regulation appears to be more complicated the more the traditional assumption of a small country in a closed economy is relaxed. Like Merrifield he finds that, due to offsetting emissions, the welfare effects of an environmental tax depend to a large extent on whether a commodity is exported or imported, and the level of tariff protection. Whalley (1991) and Felder and Rutherford (1993) have done some interesting empirical work on the impact of regulation of transnational pollution on trade patterns and the gains from trade. Whalley finds that the interregional

gains and losses of a carbon tax is highly sensitive to the type of tax implemented to reduce emissions, which is consistent with the theoretical findings of Merrifield and Krutilla. Felder and Rutherford consider six trade-regions and use simulation techniques to explore the effect of a unilateral action by one of the regions to curb global CO_2 emissions. They find significant carbon "leakages" through offsetting emissions in the non-reducing regions. They also observe that restrictions on trade in basic materials will lower the leakage rate. But at the same time they argue that this instrument has only limited effectiveness. In a setting with strategic behavior, Conrad (1993) considers theoretically the same efficiency problem as above, but he assumes strategic behavior on the part of the government. In his analysis the government sets the level of the environmental policy instruments and then firms choose the extent of their abatement levels of output given the levels of policy instruments. Conrad shows that domestic production decreases while foreign production increases by a domestic tax on inputs. He then states that most welfare will be gained if the government uses, together with the input tax, a subsidy program. In theory this is true because subsidies stimulate domestic production, which leads to an increasing share of the world market for the domestic firms and an improvement of the global environment. Ulph (1993b) shows that the choice of environmental policy instrument can have a marked impact on the incentives for producers to act strategically through their choice of investment in capital. Where strategic behavior consist in over-investment in capital in order to gain market share and abstract rents. Ulph concludes that environmental standards give less incentives for strategic behavior relative to environmental taxes or no environmental policy at all. Ulph (1993c) then explores whether the consequences of this strategic behavior of producers in turn has implications for the choice of environmental policy instruments by governments.

As with other areas of policy intervention, environmental policies also affect the distribution of income within and between countries. This in turn creates incentives for interest groups to seek to influence the policy outcome in their favour. When the influence of interest groups is taken into account we enter the domain of the "political economy of international trade". Political economy assimilates the theory of public choice into trade theory. Here, certain policies are endogenously explained in a political-economy framework as a consequence of particular conjectures of objectives, incentives and circumstances. This makes it possible to explain the creation of indirect as well as direct opportunities to erect new barriers to trade. See Moore (1990), Hillman and Ursprung (1992) and Hoekman and Leidy (1992) for an overview of "the traditional trade" literature in this area. Gawande (1992) examines the use of non-tariff barriers within a game-theoretic lobbying model, where the non-tariff barriers are supposed to possess a political component as well as a strategically retaliatory component which serves to deter foreign protectionism.

7.5 INTERNATIONAL COOPERATION

International environmental problems must be solved by voluntary multilateral agreements in order to get an efficient and effective solution. In the real world as well as in the literature the main attention is focused on agreements on reciprocal environmental problems where each country faces an incentive to cooperate but where an efficient solution depends on the behavior of the other countries concerned. These characteristics contribute to the fact that in the economic literature game theoretic models have been used for analyzing international cooperation (see for example Mäler (1989, 1991), Van der Ploeg and De Zeeuw (1992), Hoel (1992a, 1992b)).

Analyses of agreements on unidirectional environmental problems are more problematic due to the lack of incentives to cooperate for all the involved countries (Mäler, 1990). Here one faces difficulties in accomplishing an agreement. If an agreement is established, one faces difficulties in monitoring, because there exist incentives for the source country to give insufficient information.

The challenge, to come to a multilateral agreement, is to create or increase the incentives to cooperate. One way to do this is to donate side payments to enhance full cooperation. There are, however, some theoretical and practical difficulties with the implementation of such transfers. Theoretically, a side payment means that the victim rather than the polluter, should pay, which is in contradiction to for instance the Treaty of Rome. Another point is the possibility that side payments give incentives for strategic behavior; a country can receive larger side payments if it initially minimises abatement policies even below non-cooperating levels. A practical point which is mentioned by Mäler (1990) is that side payments can be avoided if the countries in question are already engaged in several areas of negotiation, for instance, trade connections. In this approach trade policy can be seen as one of the few available instruments to encourage cooperation (Anderson and Blackhurst (1992) and Folmer et al. (1993)). As an example, consider two countries, A and B, who are simultaneously involved in a transboundary pollution problem and in a trade dispute. It then is possible that country A is only willing to reduce his transfrontier emissions if country B removes his trade barriers. To be clear, there does not exist a direct or indirect relation between the environment and trade in this case, but two independent problems are connected with each other. Folmer et al. (1993) and Folmer and Van Mouche (1993) have formalized this idea in what they called 'interconnected games'.

Within a game theoretical framework they have linked environmental problems to non-environmental problems.

This idea seems especially useful if it concerns unidirectional environmental problems. Because, in that case, it is extremely difficult to achieve an agreement, due to the lack of incentives to cooperate for the source countries in the absence of clearly defined property rights.

7.6 CONCLUSIONS AND RECOMMENDATIONS FOR FURTHER RESEARCH

This chapter has reviewed the existing literature on the relation between trade and environment. Our attention was particularly focused on the theoretical issues of (i) the impacts of trade on the environment and (ii) the influence of policies on trade patterns and the environment. After reviewing four different aspects of the impact of trade on the economic and environmental welfare the following general conclusions can be given:

- the unidirectional or reciprocal character of the international externalities determine, to a large extent, the sign (positive or negative) of the impacts of trade on national environmental welfare;
- the assumed market structure is important for the magnitude of the impacts of trade on welfare;
- policies are able to influence trade patterns and therefore also the quality of the environment. The impact on both categories is very diverse, which implicates that empirical research is needed to analyze the actual impacts of policies on the welfare effects of certain trade-flows;
- trade liberalization is very likely to enhance economic welfare whereas the consequences for the environment are uncertain and need to be explored in case studies;
- unilateral domestic environmental policy is appropriate if the environmental problem occurs on a local scale. There are however impacts on international competitiveness which need international agreements on principles of environmental policy;
- if international environmental problems are involved, international environmental agreements are needed for an efficient and effective solution. The international community has, in this case, to agree about the consequences for economic as well as environmental welfare;

- the character of the international externality leads to different approaches in the literature on the issue of optimal environmental policy:
  * compensation and penalties for unidirectional environmental problems;
  * unilateral actions or full cooperation with reciprocal environmental problems;
- in the case of international unidirectional environmental problems, trade policies can be one of the few available instruments to enhance multilateral cooperation.

7.6.1 Recommendations for further research

The traditional theory of international trade is theoretically worked out very well. It is considered a general theory which, in its essence, can be used in many different situations. Disadvantages are the strict assumptions about the degree of competition in markets and the existence of indirect effects induced by various policies which affects the distribution of income within and between countries. These indirect effects are of significant importance, so they should be taken into account. Some empirical work is already done with this theory as a point of departure, but more studies are needed in order to get a better overview of the various impacts of trade and environmental policies on economic and environmental welfare.

It is not possible for the new trade theory to develop a general theory. It depends on the characteristics of the competition which concepts to follow. This results in the fact that full information about the market situation is needed in order to describe it properly. In practice, however it is not likely that it is possible to gather all the needed information. The research work assuming a market which is characterized by a Cournot oligopoly has recieved the most attention in the literature when environmental welfare is studied. A disadvantage of this approach is that it assumes non-cooperative behavior among the countries concerned. The theory of cooperative behavior in oligopolistic industries is however, not well developed. This may be an interesting subject for further research, because international cooperation is very important when international environmental problems are involved. There is certainly a need for more research involving increasing returns and imperfect competition. After all, in a lot of cases (for instance in agriculture), governments are strategically active at the national level, or rapid technical progress creates dynamic scale economies and learning-by-doing effects.

Important for follow up research is more empirical work based on different assumptions so as to improve understanding of the trade-offs between economic and environmental welfare. If these are known, the discussion can focus on the role of institutions and on the effects of policies on income distribution.

REFERENCES

Anderson, K., and R. Blackhurst (1992) *The Greening of World Trade Issues.* Harvester Wheatsheaf, New York.

Anderson, K. (1992) *The standard welfare economics of policies affecting trade and the environment.* In: K. Anderson, and R. Blackhurst. *The Greening of World Trade Issues.* Harvester Wheatsheaf, New York.

Arntzen, J., I. Hemmer, and O. Kuik (eds.) (1992) *International trade and sustainable development.* VU University Press, Amsterdam.

Asako, K. (1979) Environmental Pollution in an Open Economy. *Economic Resource.* Vol. 55, pp.359-367.

Barbier, E.B., and M. Rauscher (1993) *Trade, Tropical Deforestation and Policy Interventions.* In: C. Carraro (ed.). *Trade, Innovation, Environment.* Kluwer, Amsterdam.

Barrett, S. (1993) *The Economics of International Agreements for the Protection of Environmental and Agricultural Resources.* Paper prepared for the FAO, Rome. january 1993.

Baumol, W.J., and W.E. Oats (1988) *The theory of environmental policy.* Second edition. Cambridge University Press, Cambridge.

Birdsall, N., and D. Wheeler (1992) *Trade Policy and industrial Pollution in Latin America: Where are the Pollution Havens?.* In: P. Low (ed.). *International Trade and the Environment.* World Bank Discussion Paper, no.159. The World Bank, Washington, D.C.

Blackhurst, R. (1977) *International Trade and Domestic Environmental Policies in a Growing World Economy.* In: R. Blackhurst, et.al. *International Relations in a Changing World.* Geneva: Sythoff-Leiden.

Braga, C.A.P. (1992) *Tropical Forests and Trade Policy: The Case of Indonesia and Brazil.* In: P. Low (ed.). *International Trade and the Environment.* World Bank Discussion Paper, no.159. The World Bank, Washington, D.C.

Brander, J.A., and B.J. Spencer (1983) Strategic commitment with R and D: The symmetric case. *Bell J. Econom.* 14, 225-235.

Brander, J.A., and B.J. Spencer (1985) Export subsidies and international market share rivalry. *Journal of International Economy.* 18, 83-100.

Chichilnisky, G. (1993) *Property Rights and the Dynamics of Renewable Resources in North-South Trade.* In: C. Carraro (ed.). *Trade, Innovation, Environment.* Kluwer, Amsterdam.

Conrad, K. (1993) Taxes and Subsidies for Pollution-Intensive Industries as Trade Policy. *Journal of Environmental Economics and Management.* 25, 121-135.

Dixit, A.K. (1984) International trade policy for oligopolistic industries. *Economic Journal*. Vol. 94 (suppl.), pp.1-16.

Felder, S., and T.H. Rutherford (1993) Unilateral CO_2 Reductions and Carbon Leakage: The Consequences of International Trade in Oil and Basic Materials. *Journal of Environmental Economics and Management*. Vol. 25, pp.162-176.

Folmer, H., P van Mouche and S. Ragland (1993) Interconnected Games and International Environmental Problems. *Environmental and Resource Economics* (forthcoming).

Folmer, H. and P. van Mouche (1993) Tensor Games. *Annals of Operations Research* (forthcoming).

Gawande, K. (1992) *Strategically Deterrent Non-tariff barriers Within a Lobbying Model: Theory and Evidence*. University of New Mexico.

Hardin, G. (1968) The Tragedy of the Commons. *Science*, 162, 1243-8.

Heerink, N.B.M., J.F.M. Helming, O.J. Kuik, A. Kuyvenhoven and H. Verbruggen (1993) *International trade and the environment; Theory and policy issues*. Wageningen Economic studies 30, Agricultural University, Wageningen.

Heerink, N.B.M., O.J. Kuik, A. Kuyvenhoven en H. Verbruggen (1994) Is internationale handel goed of slecht voor het milieu? *Economisch Statistische Berichten*, nr.3946, pp.77-81.

Helpman, E., and P.R. Krugman (1991) *Market Structure and Foreign Trade; Increasing returns, Imperfect Competition and the International Economy*. MIT-Press, Cambridge.

Hillman, A.L., and H.W. Ursprung (1992) *The influence of environmental concerns on the political determination of trade policy*. In: K. Anderson, and R. Blackhurst. *The Greening of World Trade Issues*. Harvester Wheatsheaf, New York.

Hoekman, B., and M. Leidy (1992) *Environmental policy formation in a trading economy: a public choice perspective*. In: K. Anderson, and R. Blackhurst. *The Greening of World Trade Issues*. Harvester Wheatsheaf, New York.

Hoel, M. (1992a) *Emission taxes in a Dynamic Game of CO_2 Emissions*. In: R. Pethig (ed). *Conflict and Cooperation in Managing Environmental Resources*. Springer, Berlin.

Hoel, M. (1992b) International Environment Conventions: The Case of Uniform Reductions of Emissions. *Environment and Resource Economics*. Vol.2, no.2, pp.141-159

Ierland, E.C. van, S. Kruitwagen and E. Hendrix (1994) *Guided Bilateral Trade for Acidification in Europe*. In: E.C. van Ierland (ed.). *International Environmental Economics; Theories and applications to climate change, acidification and international trade*. Elsevier, Amsterdam (forthcoming)

Krutilla, K. (1991) Environmental Regulation in an Open Economy. *Journal of Environmental Economics and Management*. Vol.20, pp.127-142.

Low, P. (ed.) (1992) *International Trade and the Environment*. World Bank Discussion Paper, no.159. The World Bank, Washington, D.C.

Low, P., and A. Yeats (1992) *Do "Dirty" Industries Migrate?* In: P.Low (ed.). *International Trade and the Environment*. World Bank Discussion Paper, no.159. The World Bank, Washington, D.C.

Lucas, R.E.B., D. Wheeler, and H. Hettige (1992) *Economic Development, Environmental Regulation and the International Migration of Toxic Industrial Pollution: 1960-1988*. In: P.Low (ed.). *International Trade and the Environment*. World Bank Discussion Paper, no.159. The World Bank, Washington, D.C.

Mäler, K-G. (1989) *The Acid Rain Game*. In: H. Folmer and E.C. van Ierland (eds.) *Valuation Methods and Policy making in Environmental Economics*. Elsevier, Amsterdam.

Mäler, K-G. (1990) International Environmental Problems. *Oxford Review of Economic Policy*. Vol.6, no.1.

Mäler, K-G. (1991) *The Acid rain Game II*. Mimeo. Stockholm School of Economics, Stockholm.

Markusen, J.R., E.R. Morey, and N.D. Olewiler (1993) Environmental Policy when Market Structure and Plant Locations Are Endogenous. *Journal of Environmental Economics and Management*. 24, 69-86.

McGuire, M. (1982) Regulation, Factor Rewards, and International Trade. *Journal of Public Economics*. Vol. 17, pp.335-354.

Merrifield, J.D. (1988) The impact of selected abatement strategies on transnational pollution, the terms of trade and factor rewards: A general equilibrium approach. *Journal of Environmental Economics and Management*. Vol.15, 259-284.

Moore, M.O. (1990) *New Developments in the Political Economy of Protection*. In: C.A. Carter, A.F. McCalla, and J.A. Sharples (eds.). *Imperfect Competition and Political Economy; The New Trade Theory in Agricultural Research*. Westview Press, Boulder.

OECD, (1991) *The State of the Environment*. OECD, Paris.

Pethig, R. (1976) Pollution, Welfare and Environmental Policy in the Theory of Comparative Advantage. *Journal of Environmental Economics and Management*. Vol.2, pp.160-169.

Ploeg, F. van der, and A. de Zeeuw (1992) International Aspects of Pollution Control. *Environment and Resource Economics*. Vol. 2, no.2, pp.117-139.

Ploeg, F. van der, J. Verbeek, and C. van Marrewijk (1993) *Pollution, Abatement and Endogenous Growth*.

Ploeg, F. van der, and J.E. Lighthart (1993) *Sustainable Growth and Renewable Resources in the Global Economy*. In: C. Carraro. Trade, Innovation, Environment. Kluwer, Amsterdam.

Raucher, M. (1989) Foreign Debt and Renewable Resources. *Metro economica*. Vol.40, no.1, pp.57-66.

Richardson, J.D. (1989) Empirical Research on Trade Liberalization with Imperfect Competition: A Survey. *OECD Economic Studies*. 12:7-50.

Richardson, J.D. (1990) *International Trade, National Welfare, and the Workability of Competition: A Survey of Empirical Estimates*. In: C.A. Carter, A.F. McCalla, and J.A. Sharples. *Imperfect Competition and Political Economy; The New Trade Theory in Agricultural Trade Research*. Westview Press, Boulder.

Runge, C.F. (1993) *Economic Trade and Environmental Protection*. In: D.W. Bromley (ed.). *Handbook of Environmental Economics*.

Siebert, H. (1977) Environmental Quality and the Gains from Trade. *Kyklos*. Vol.30, no.4, pp.657-673.

Siebert. H. (1985) *Spatial Aspects of Environmental Economics*. In: A. Kneese, and J. Sweeney (eds.). *Handbook of Natural Resource and Energy Economics*. North-Holland, New York.

Strand, J. (1990) *Lending Terms, Debt Concessions, and Developing Countries' Resource Extraction*. Memorandum from Department of Economics, University of Oslo, no.5.

Strand, J. (1991) Foreign Aid, Capital Accumulation and Developing Countries' Resource Extraction. *Journal of Development Economics*.

Thursby, M.C., and J.G. Thursby. (1990) *Strategic Trade Theory and Agricultural Markets: An Application to Canadian and U.S. Wheat Exports to Japan*. In: C.A.

Carter, A.F. McCalla, and J.A. Sharples. *Imperfect Competition and Political Economy; The New Trade Theory in Agricultural Trade Research*. Westview Press, Boulder.

Ulph, A. (1993a) *Environmental Policy, Plant Location and Government Protection*. Discussion Papers in Economics and Econometrics, no.9305. University of Southampton, Southampton.

Ulph, A. (1993b) *Environmental Policy and Strategic International Trade*. Discussion Papers in Economics and Econometrics, no.9304. University of Southampton, Southampton.

Ulph, A. (1993c) *Environmental Policy and International Trade when Governments and Producers act Strategically*. Paper written for a project funded by the ESRC grant Y320-25-3038.

Whalley, J. (1991) The Interface Between Environmental and Trade Policies. *The Economic Journal*. Vol.101, no.405, pp.180-189.

WCED (The World Commission on Environment and Development) (1988) *Our Common Future*. Oxford University Press, Oxford.

Chapter 8 EMISSION TAXES AND INTERNATIONAL MARKET SHARE RIVALRY

Klaus Conrad

ABSTRACT

Integration of environmental objectives into economic policy is going to become essential. However, as pollution control and abatement expenditures are likely to affect production cost structures and the international competitiveness of industrial firms, strategic thinking and rent capturing aspects will be part of an environmental policy. The chapter analyses models of strategic environmental policy in international competition, and characterizes optimal intervention using different instruments. Production will be either for a third-country market or there will be intraindustry trade explained by differences in cost and in consumer preferences. The behavior of the governments and firms is modeled as a two stage subgame perfect Nash equilibrium. The government makes the first move by committing itself to an emission tax rate and/or subsidies for abated emissions before abatement and output decisions are made by private firms. The practical implications for environmental policy-making which emerge from the theoretical results is that national welfare can be raised by increasing the deviation of the emission tax from marginal damage.

8.1 INTRODUCTION

The past years have seen the emergence of global threats - ozone depletion, greenhouse effect and global warming - but they haven't seen new forms of international cooperation to cope with these threats. Even the member countries of the European Community couldn't agree to introduce a carbon dioxide tax beginning January 1, 1993. There still seems to be a growing belief that economic growth and environmental protection might be substitutable. Pollution control and abatement expenditures are likely to affect production cost structures and the international competitiveness of industrial firms. As a result of requirements for a higher degree of environmental protection, production costs are likely to increase. Consumer prices of many products will increase as well, especially for those products that involve a relatively polluting production process. Although all the economic agents (including firms and consumers) can contribute to a cleaner environment, a high degree of government intervention is required to make the polluter pay the full cost of polluting.

However, if a member country of, let's say the European internal market, takes initiatives to set up higher standards than required by the European Commission, such moves may be considered as protectionist measures by other member countries or as a strategic environmental policy such as a subsidy for abatement activities.

In this chapter we will evaluate the impact on welfare of an environmental policy with no environmental tax versus a policy with a Pigouvian tax rate equal to marginal damage or equal to a strategic tax rate under noncooperative as well as cooperative behavior. If strategic taxation or a subsidization of abatement measures dominates all other alternatives, we do not have to wonder why environmental objectives are not valued properly in environmental policy.

The methodological background of this chapter is the strategic industrial policy originated by Brander and Spencer (1983, 1985) and extended by Eaton and Grossman (1986). With respect to international externalities, trade policy and (emission) tax competition, it parallels the work by Markusen (1975), Krutilla (1991) and Mintz and Tulkens (1986). In our model the industry in question (e.g. steel or chemicals) is a homogeneous product duopoly located in two countries. There is one producer of the good in each country, interacting as Cournot duopolists. Production will be either for a third-country market or there will be intra-industry trade explained by differences in marginal cost and in consumer preferences. If national governments levy emission taxes, this policy might change the international competitiveness of their firms. Therefore, it is reasonable to expect that governments will act strategically when setting their national environmental policy instruments. With industrial policy and these international strategic interactions deal also the papers by Barrett (1992), Conrad (1993 a, 1993 b), Kennedy (1993) and Ulph (1992). The paper by Barrett is similar in spirit to our homogeneous duopoly with production for a third country market, developed in Conrad (1993 a) and in section 8.2 and 8.4. Whereas we deal with emission taxes and subsidies, Barrett deals with standards. In Ulph production is also for a third market but the governments do not act strategically in terms of a two stage game played by firms and governments. Strategic incentives to distort pollution taxes in free trading economies are considered in Conrad (1993 b) and Kennedy (1993). There is imperfect competition among producers with trade occuring between countries and a tax choice game between competing governments. The conclusions from the models are similar, but the components of the tax rate operate in Conrad (see also section 8.3) in the same direction whereas in Kennedy they can operate in opposite directions. Since there are two types of market imperfections - imperfect competition and negative externality -, we introduce two instruments to achieve cost-effectiveness in environmental policy (see section 8.3 and 8.5).

8.2 ENVIRONMENTAL TAXES IF OUTPUT OF THE INTERNATIONAL DUOPOLY IS FOR A THIRD-COUNTRY MARKET

8.2.1 The Duopolists' Response in Supply to Environmental Taxes

Let us denote by small letters the domestic variables and by capital letters the foreign variables. I begin by analyzing the last stage (the choice of output) in a duopolist's rivalry for greater output shares. The domestic firm produces output x at cost c(x, q(t)), where q(t) is the price of the polluting input (say coal). All other input prices are constant and have been omitted as arguments in the cost function. The input price q(t) consists of the basic price, the cost of abatement and the costs from taxing nonabated emissions:

$$q(t) = q_0 + ca \cdot a \cdot e + t(1-a) \cdot e \tag{8.1}$$

q_0 is the basic price of the input, ca = ca(a) is the unit cost of abatement which depends on the degree of the abatement activity a($0<a<1$), e is an emission coefficient of the input (e.g. tons of SO_2 per ton of input), and t is an emission tax rate. We assume $ca_a>0$ and $ca_{aa}>0$ (using subscripts to denote derivatives).[1] With r for revenue, profit π of the domestic firm is:

$$\pi(x,X;t) = r(x,X) - c(x,q(t)) \tag{8.2}$$

Without loss of generality we examine only the duopoly case with revenue as

$$r(x,X) = p(x + X) \cdot x$$

Similarly, profit Π of the foreign firm is given by:

$$\Pi(x,X;T) = R(x,X) - C(X,Q(T)) \tag{8.3}$$

where[2]

$$Q(T) = Q_0 + CA \cdot A \cdot e + T(1-A) \cdot e \tag{8.4}$$

and $$R(x,X) = p(x + X) . X$$

The Nash equilibrium in outputs is characterized by first order conditions

[1] We therefore assume that abatement costs are proportional to the quantity of emissions to be abated, given the degree of abatement. Hence a firm's aggregated abatement technology has constant returns to scale in abated emissions but costs per unit of abated emissions are to be convex in the degree of abated emissions.

[2] The emission coefficient e of the input is assumed to be the same in the two countries.

$$\pi_x = r_x(x,X) - c_x(x,q(t)) = 0 \tag{8.5}$$

$$\Pi_X = R_X(x,X) - C_X(X,Q(T)) = 0 \tag{8.6}$$

and second order conditions

$$\pi_{xx} < 0 \; ; \; \Pi_{XX} < 0 \, . \tag{8.7}$$

Following the literature (see e.g. Spencer and Brander (1983)) we assume that own effects of output on marginal profit dominate cross effects, giving rise to the following condition:

$$\Omega = \pi_{xx} \Pi_{XX} - \pi_{xX} \Pi_{Xx} > 0 \tag{8.8}$$

This condition is fairly general, and if it holds globally, it ensures uniqueness and global stability of the equilibrium in noncooperative models.

The solutions to (8.5) and (8.6) depend on domestic and foreign emission taxes t and T and can be written as

$$x = x(t,T); \quad X = X(t,T) \tag{8.9}$$

Outputs depend on marginal cost which in turn depend on the price of the pollution-intensive input. This price in turn depends on the degree of abatement a and on the tax rate t (A and T for the foreign firm respectively).

The degree of abatement a(t) is a function of t and its level is chosen prior to production. In order to determine a(t), the firm minimizes the unit cost q of the pollution intensive input in (8.1) with respect to a, given t:

$$\min_a q\,(a;\, t)$$

The first order condition is

$$\frac{dq}{da} = ca_a \cdot a + ca - t = 0 \tag{8.10}$$

i.e. the marginal cost of abatement is equal to the tax rate.
Total differentiation of (8.10) yields da/dt > 0 since $ca_{aa} > 0$ by assumption; the degree of abatement increases with t. Similarly, for the foreign firm:

$$\frac{dQ}{dA} = CA_A \cdot A + CA - T = 0 \tag{8.11}$$

The abatement parameters a and A, determined from (8.10) and (8.11), are treated as fixed in the subsequent output game.

If in the first stage of the game the national governments change t, or T respectively, then the reaction functions x = f(X,t) and X = F(x,T), derived from (8.5) and (8.6), will shift and output and market shares in (8.9) will change. If the domestic government raises the emission tax rate t, then the domestic firm will produce less and the foreign firm will gain market shares. An algebraic analysis requires total differentiation of (8.5) and (8.6) with respect to x, X, t and T. Using (8.7), (8.8) and Shepard's lemma (i.e. $c_q(x,q) = v$ where v is the quantity of the pollution intensive input), we obtain:[3]

$$\frac{dx}{dt} < 0 \,, \quad \frac{dx}{dT} > 0 \,, \quad \frac{dX}{dT} < 0 \,, \quad \frac{dX}{dt} > 0 \tag{8.12}$$

A firm's Nash equilibrium level of output is decreasing in the domestic tax and increasing in the foreign tax. It is straightforward to show that the supply of goods on the third-country market is declining in the tax rate, i.e.

$$\frac{d(x + X)}{dt} < 0 \quad \text{and} \quad \frac{d(x + X)}{dT} < 0. \tag{8.13}$$

In the next section the government is introduced. Its goal is to maximize national welfare in terms of industry profit less damage from pollution, by setting an appropriate emission tax level. Since the two countries are competitors in third-country markets, all output is for export and home consumer surplus is not at issue. In section 8.3 we will extend our model by including home consumer surplus. The governments know the effect of environmental taxes on market shares (i.e. (8.12)) and on total supply i.e. that taxes will increase the world market price (see (8.13)).

8.2.2 Optimal Emission Taxes under Noncooperative Behavior

The domestic government wishes to maximize profit of the domestic firm less damage from global emissions plus the revenue (or minus the cost) of any policies carried out. A global pollutant implies that non-abated foreign emissions also have an impact on national damage and on marginal damage. With d (U) as the convex damage function from total pollution U = (1-a) e .v + (1-A) e .V, the objective function is:

[3] An appendix containing the mathematical details in the paper will be sent by the author on request.

$$\max_{t} \quad w(t;T) = r(x,X) - c(x,q(t)) + t \cdot (1-a) \cdot e \cdot v - d(U) \tag{8.14}$$

Similarly, the objective function of the foreign government is:

$$\max_{T} \quad W(T;t) = R(x,X) - C(X,Q(T)) + T \cdot (1-A) \cdot e \cdot V - D(U) \tag{8.15}$$

Taxes affect the degree of abatement (see (8.10) and (8.11)) but not the resolution of the output game given abatement levels which is represented by (8.9); i.e. x = x(t,T) and X = X(t,T). The first order condition for (8.14) is an implicit reaction function of the domestic government:

$$r_X \frac{dX}{dt} + (1-a)e \cdot v_x \frac{dx}{dt} (t-md) = 0 \tag{8.16}$$

where md (respectively MD) is marginal damage of the domestic (foreign) country. Rewriting (8.16) reveals the structure of the optimal tax on a global pollutant ($\hat{t}_G$), given the foreign tax T:

$$\hat{t}_G = md + \frac{1}{(1-a)e \; v_x \dfrac{dx}{dt}} \left[-\frac{p}{\eta} \frac{x}{x+X} \frac{dX}{dt} + md \cdot (1-A)e \; V_X \frac{dX}{dt} \right] \tag{8.17}$$

where $\eta = \dfrac{d(x+X)}{dp} \dfrac{p}{x+X} < 0$ is the price elasticity of demand on the world market.

Equation (8.17) recommends an optimal tax in terms of marginal damage, adjusted by a rent capture effect and by a transboundary effect, weighted by a technology effect.

Since dX/dt > 0, the two terms in the bracket will be subtracted from the Pigouvian tax rate; i.e. from marginal damage (note that dx/dt < 0). The first term shows the impact of market structure (export share) and of conduct on performance (p-mc) and reflects the effort of the government to capture rent from the world market lost otherwise under pure md-taxation. The downward adjustment of md is enforced by the transboundary externality effect md . (1-A)e V_X. Since under an emission tax, output X of the foreign country increases, it will produce (1-A)e V_X additional emissions which add to global pollution. If however the competing country now contributes more to global pollution, then it makes no sence to restrict domestic production according to a md-taxation. The technology effect in terms of (1-a)e v_x represents the impact of the production technology on the environment.

8.2.3 Optimal Emission Taxes under Cooperative Behavior

Next we discuss the cooperative equilibrium if both countries jointly maximize international welfare. In this case they maximize total welfare with respect to t and T:

$$\max_{t,T} TW(t,T) = r(x,X) - c(x,q(t)) + R(x,X) - C(X,Q(T)) + t(1-a)e \cdot v + T(1-A)e \cdot V - d(U) - D(U). \tag{8.18}$$

If we solve the first order conditions

$$\frac{\partial TW}{\partial t} = 0 \quad and \quad \frac{\partial TW}{\partial T} = 0$$

simultaneously for t and T, we obtain:

$$\hat{t}_{co} = md + MD - \frac{p}{(1-a)e \cdot v_x} \frac{1}{\eta} \frac{X}{x+X}$$
$$\hat{T}_{co} = md + MD - \frac{p}{(1-A)e \cdot V_X} \frac{1}{\eta} \frac{x}{x+X} \tag{8.19}$$

Tax rates are now rather different in size and in their interpretation. They are higher because they reflect marginal damage in both countries. And as $\eta < 0$, they should be even higher than marginal damage. Any consideration of the kind "if I produce less he will pollute more" does not enter into the tax rates. The country with the higher share on the world market should levy a higher tax rate in order to capture more rent in terms of foreign consumer surplus.

We conclude that the tax rate is set below marginal damage in the noncooperative game because this improves firms' strategic positions in the following stage. With government support, they commit themselves to lower costs. When there is collusion in tax rates, firms will improve their strategic position since the governments agree to jointly raise costs.

8.3 ENVIRONMENTAL TAXES UNDER INTRA INDUSTRY TRADE

8.3.1 The Duopolists' Response in Domestic and Foreign Supply to Environmental Taxes

We again assume a duopoly on the world market but now each monopolist produces for the domestic as well as for the foreign market. We again analyze first the last stage of the game, the choice of output for the domestic and for the competing foreign market. The domestic firm produces output x for the domestic market and output x_E for the foreign market. Profit then is

$$\pi(x,x_E,X,X_E;t) = r(x,x_E,X,X_E) - c(x+x_E,q(t)) \tag{8.20}$$

Revenue r is the sum of revenue in the domestic market and of revenue in the foreign market:

$$r(x,x_E,X,X_E) = p(x+X_E)x + P(X+x_E)x_E \tag{8.21}$$

$x+X_E$ is total sales in the domestic country's market and $p(x+X_E)$ is the inverse domestic demand function. $X+x_E$ is total sales in the foreign country's market and $P(X+x_E)$ is the inverse foreign demand function. Similarly, profit Π of the foreign firm is given by:

$$\Pi(X,X_E,x,x_E;T) = R(X,X_E,x,x_E) - C(X+X_E,Q(T)) \tag{8.22}$$

with q(t) and Q(T) as defined in (8.1) and (8.4). The Nash equilibrium in the four outputs is characterized by the first-order conditions:

$$\pi_x = p + xp' - c_{\bar{x}} = 0 \tag{8.23}$$

$$\pi_{x_E} = P + x_E P' - c_{\bar{x}} = 0 \tag{8.24}$$

$$\Pi_X = P + XP' - C_{\bar{X}} = 0 \tag{8.25}$$

$$\Pi_{X_E} = p + X_E\, p' - C_{\bar{X}} = 0 \tag{8.26}$$

where $\bar{x} = x + x_E$, $\bar{X} = X + X_E$. The second-order conditions are:

$$\pi_{xx}<0, \quad \pi_{x_E x_E}<0; \quad \Pi_{XX}<0, \quad \Pi_{X_E X_E}<0$$

and the condition for uniqueness and global stability of the equilibrium is

$$\Omega = (\pi_{xx}+\pi_{xx_E}\frac{p'}{P'})(\Pi_{XX}+\Pi_{XX_E}\frac{P'}{p'}) - \pi_{xX_E}\cdot\Pi_{Xx_E} > 0. \qquad (8.27)$$

The solution of (8.23) - (8.26) depends on domestic and foreign emission taxes t and T and can be written as:

$$x = x(t,T),\ x_E = x_E(t,T),\ X = X(t,T),\ X_E = X_E(t,T) \qquad (8.28)$$

If in the first stage of the game the national governments change t, or T respectively, then the reaction functions $x = f(X,X_E,t)$ and $x_E = g(X,X_E,t)$, derived from (8.23) and (8.24), and $X = F(x,x_E,T)$, $X_E = G(x,x_E,T)$, derived from (8.25) and (8.26), will shift and outputs in (8.28) and hence foreign market shares will change. If the dome stic government raises the emission tax rate t, then the domestic firm will produce and export less and the foreign firm will gain market shares. An algebraic analysis requires total differentiation of (8.23) to (8.26) with respect to x,x_E,X,X_E,t and T. We obtain:[5]

$$\frac{dx}{dt} < 0,\ \frac{dx_E}{dt} < 0,\ \frac{dX}{dt} > 0,\ \frac{dX_E}{dt} > 0 \qquad (8.29)$$

$$\frac{dx}{dT} > 0,\ \frac{dx_E}{dT} > 0,\ \frac{dX}{dT} < 0,\ \frac{dX_E}{dT} < 0 \qquad (8.30)$$

A firm's Nash equilibrium levels of output for the domestic and for the foreign market is decreasing in the domestic tax and increasing in the foreign tax. To simplify the computations we assume the demand functions to be linear, which implies

$$\begin{aligned}\frac{dx_E}{dt} &= \frac{p'}{P'}\frac{dx}{dt}, \quad \frac{dX_E}{dt} = \frac{P'}{p'}\frac{dX}{dt}\\ \frac{dx_E}{dT} &= \frac{p'}{P'}\frac{dx}{dT}, \quad \frac{dX_E}{dT} = \frac{P'}{p'}\frac{dX}{dT}\end{aligned} \qquad (8.31)$$

It can also be shown that the supply of goods on both markets is declining in the tax rate, i.e.

$$\frac{d}{dt}(x + X_E) < 0, \qquad \frac{d}{dt}(X + x_E) < 0 \qquad (8.32)$$

[5] Again, proofs will be sent on request.

$$\frac{d}{dt}(x + X_E) < 0, \qquad \frac{d}{dt}(X + x_E) < 0 \tag{8.32}$$

and similarly for T. Thus the governments know that environmental taxes will raise prices in both markets.

8.3.2 Optimal Emission Taxes and Intra Industry Trade the Noncooperative Case

The goal of the governments is now to maximize national welfare in terms of the sum of home consumer surplus, industry profit, and tax revenue less damage from pollution. Hence the objective function, given the foreign tax rate, is:

$$\max_t w(t;T) = \left[\int_0^{x+X_E} p(\xi)\, d(\xi) - p(x + X_E)(x + X_E) \right] \tag{8.33}$$

$$+ \pi(x,x_E,X,X_E,t) + t(1 - a)\, e \cdot v - d(U)$$

The first-order condition for (8.33) as an implicit reaction function of the domestic government can be rewritten as marginal damage less four strategic tax reducing effects:

$$t_G^* = md + \frac{1}{(1-a)ev_{\bar{x}}\frac{d\bar{x}}{dt}} \left[\frac{p}{\eta_p} \frac{d(x + X_E)}{dt} - \frac{p}{\eta_p} \frac{x}{x + X_E} \frac{dX_E}{dt} \right. \tag{8.34}$$

$$\left. - \frac{P}{\eta_P} \frac{x_E}{X+x_E} \frac{dX}{dt} + md \cdot (1-A)e\, V_{\bar{X}} \frac{d\bar{X}}{dt} \right]$$

where $\eta_p = \dfrac{d(x+X_E)}{d\,p} \dfrac{p}{x+X_E} < 0$ is the domestic price elasticity of demand

and $\eta_P = \dfrac{d(X+x_E)}{d\,P} \dfrac{P}{X+x_E} < 0$ is the foreign price elasticity.

Since $d(x+X_E)/dt < 0$, $d\,X_E/dt > 0$, $dX/dt > 0$ and $d\bar{X}/dt > 0$, all four terms in the bracket are positive. Divided by $d\bar{x}/dt < 0$, four terms will be subtracted from marginal damage. The first term represents the loss in consumer surplus due to the reduced availability of goods on the domestic market. We label it the deadweight-loss-effect;[6] it shows that the government is concerned about higher prices under monopolistic domestic supply and is hence motivated to set the tax rate below md. The second term we label the import-effect; it represents the loss in market power on the domestic market due to import pressure from higher domestic emission taxation. The third term we label the export-effect; it shows the reduced competitiveness of the domestic firm in the foreign market due to a higher production level of the foreign firm thanks to the domestic emission tax. The fourth term is the transboundary-externality-effect which we know already from section 8.2. Finally, the four rent-capturing effects are weighted by the technology-effect which balances the importance of the emission intensive input in production with its impact on the environment. The practical implication from our theoretical result is clear: in an open economy with imperfect competition, policy makers are inclined to set emission taxes below marginal damages. The four tax reducing strategic effects might be so high that there will be no tax at all.

The foreign government can act in a similar way by imposing an emission tax $T_{\hat{G}}$, given the other country is passive. A noncooperative Nash equilibrium in emission taxes occurs if each government is assumed to choose its tax level given the tax level of the other government. The optimal values of the tax rates in equilibrium are the solution of the two implicit reaction functions $\partial w/\partial t = 0$ and $\partial W/\partial T = 0$. Since the tax rate of the foreign government affects welfare of the other government only indirectly through its impact on output levels, the solution implies marginal damage taxation with four rent-capturing deductions. Although the tax rates have the same form as the single country tax rate (8.34), their values will be different at the noncooperative equilibrium. If we compare our tax rate t_G^* in (8.34) with the tax rate $\hat{t}_G$ in (8.17) under third-market production, we recognize that the export-effect, the transboundary-externality-effect and the technology-effect are the same in both formulas. If there is perfect competition in the domestic market ($\eta_p = -\infty$), then they are alike. If there are no abatement measures available at reasonable costs (e.g. CO_2 emissions), then a = A = 0 and the four strategic effects will become a lower weight. Finally, under perfect competition and symmetric firms, t = ½ md and T = ½ MD is the solution of the Nash equilibrium (if the damage functions are the same and marginal costs are constant).

[6] This effect is wellknown from the literature and Baumol and Oates (1989, Chapter 6) is a source of reference on the Pigouvian tax rate under market imperfections.

8.3.3 Optimal Emission Taxes and Intra Industry Trade: the Cooperative Case

Now both governments jointly maximize international welfare with respect to t and T:

$$\max_{t,T} TW(t,T) = \int_0^{x+X_E} p(\xi)d\xi - c(x+x_E,q(t)) + t(1-a)e\ v - d(P) + \int_0^{X+x_E} P(\xi)d\xi - C(X+X_E,Q(T)) + T(1-A)e\ V - D(U) \tag{8.35}$$

with D(P) as the damage function of the foreign country. Since revenue from export is outlay for import by the other country, these trade values cancel out. If we solve the first order conditions

$$\frac{\partial TW}{\partial t} = 0 \quad and \quad \frac{\partial TW}{\partial T} = 0$$

simultaneously for t and T, we obtain:

$$t_{co}^* = (md + MD) + \frac{p}{(1-a)e\ v_{\bar{x}}} \frac{1}{\eta_p} \frac{x + x_E}{x + X_E}$$
$$T_{co}^* = (md + MD) + \frac{P}{(1-A)e\ V_{\bar{X}}} \frac{1}{\eta_P} \frac{X + X_E}{X + x_E} \tag{8.36}$$

Compared to the non-cooperative case, the tax rates now reflect marginal damage in both countries and there is only the deadweight-loss-effect which suggests to lower the tax rates below marginal damage. If there is perfect competition in one country, then the tax rate will be equal to marginal damage in that country. We also note that the cooperative solution for the third-country production case suggested tax rates higher than the sum of marginal damages to capture consumer surplus from the third market. However, (8.36) shows the concern of the governments about their own consumers and hence the tax rates are below md + MD.

8.4 TAXES AND SUBSIDIES UNDER COMPETITION ON THE THIRD-COUNTRY MARKET

8.4.1 The Noncooperative Case

Since there are two types of market failure - negative externalities and imperfect competition - one instrument can not correct both of them. We therefore introduce a subsidy for pollution abatement efforts in addition to the emission tax. If the purpose of the subsidy is to reduce global pollution it should be allowed by GATT codes.

The effect of a subsidy s for abatement activities on the abatement decision and on the price of the pollution intensive input can be specified by modifying (8.1) to

$$q(s,t) = q_0 + (ca - s)\, a \cdot e + t(1 - a)\, e \tag{8.37}$$

First of all, we have to show that a firm's Nash equilibrium level of output is increasing in its own abatement subsidy s and decreasing in the foreign firm's abatement subsidy S. Total differentiation of (8.5) - (8.6) with respect to x, X, s and S given t and T yields:

$$\frac{dx}{ds} > 0 \ , \qquad \frac{dx}{dS} < 0, \tag{8.38}$$

and symmetrically for dX/dS > 0, dX/ds < 0. Of course, the subsidy has the opposite effect in sign on output quantities than the effects derived for taxes on emissions (see (8.12)). The optimal subsidy and tax rates are determined by maximizing national welfare which is profit plus tax revenue minus the cost of the subsidy and damage from emission.

$$\max_{s,t} w = r(x,X) - c(x,q(s,t)) + t(1 - a)e \cdot v - s \cdot a \cdot e \cdot v - d(U) \tag{8.39}$$

where x = f(t,T,s,S) and X = F(t,T,s,S). With T and S fixed by the foreign country, we first examine the incentive facing a single country to subsidize. If the firm knows about the level of s and t, it responds by minimizing q(a;s,t) with respect to the degree of abatement a. The first order condition following from (8.37)is given by:

$$ca_a \cdot a + ca = s + t \tag{8.40}$$

The government, when maximizing the welfare function (8.39) will take into account the change in the abatement coefficient resulting from cost-minimizing behavior of the firms. Rearranging the two first order conditions with respect to t and s yields a positive subsidy rate and a downwards adjusted marginal damage as the emission tax rate:

$$\hat{t}_s = md + \frac{1}{ev_x \dfrac{dx}{dt}} \left[- \frac{p}{\eta} \frac{x}{x+X} \frac{dX}{dt} + md \cdot (1 - A)eV_X \frac{dX}{dt} \right] \tag{8.41}$$

$$\hat{s} = \frac{-1}{ev_x \dfrac{dx}{dt}} \left[- \frac{p}{\eta} \frac{x}{x+X} \frac{dX}{dt} + md\,(1 - A)\, eV_X \frac{dX}{dt} \right] \tag{8.42}$$

We observe a positive subsidy rate (see (8.12)) and its size is equal to the term by which marginal damage is adjusted downwards when taxing emissions. Compared to the non-subsidy tax rate in (8.17), adjustment is now weaker because (1-a) < 1 in the denominator of (8.17) implies a stronger downward adjustment. The tax rate as well as the subsidy rate will be affected by the relevance of the export effect, the transboundary externality effect and by the technology effect. If the export effect matters, the subsidy rate is high and the tax rate will be significantly reduced. Since $\hat{t}_s$ and $\hat{s}$ add to md, firms choose the degree of abatement such as to equate marginal abatement cost and marginal damage by observing $\hat{s} + \hat{t}_s$ (see (8.40)). Hence the tax and subsidy program achieves cost-effectiveness in environmental policy.

Since abatement subsidies increase national welfare, there might be rivalry in abatement subsidies. Abatement subsidies shift out the reaction functions and by imposing such a subsidy, the domestic country gains market shares at the expense of the foreign country. Since the tax-and-subsidy program improves the global environment, it is difficult for a country without such a subsidy program to complain about it. Hence the actions of both governments must be considered. The noncooperative equilibrium which would occur if each government acted independently given the subsidy and tax rates set by the other can be obtained by solving four first order conditions: two from maximizing (8.39) ($\partial w/\partial t = 0, \partial w/\partial s = 0$) and two from maximizing welfare by the foreign government ($\partial W/\partial T = 0, \partial W/\partial S = 0$). Since the tax and subsidy program of each government affects the welfare of the other government only indirectly, i.e. through its impact on the output levels, the tax-and-subsidy programs will have the same form as (8.41) and (8.42).

8.4.2 The Cooperative Case

It is obvious that a noncooperative equilibrium is suboptimal and that total welfare can be raised by cooperative behavior. This implies that the two governments maximize international welfare with respect to s, t, S and T:

$$\max_{s,t,S,T} TW = r(x,X) - c(x,q(s,t)) + R(x,X) - C(X,Q(S,T)) + (t(1-a) - s \cdot a)e \cdot v + (T(1-A) - S \cdot A)e \cdot V - d(U) - D(U) \tag{8.43}$$

where Q(S,T) is specified as q(s,t) in (8.37). In order to derive optimal tax and subsidy rates we have to solve four first order conditions simultaneously. The result is:

$$\hat{t}_{s,co} = md + MD - \frac{1}{e\, v_x}\frac{p}{\eta}\frac{X}{x+X}, \qquad \hat{s}_{co} = \frac{1}{e\, v_x}\frac{p}{\eta}\frac{X}{x+X} < 0 \tag{8.44}$$

$$\hat{T}_{S,co} = md + MD - \frac{1}{e\, V_X}\frac{p}{\eta}\frac{x}{x+X}, \qquad \hat{S}_{co} = \frac{1}{e\, V_X}\frac{p}{\eta}\frac{x}{x+X} < 0 \tag{8.45}$$

Thus, the optimal policy is to tax abatement activities.[7] Since own exports exert a negative externality on the other country's revenue, the tax rate is higher than total marginal damage. Besides marginal damage taxation both instruments are used to raise marginal cost and hence prices for rent-capturing purposes.

8.5 A TAX AND SUBSIDY PROGRAM UNDER INTRA-INDUSTRY TRADE

8.5.1 The Noncooperative Case

To show the effect of a subsidy program on exports and imports we have to differentiate (8.23) - (8.26) totally with respect to x, x_E, X, X_E, s and S given t and T. We obtain:

[7] This result is similar in spirit to the taxation of R&D rivalry developed by Spencer and Brander (1983).

$$\frac{dx}{ds} = \left[- \frac{a}{1-a} \right] \frac{dx}{dt} > 0, \qquad \frac{dx_E}{ds} = \frac{p'}{P'} \frac{dx}{ds} > 0$$

$$\frac{dx}{dS} = \left[- \frac{A}{1-A} \right] \frac{dx}{dT} < 0, \qquad \frac{dx_E}{dS} = \frac{p'}{P'} \frac{dx}{dS} < 0. \tag{8.46}$$

Similar relations hold for the foreign country. The optimal subsidy and tax rate is determined by maximizing national welfare which is the sum of consumer surplus, profit of the domestic firm, net revenue from the tax-and-subsidy program minus damage from emissions:

$$\max_{s,t} w(s,t;S,T) = \left[\int_0^{x+X_E} p(\xi)d\xi - p(x + X_E)(x + X_E) \right] + \pi(x,x_E,X,X_E,t,s) + t(1 - a)e \cdot v - s \cdot a \cdot e \cdot v - d(U) \tag{8.47}$$

where again all quantities are functions of t, T, s and S. With T and S fixed by the foreign country we find a tax rate as in (8.34), but without (1-a) in the denominator, and a subsidy rate with the four rent-capturing effects:

$$t_s^* = md + \frac{1}{e\, v_{\bar{x}} \frac{d\bar{x}}{dt}} \tag{8.48}$$

[see the four strategic effects in (8.34)]

$$s^* = \frac{-1}{e\, v_{\bar{x}} \frac{d\bar{x}}{dt}} \tag{8.49}$$

[see the four strategic effects in (8.34)]>0

Again the instruments add to md, so that the firm equates marginal abatement costs to marginal damage. The dead weight-loss-effect, the import effect, the export effect and the transboundary externality effect have a positive impact on the size of the subsidy rate and a negative impact on the emission tax rate. Again the noncooperative equilibrium in the tax and subsidy rate could be derived by solving four first order conditions. The structure of the equations for the four instruments will be the same as in (8.48) and (8.49), but the rates will be different at the noncooperative equilibrium.

8.5.2 The Cooperative Case

Let us finally look at the set of instruments if the two government maximize international welfare with respect to s, t, S and T:

$$\begin{aligned}\max_{s,t,S,T} TW = & \left[\int_0^{x+X_E} p(\xi)d\xi - p(x+X_E)(x+X_E) \right] \\ & + \left[\int_0^{X+x_E} P(\xi)\, d\xi - P(X+x_E)(X+x_E) \right] \\ & + \pi(x,x_E,X,X_E;s,t,S,T) + \Pi(x,x_E,X,X_E;s,t,S,T) \\ & + [t(1-a) - s \cdot a]e \cdot v + [T(1-A) - S \cdot A]e \cdot V \\ & - d(U) - D(U)\end{aligned} \tag{8.50}$$

Since the export-import relations will cancel out, the governments maximize consumers' surplus, profits, net revenue from the program less damages from emissions with strategic trade considerations. In order to derive optimal tax and subsidy rates we have to solve four first order conditions simultaneously. The result is:

$$t^*_{S,co} = md + MD + \frac{1}{e \cdot v_{\bar{x}}} \left[\frac{x + x_E}{p^{-1}\eta_p(x+X_E) + P^{-1}\eta_P(X+x_E)} \right]$$

$$s^*_{co} = -\frac{1}{ev_{\bar{x}}} \text{ [see the bracket above] } > 0$$

$$T^*_{S,co} = md + MD + \frac{1}{e \cdot V_{\bar{X}}} \left[\frac{X + X_E}{p^{-1}\eta_p(x+X_E) + P^{-1}\eta_P(X+x_E)} \right] \tag{8.51}$$

$$S^*_{co} = -\frac{1}{e \cdot V_{\bar{X}}} \text{ [see the bracket above] } > 0$$

Thus, the optimal tax policy is to tax emissions below marginal damage and to subsidize abatement efforts. Compared to the cooperative tax-and subsidy program when production is for a third-country market (see (8.44) and (8.45)), the most important difference is that abatement activities will now be subsidized and not taxed.

The reason is obvious; whereas the governments are not concerned about deadweight loss due to inflation on the third country market, they are concerned about there own consumers' deadweight loss. Therefore the price elasticities in both countries have an influence on the optimal deductions from the sum of marginal damages and on the size of the subsidy rates. In section 8.3.3 the government used only taxes under cooperative behaviour and their rates differed if there was perfect competition in one country (see (8.36)). In (8.51) however, perfect competition in one country implies t = md + MD in both countries and no subsidies. Again, compared to a cooperative tax program a cooperative tax-and-subsidy program is cost-effective as the two instruments add to md + MD which is equal to marginal abatement costs in either country.

8.6 CONCLUDING REMARKS

Our purpose has been to show that there are vested interests for low emission taxes for own selfish reasons given the level of the policy instruments in the competing country. Our approach permits to integrate international aspects of the environmental issue such as national environmental policy and its influence on competitiveness and trade, the specific issue of transfrontier pollution and the global environmental problem. Given the results derived in the chapter, there will be no prior harmonization of environmental instruments after the completion of the European Market. Maybe a competitive process among national environmental authorities will bring about a harmonization ex post. The conclusion from our model is that no environmental taxes are an inferior strategy dominated by other strategies. Also a Pigouvian marginal damage tax will be dominated by other tax strategies. However, whereas a Pigouvian tax is never an optimal tax in a strategic setting, a zero emission tax is if marginal damage taxation will be exactly compensated by rent-capturing effects. Even under a cooperative environmental policy we can not expect a Pigouvian tax as an optimal policy.

REFERENCES

Barrett, S. (1992) *Strategic environmental policy and international trade*. CSERGE Working Paper, GEC 92-19.

Baumol, W. J. and Oates, W.E. (1988) *The theory of environmental policy*. 2. ed., Cambridge Univ. Press.

Brander, J. A. and Spencer, B. J. (1983) Strategic commitment with R&D: The symmetric case. *Bell Journal of Economics*, 14, 225-235.

Brander, J. A. and Spencer, B. J. (1985) Export subsidies and international market share rivalry. *Journal of International Economics*, 18, 83-100

Butler, A. (1992) *Environmental protection and free trade: Are they mutually exclusive?* Federal Reserve Bank of St. Louis, 3-16.

Conrad, K. (1993) Taxes and subsidies for pollution intensive industries as trade policy. *J. Environ. Econom. Management*, 25, 121-135 (a).

Conrad, K. (1993) *Optimal environmental policy for oligopolistic industries in an open economy*. Disc. paper No. 476-93. Mannheim Univ.

Eaton, J. and Grossman, G. M. (1986) Optimal trade and industrial policy under oligopoly. *Quarterly Journal of Economics*, 100, 383-406.

Kennedy, P. *Equilibrium pollution taxes in open economies with imperfect competition*, Disc. paper 93-01, Department of Economics, Univ. of Victoria, British Columbia.

Krutilla, K. (1991) Environmental regulation in an open economy. *J. Environ. Econom. Management* 20, 127-142.

Markusen, J.R. (1975) International externalities and optimal tax structure. *J. International Econom*. 5, 15-29.

Mintz, J. and Tulkens, M. (1986) Commodity tax competition between member states of a federation: Equilibrium and efficiency. *Journal of Public Economics* 29, 133 - 172.

Spencer, B.J. and Brander, J. A. (1983) International R&D rivalry and industrial strategy. *Review of Economic Studies*, 50, 707 - 722.

Ulph, A. (1992) *The choice of environmental policy instruments and strategic international trade*. In Pethig, R. (ed.): Conflicts and Cooperation in Managing Environmental Resources, Springer.

Chapter 9 ENVIRONMENTAL POLICY AND TRADE LIBERALIZATION UNDER IMPERFECT COMPETITION

Peter W. Kennedy[1]

ABSTRACT

This chapter examines the strategic incentives to distort pollution taxes in open economies with imperfect competition. I first examine a world economy in which a pollution tax is the only instrument available. The associated free-trading equilibrium pollution taxes are distorted away from the globally efficient taxes by three separate effects: the usual transboundary externality effect, a rent capture effect and a pollution-shifting effect. I then examine a setting in which both a pollution tax and a production subsidy are available. The equilibrium in which taxes and subsidies are set non-cooperatively is globally efficient in the absence of transboundary pollution. When pollution is at least partly transboundary the equilibrium subsidy is positively distorted by a strategic incentive to capture foreign rents and the equilibrium tax is negatively distorted by the usual transboundary externality. Finally, I examine a restricted equilibrium in which countries agree not to use subsidies for trade-related goals. In the absence of transboundary pollution the pre- and post agreement equilibria coincide and both are efficient. When pollution is transboundary the pollution tax takes on the strategic trade role previously played by the subsidy and is negatively distorted relative to the pre-agreement equilibrium and relative to the globally efficient solution.

9.1 INTRODUCTION

One of the most contentious issues raised in the debate over the North American Free Trade Agreement (NAFTA) was the likely impact that trade liberalization would have on environmental quality. Opponents of the deal argued that the United States and Canada would be forced to weaken their environmental standards in an attempt to compete with Mexico. Proponents of NAFTA counter-argued that freer trade would make Mexicans richer and thus more demanding of higher environmental quality. Free trade would therefore lead to higher environmental standards in North America. This type of debate is likely to arise increasingly around the world. The

[1] Department of Economics University of Victoria Victoria, British Columbia Canada V8W 3P5 December 1993.

latest round of GATT talks for the most part excluded environmental issues, but the environment is almost surely going to be a sticking point in the next round of talks. Similarly, any move towards a Pacific free-trade zone or an expansion of the European Union to include Eastern European countries will likely face resistance based at least in part on fears about the impact of freer trade on environmental quality. Moreover, concerns about trade and the environment are not limited to instances of trade between rich countries and relatively poor countries. The same sort of concerns were raised in the debate over the Canada-United States free trade agreement (the precursor to the NAFTA) and the issue continues to generate disputes within the existing European Union.

There is undoubtedly some truth in both sides of the argument over trade and the environment. Gains from trade will make countries richer and this may tend to produce higher environmental standards (although possibly still with lower environmental quality in the aggregate).[2] On the other hand, there surely exist strategic incentives to relax environmental standards for trade-related goals. Which effect will dominate in the end is an empirical question; no amount of theory is likely to resolve this debate definitively. However, there is a clear role for theoretical analysis in clarifying the debate and focusing attention on key aspects of the relationship between trade liberalization and environmental policy.

The purpose of this chapter is to provide an analysis of the strategic incentives to distort environmental policy in open economies in the presence of imperfect competition. The framework for the analysis has three main features. First, global trade is driven by imperfect competition among producers. Trade occurs between countries because the multinational firms based in those countries find it profitable to sell in foreign markets. This oligopolistic framework allows me to abstract from the traditional bases for trade (such as comparative advantage) and focus on strategic effects. Second, the tax choice game between competing countries is modeled explicitly. This permits an examination of strategic policy distortions in equilibrium. Third, the model allows for transboundary pollution. A single parameter captures the extent to which pollution is transboundary and this permits an examination of a continuum of cases, from purely local pollution to perfectly transboundary pollution. The chapter has two main parts. In the first part I examine a world in which a pollution tax is the only instrument available. I compare the free-trading Nash equilibrium taxes with the globally efficient taxes in this setting. My purpose is to highlight the nature of the strategic distortions to pollution taxes that may arise in

[2] The higher world output associated with freer trade may produce more pollution even if production techniques become cleaner.

free-trading economies. The second part of the chapter extends consideration to a setting in which both pollution taxes and production subsidies are available. I focus on production subsidies because the issue of "unfair" production subsidies has created so much conflict in recent trade talks surrounding both the GATT and the NAFTA.[3] Production subsidies can have a legitimate role in an imperfectly competitive setting as a measure to correct for the associated under-production problem. However, there exist incentives to misuse (distort) production subsidies for trade-related goals. I examine these incentives in the context of a non-cooperative Nash equilibrium in pollution taxes and unrestricted subsidies. I then examine the implications of a trade liberalization agreement that prohibits the use of subsidies for trade-related goals. I focus on the implications for equilibrium pollution taxes.

The rest of the chapter is organized as follows. Section 9.2 presents the model. Section 9.3 compares the Nash equilibrium taxes with the globally efficient taxes when a pollution tax is the only instrument available. Section 9.4 then derives the Nash equilibrium in pollution taxes and unrestricted subsidies. Section 9.5 examines the Nash equilibrium under a trade-liberalization agreement. Section 9.6 concludes. For a review of related literature the reader is referred to Kennedy (1994).

9.2 THE MODEL

There are two identical countries each producing a polluting homogeneous good x. Production in each country occurs within a symmetric oligopolistic industry with n firms. The entry of new firms is prevented by some sunk cost already incurred by established firms. The 2n firms compete openly in the two markets. Marginal production cost is constant and equal to the chosen level of pollution abatement θ. Output by a representative firm in country i is denoted y_i. This output is divided between home country sales y_i^H and foreign country sales y_i^F. Total production by country i firms is denoted Y_i. Production in country i generates pollution $Z_i=(Y_i/\theta_i)$ in country i, and a fraction $\alpha \in [0,1]$ of this pollution also affects the other country. If $\alpha=0$ then pollution is purely local. If $\alpha=1$ then pollution is perfectly transboundary.[4] Pollution generates environmental damage in country i according to the function $e_i=e(Z_i+\alpha Z_{-i})$ where a "-i" subscript denotes the other country. Damage is increasing and convex in pollution. The inverse demand for x in each country is $p(X_i)$ where X_i

[3] For example, alleged subsidies in aircraft and automobile manufacturing were a major source of difficulty in the last round of GATT talks. Similarly, the issue of subsidies in timber production continues to generate considerable conflict between the United States and Canada.

[4] Note that the degree to which pollution is transboundary is entirely independent of the extent of trade. It is determined only by chemical and geographical factors.

is the amount sold in country i.[5] The notational distinction between X_i and Y_i is needed to allow for imports and exports. Welfare in each country is measured as the sum of consumer surplus and profit, less environmental damage. Country i levies a tax τ_i on pollution and pays a subsidy s_i on domestic production.

9.3 EQUILIBRIUM WITHOUT SUBSIDIES

In this section I compare the Nash equilibrium taxes with the globally efficient taxes when a pollution tax is the only instrument available. This section is based on Kennedy (1994) and the reader is referred to that source for a more extensive analysis.[6]

9.3.1 Efficiency

It should be noted at the outset that the efficient taxes I present here are second-best. The first-best solution can only be achieved with the use of a pollution tax *and* a production subsidy to address the under-production associated with the imperfect competition. I consider the role of the production subsidy in section 9.4.

The planning problem is to set a uniform tax on pollution to maximize the welfare of a representative country, given the equilibrium behaviour of firms.[7] Welfare is defined as the sum of consumer surplus and profit, less environmental damage. The first-order condition to this planning problem is

$$(p - t)(\partial X/\partial t) - X = (1+\alpha)e'[t(\partial X/\partial t) - X]t^2 \tag{9.1}$$

where $e' \equiv \partial e/\partial Z$ and $t \equiv \tau^{1/2}$. The LHS is marginal abatement cost. The RHS is marginal global damage. Note that marginal abatement cost has two components: the first term reflects the welfare cost of the reduced output associated with the tax; the second term reflects the increased marginal cost of production.

[5] Note that this inverse demand function implies the absence of income effects.

[6] Related papers of interest include Barrett (1994), Cumberland (1979), Markusen et.al. (1993), Mintz and Tulkens (1986), Oates and Schwab (1988), Ulph (1992) and Wildason (1988).

[7] It is sufficient to consider a uniform tax because countries are identical, and because damage is strictly convex in pollution.

In a perfectly competitive market in which price is equal to the tax-inclusive marginal cost, equation (9.1) would reduce to $t^2 = e'(1+\alpha)$ or $\tau = e'(1+\alpha)$. That is, the tax would be set equal to marginal damage; this is the standard Pigouvian rule. In contrast, in an imperfectly competitive equilibrium, price is distorted above social marginal cost, and so there is a strictly positive welfare cost associated with a reduction in output. This means that the efficient tax is set below marginal damage.[8] This result can be seen more clearly by rewriting (9.1) as

$$(t - e'(1+\alpha)/t) = Xp'\epsilon/(\epsilon-1)2n \tag{9.2}$$

where $\epsilon \equiv (\partial Y/\partial t)t/Y$ is the elasticity of equilibrium output with respect to t. The RHS of (9.2) is negative for finite n so the optimal t^2 is less than $e'(1+\alpha)$.

9.3.2 Nash equilibrium

The governments in both countries will choose their tax rates knowing that the choice of the domestic tax rate will affect the world price and therefore affect production both in the home country and in the foreign country. This is the source of the strategic interaction between the two countries. Each country is able to affect the aggregate reaction function for the firms based there by changing its tax rate. Manipulation of the reaction function in turn has the potential to alter the equilibrium aggregate production level and the equilibrium mix of production. Each country has an incentive to exploit this to its advantage.

The policy problem for each national government is to set its tax rate to maximize domestic welfare, given the equilibrium behaviour of firms and given the tax rate of the other country. The game between governments gives rise to the following symmetric equilibrium condition:

$$(t - e'/t) = Xp'\epsilon_i/(\epsilon_i-1)2n + \alpha e'\epsilon_{-i}/(\epsilon_i-1)t \tag{9.3}$$

where $\epsilon_i \equiv (\partial Y_i/\partial t_i)t_i/Y_i$ and $\epsilon_{-i} \equiv (\partial Y_{-i}/\partial t_i)t_i/Y_{-i}$. Comparing (9.3) with the efficiency condition in (9.2) clearly indicates that the Nash equilibrium will generally not coincide with the globally efficient solution. At the globally efficient taxes there is an incentive for each country to deviate.

The equilibrium incentive to deviate from the efficient tax can be thought of as comprising three separate effects. The first is the usual *transboundary externality effect*.

[8] For an analogous result in the monopoly case see Barnett (1980) and Lee (1975).

Each country ignores the impact of the pollution created within their boundaries on the environment outside their boundaries, and tax rates therefore tend to be set lower than is globally efficient.[9] If pollution is purely local then this effect vanishes. It is important to stress that this source of inefficiency does not stem from the openness of the economies per se. The same transboundary pollution problem would persist in a non-cooperative equilibrium between closed economies.

The second effect is a *rent capture effect*. A unilateral change in a country's tax rate has a bigger effect on equilibrium domestic production than it does on equilibrium domestic consumption (since foreign costs are not affected). The difference is reflected in a change in net exports. A unilateral reduction in the domestic tax rate therefore has the potential to raise net exports and thereby permit the capture of rent from foreigners. This rent comprises the profits from net exports plus the tax revenue earned on net exports. Both contribute directly to the domestic surplus. The rent capture effect is negative; that is, it tends to negatively distort the equilibrium tax rates from their efficient levels. The reason is as follows. The pollution tax raises the marginal cost of production for firms in the taxing country and so adversely affects the ability of those firms to compete for net exports. Each country attempts to gain a competitive advantage over its trading partner by reducing the tax. Note that this distortion is a purely strategic effect arising directly from the openness of the economies. Moreover, no rents are actually captured in equilibrium because both countries act symmetrically, but tax rates are distorted nonetheless. The distortions are therefore purely destructive.

The third effect is a *pollution shifting* effect and it works in the opposite direction to the rent capture effect. A unilateral tax increase tends to shift production and its associated pollution to the other country. It is therefore possible to achieve a reduction in domestic pollution with a lower adverse impact on domestic consumption than would be possible in a closed economy. Of course the strategy is potentially effective only if pollution is not perfectly transboundary. If pollution is perfectly transboundary then shifting production to the other country provides no relief for the domestic environment. In that case the pollution shifting effect vanishes. When pollution is less than perfectly transboundary the pollution shifting effect is strictly positive; it tends to positively distort the equilibrium tax rates away from their efficient levels. The effect is strongest when pollution is purely local because none of the shifted pollution returns to damage the domestic environment. Like the rent capture effect, the pollution shifting effect is a purely strategic effect that arises

[9] It should be noted that the relevant "boundaries" here are not necessarily geographical boundaries. The relevant boundaries are the boundaries of the interests of domestic residents.

directly from the openness of the economies. It is also purely destructive. Pollution shifting yields no benefits in equilibrium because each country acts symmetrically, but equilibrium taxes are nonetheless distorted in the attempt to shift pollution.

The rent capture effect and the pollution shifting effect operate in opposite directions, so the net strategic effect could potentially be either positive or negative. It turns out that the rent capture effect dominates the pollution shifting effect except in the case of perfect competition with no transboundary pollution.[10] In that special case the two effects are exactly offsetting. The reason is the following. Diverting production to the foreign country through a unilateral increase in the tax rate has a cost in foregone rent, and a benefit in shifted pollution. Under perfect competition the foregone rent is simply the tax revenue. When pollution is purely local the private benefit from shifting pollution is equal to the domestic social benefit of reducing production. This cost and benefit are necessarily equal at a social optimum implemented by a Pigouvian tax. Note that the usual transboundary externality also vanishes when pollution is purely local so the Nash equilibrium is efficient in this special case. In all other cases the Nash equilibrium is inefficient and pollution taxes are too low relative to the global optimum. In particular, imperfect competition leads to a strictly negative net strategic effect, even when pollution is purely local: the rent captured under imperfect competition exceeds the tax revenue, and so must also exceed the benefits from shifting pollution. When pollution is at least partly transboundary then the benefits of shifting pollution are even smaller and so the rent capture effect must continue to dominate. This net negative effect reinforces the usual transboundary externality giving rise to equilibrium taxes that are lower than is globally efficient.[11]

9.4 EQUILIBRIUM WITH SUBSIDIES

In this section I introduce production subsidies. I examine the Nash equilibrium taxes and subsidies before the introduction of any agreement to restrict the use of subsidies. It will be useful to begin with a characterization of the globally efficient taxes and subsidies.

[10] See Kennedy (1994) for the proof.

[11] The net strategic effect also tends to compound the environmental damage associated with the usual transboundary externality because the lower taxes lead to higher output.

9.4.1 Efficiency

The tendency towards under-production in an industry characterized by imperfect competition implies a potential role for a production subsidy in implementing an efficient allocation. The actual role for a production subsidy depends on the underlying reason for the imperfect competition and on whether or not direct competition policy can better enhance competition. I abstract from this issue of choosing between alternative policies and simply assume that a production subsidy is the instrument of choice for boosting production. This is justified by the fact that production subsidies are widely observed in real economies and have been one of the main items of concern in recent world trade negotiations.

The planning problem is to set the taxes and subsidies to maximize global welfare. The first step towards solving this problem is to characterize the industry equilibrium when all firms face the same tax and subsidy rates.

Industry equilibrium

The problem for the representative firm based in country i is

$$\max_{y^H_i, y^F_i, \theta_i} p(X_i)y^H_i + p(X_{-i})y^F_i - (\theta_i - s)y_i - \tau(y_i/\theta_i) \tag{9.4}$$

where τ is the tax rate on pollution and s is the production subsidy. Demand conditions are identical in the two markets so the firm will sell half its output in each country. The conditions for profit-maximization therefore reduce to

$$p + s + yp'/2 = (\theta_i + \tau/\theta_i) \tag{9.5}$$

$$\theta^2_i = \tau \tag{9.6}$$

where the country subscripts have been omitted because τ and s are independent of i, and where $p \equiv p(X)$ and $p' \equiv p'(X)$. The second-order conditions are satisfied if $3p' + yp'' < 0$. This condition is henceforth assumed. Recalling that $t \equiv \tau^{1/2}$, these conditions can be rewritten as $\theta = t$ and

$$p + s + yp'/2 = 2t \tag{9.7}$$

where 2t is the after-tax marginal cost faced by the firm. The Nash equilibrium can be found by multiplying (9.7) by 2n and setting $ny = X$ to obtain

$$2np + 2ns + p'X - 4nt = 0 \tag{9.8}$$

The stability of this equilibrium is guaranteed by the constancy of marginal cost and the second-order conditions for (9.4). The effects of symmetric policy changes on domestic consumption in both countries can be derived from (9.8):

$$\partial X/\partial t = 4n/[(2n+1)p' + Xp''] \tag{9.9}$$

$$\partial X/\partial s = -2n/[(2n+1)p' + Xp''] \tag{9.10}$$

It follows from the second-order conditions for profit-maximization that $\partial X/\partial t<0$ and $\partial X/\partial s>0$.

The planning problem

The planning problem can be formulated as maximizing the welfare of a representative country, given the equilibrium behaviour of firms embodied in (9.9) and (9.10). Throughout I assume that subsidies can be financed by lump-sum taxes and that pollution tax revenues are returned to citizens in a lump-sum fashion. I abstract from any income effects associated with this intervention. Welfare is defined as the sum of consumer surplus and profits (gross of tax payments and net of subsidy receipts), less environmental damage:

$$W = [\int_0^x p(\tilde{X})d\tilde{X} - pX] + (pX - 2tX + \tau X/t) - e((1+\alpha)X/t) \tag{9.11}$$

$$= [\int_0^x p(\tilde{X})d\tilde{X} - tX - e((1+\alpha)X/t)$$

since tax revenue is $\tau X/t = tX$. The second-order condition for the problem in (9.4) and the convexity of e(.) are sufficient to ensure that W is concave in t and s. The first-order conditions to the planning problem are

$$(p - t)(\partial X/\partial t) - X = (1+\alpha)e'[t(\partial X/\partial t) - X]/t^2 \tag{9.12}$$

$$(p - t)(\partial X/\partial s) = (1+\alpha)e'(\partial X/\partial s)/t \tag{9.13}$$

Joint solution of (9.12) and (9.13) yields conditions for the efficient tax and subsidy rates respectively:

$$t^2 = (1+\alpha)e' \tag{9.14}$$

$$p = t + (1+\alpha)e'/t \tag{9.15}$$

Consider condition (9.15) first. This condition states that the optimal subsidy is chosen to equate price with the marginal global social cost of production. Marginal global social cost comprises the private marginal cost of production t (since $\theta = t$ in equilibrium), plus the marginal cost of global environmental damage. The planner uses the subsidy to correct for the inefficiency associated with under-production due to imperfect competition.
Now consider condition (9.14). This condition states that the pollution tax is optimally set equal to marginal global environmental damage. This is the standard Pigouvian taxing rule. This result contrasts with the planning solution in section 9.3 where a subsidy is unavailable. In that case the pollution tax must be moderated to take account of the under-production associated with the imperfect competition. However, when the planner has two instruments available, the subsidy can be used to fully correct for the under-production and so the pollution tax can be freely set at its Pigouvian level.

9.4.2 Nash equilibrium

Now consider the Nash equilibrium in taxes and subsidies. Each country chooses its tax and subsidy rates, taking as given the other country's policies, and the equilibrium behaviour of firms. To derive the equilibrium policies, I begin by characterizing the industry equilibrium when tax and subsidy rates can differ across countries.

Industry equilibrium

The problem for the representative firm based in country i is the same as the problem in (4) except that τ is replaced by τ_i and s is replaced by s_i. The first-order conditions are given by $\theta = t$ and

$$p + s_i + y_i p'/2 = 2t_i \tag{9.16}$$

Multiplying (9.16) by n and setting $ny_i = Y_i$ then yields the aggregate reaction function for firms based in country i:

$$2np + 2ns_i + Y_i p' = 4nt_i \tag{9.17}$$

An analogous reaction function can be derived for firms based in the other country. Solving for the equilibrium yields

$$2np + p'X = 2n(t_i + t_{-i}) - n(s_i + s_{-i}) \tag{9.18}$$

where $X=(Y_i + Y_{-i})/2$ is the equilibrium consumption in each country. Substituting (9.18) into (9.17) then yields

$$Y_i = X + 2n(t_i - t_{-i})/p' - n(s_i - s_{-i})/p' \tag{9.19}$$

It will soon be shown that the relationship between market share and the relative policy rates reflected in (9.19) creates a strategic incentive for each country to distort its policies so as to capture rent from the foreign country. But it is first necessary to state some comparative static results. The effects of unilateral policy changes in country i on domestic consumption in both countries can be derived from (9.18):

$$\partial X/\partial t_i = 2n/[(2n+1)p' + Xp''] \tag{9.20}$$
$$\partial X/\partial s_i = -n/[(2n+1)p' + Xp''] \tag{9.21}$$

It follows from the second-order conditions for profit-maximization that $\partial X/\partial t_i < 0$ and $\partial X/\partial s_i > 0$. Furthermore, it is straightforward to show that $\partial Y_i/\partial t_i < 0$ and $\partial Y_i/\partial s_i > 0$, and $\partial Y_{-i}/\partial t_i > 0$ and $\partial Y_{-i}/\partial s_i < 0$.[12]

Nash equilibrium taxes and subsidies

Now consider the game between national governments. Each country chooses its policies to maximize domestic welfare taking as given the equilibrium behaviour of firms and the policies of the other country. Domestic welfare is defined as domestic consumer surplus, plus profits to domestic firms (gross of tax payments and net of subsidy receipts), less domestic environmental damage:

$$W_i = [\int_0^x p(\tilde{X})d\tilde{X} - pX] + (pY_i - t_iY_i) - e(\alpha Y_{-i}/t_{-i} + Y_i/t_i) \tag{9.22}$$

[12] These comparative static results follow directly from the stability of the equilibrium. See Dixit [7] for further discussion of comparative statics in oligopoly.

Differentiating (9.22) with respect to t_i and s_i, and setting $t_i=t_{-i}=t$, $s_i=s_{-i}=s$ and $Y_i=Y_{-i}=X$, yields the symmetric Nash equilibrium conditions:[13]

$$(p - t)(\partial Y_i/\partial t_i) - X - \alpha e'(\partial Y_{-i}/\partial t_i)/t - e'[t(\partial Y_i/\partial t_i) - Y_i]/t^2 = 0 \qquad (9.23)$$

$$(p - t)(\partial Y_i/\partial s_i) - e'[\alpha(\partial Y_{-i}/\partial s_i) + (\partial Y_i/\partial s_i)]/t = 0 \qquad (9.24)$$

It is straightforward to show, using (9.17), that $(\partial Y_i/\partial s_i)=-(\partial Y_i/\partial t_i)/2$ and $(\partial Y_{-i}/\partial s_i)=-(\partial Y_{-i}/\partial t_i)/2$. Making these substitutions in (9.24) and substituting the resulting expression into (9.23) yields

$$t^2 = e' \qquad (9.25)$$

This condition states that in Nash equilibrium the pollution tax is set equal to marginal domestic damage. Two points are noteworthy about this equilibrium tax-setting rule. First, it is inefficient from a global perspective if pollution is transboundary. Recall from (9.14) that global efficiency requires that the tax be set equal to marginal global damage. For a given level of pollution, the equilibrium tax rate is too low from a global perspective when pollution is transboundary. The second point to note is that this inefficiency has nothing to do with trade per se. It is simply a reflection of the usual transboundary externality that would persist even without trade.

Now consider condition (9.24) for the equilibrium subsidy rate. By adding and subtracting $t(dX/ds\ )$, and noting that $t=e'/t$, this condition can be usefully rewritten as

$$\begin{aligned}(p - 2t)(\partial X/\partial s_i) &+ (p - t)[(\partial Y_i/\partial s_i) - (\partial X/\partial s_i)] \\ &- e'[(\partial Y_i/\partial s_i) - (\partial X/\partial s_i) + \alpha(\partial Y_{-i}/\partial s_i)]/t = 0\end{aligned} \qquad (9.26)$$

The three terms in (9.26) correspond to three separate incentives underlying the subsidy choice. Consider each term in turn. The first term represents the gain in social surplus (price less marginal domestic social cost) from a marginal increase in domestic consumption. This term reflects the role of the subsidy in correcting for the

[13] Existence of equilibrium is only assured if the subsidy rate cannot be set arbitrarily large. Political consideration of domestic surplus distribution is likely to provide the necessary constraint on policy.

under-production associated with imperfect competition. This is the same role that the subsidy plays in the global planning problem. It has nothing to do with trade per se. Note also that this first term would vanish under perfect competition, when equilibrium price would be equal to the after-tax marginal cost. There are no non-strategic welfare gains from boosting production in the perfectly competitive case.

The second term in (9.26) reflects the role of the subsidy in affecting the distribution of monopoly rents between the two countries. The subsidy tends to boost net exports (Y_i-X) for the home country and thereby transfer rents from the foreign country. Equivalently, it serves to protect domestic firms from foreign firms that might otherwise capture the rents available in the domestic market. This is a purely strategic use of the production subsidy designed to distort trade flows.[14] It is to this use of production subsidies that trade liberalization negotiations have in principle been directed. The potential welfare gains from eliminating this strategic use of the production subsidy are clear. The attempted distortion of trade flows is purely destructive in equilibrium. Both countries attempt to distort trade flows in their favour but these attempts are mutually offsetting in equilibrium. All that remains in equilibrium is the welfare cost of the distorted production decisions.

The subsidy-induced boost to net exports comes at the cost of higher domestic pollution and this tends to moderate the incentive to capture rents through setting a high subsidy. This moderating effect is reflected in the third term in (9.26). It is clearly reminiscent of the pollution shifting effect identified in section 9.3. It tends to reduce the equilibrium subsidy. The magnitude of this moderating effect depends on the degree to which pollution is transboundary. If pollution is purely local ($\alpha=0$) then domestic pollution rises by the entire amount associated with the increase in net exports. The moderating effect is strongest in this case. However, if pollution is to some extent transboundary then there is an offsetting effect: the subsidy-induced reduction in foreign production causes a reduction in domestic pollution. Domestic damage does not increase commensurately with the increase in domestic production because the transboundary element of the pollution associated with the displaced foreign production is eliminated from the domestic environment. If pollution is perfectly transboundary ($\alpha=1$) then the substitution of domestic production for foreign production has no impact at all on the domestic environment. Domestic pollution rises only by the amount due to the subsidy-induced increase in foreign consumption.[15] The incentive to set a lower subsidy is weakest in this case.

[14] The production subsidy plays the same strategic role as of an export subsidy. In fact, in this symmetric setting the two are substitute instruments.

[15] To see this in (9.26), note that $Y_i+Y_{-i}=2X$, so $(\partial Y_i/\partial s_i) + \alpha(\partial Y_{-i}/\partial s_i) = 2(\partial X/\partial s_i)$ when $\alpha=1$.

If pollution is purely local then the conflicting incentives associated with rent capture and pollution shifting are exactly offsetting. To see this, rewrite (9.26) as

$$(p - 2t)(\partial Y_i/\partial s_i) - e'\alpha(\partial Y_{-i}/\partial s_i)/t = 0 \tag{9.27}$$

When $\alpha=0$ the only incentive remaining is to correct for the under-production: the subsidy is chosen to equate price with marginal domestic social cost. The strategic incentive to distort the subsidy is neutralized for the following reason. When pollution is purely local the entire increase in net exports contributes to new domestic pollution with no transboundary offset. These net exports therefore have a domestic environmental cost of $e'/t=t$. But this is precisely the amount of marginal rent captured on net exports when price is set at marginal domestic social cost. There is therefore no net gain to boosting net exports by setting a higher subsidy than is needed to correct for the under-production problem. In contrast, when pollution is at least partly transboundary ($\alpha>0$) then boosting net exports can capture rents in excess of the domestic environmental cost incurred because the substitution of domestic production for foreign production does not have a commensurate effect on domestic damage. In this case the equilibrium subsidy is set too high relative to what is globally efficient and gives rise to an equilibrium price below marginal domestic social cost.

This is clear from (9.27) since $(\partial Y_{-i}/\partial s_i)<0$. This distortion imposes a welfare cost on both countries. When pollution is transboundary there are two additional sources of inefficiency in equilibrium. The first is that associated with the equilibrium tax-setting rule discussed earlier. The equilibrium tax is set equal to marginal domestic damage rather than marginal global damage. The second source of inefficiency relates to the role of the subsidy in correcting for under-production. Recall from (9.15) that global efficiency requires price to be equated with marginal global social cost. However, in the non-cooperative equilibrium, each country is concerned only with marginal domestic social cost; this excludes the cost of production associated with damage done outside its boundaries. Thus, strategic trade considerations aside, each country will tend to over-subsidize production in correcting for the perceived under-production associated with imperfect competition. It is important to reiterate that neither of these inefficiencies are due to strategic distortions associated with trade. They are due to the usual transboundary externality problem and would persist in a non-cooperative equilibrium between closed countries when pollution is transboundary.

When pollution is purely local there are no transboundary externalities and there are no strategic trade distortions. In this case the Nash equilibrium is efficient. Note that this equilibrium coincides with the equilibrium in section 9.3 without subsidies when there is perfect competition and no transboundary pollution. The reason is straightforward: the equilibrium subsidy effectively implements a competitive industry equilibrium allocation.

9.5. EQUILIBRIUM WITH RESTRICTED SUBSIDIES

I now turn to the Nash equilibrium when subsidies are restricted by a trade liberalization agreement. I confine consideration to an agreement which prohibits the use of subsidies for trade-related goals. In particular, the agreement allows the use of subsidies to correct for the under-production associated with imperfect competition but prohibits the use of subsidies for the purely strategic goal of distorting trade flows. I assume that the permitted subsidy is that which equates equilibrium price with marginal global social cost. That is, the trade agreement explicitly recognizes the domestic and transboundary costs of pollution. We have seen from the previous section that there is no equilibrium incentive for either country to set a subsidy lower than that which equates price and marginal global social cost, so the agreed restriction on subsidies will be binding in equilibrium.[16] It follows that in the post-agreement equilibrium:

$$p = t + (1+\alpha)e'/t \tag{9.28}$$

What does the agreement imply for equilibrium pollution taxes? Expression (9.23) specifies the Nash equilibrium tax rate choice for a given subsidy level. Substituting (9.28) into (9.23) therefore yields an expression for the Nash equilibrium tax rate when the subsidy is set at the restricted level:

$$(1 - e'/t^2) = \alpha e'[(\partial Y_i/\partial t_i) - (\partial Y_{-i}/\partial t_i)]/tX \tag{9.29}$$

It is clear from (9.29) that if there is no transboundary pollution ($\alpha=0$) then the equilibrium tax is set equal to marginal domestic (and global) damage. That is, the agreement does not distort the tax-setting rule relative to the pre-agreement equilibrium, and moreover, the post-agreement equilibrium is globally efficient. This follows from the fact that there is no strategic distortion of trade flows in the pre-agreement equilibrium when pollution is not transboundary. The agreement simply binds countries to behave in a manner that is privately optimal anyway.

[16] In the case of no transboundary pollution it will weakly binding.

On the other hand, if pollution is at least partly transboundary then the agreement causes a negative distortion of the tax setting-rule relative to the pre-agreement equilibrium. When $\alpha>0$ the RHS of (9.29) is negative, and so $t^2<e'$. That is, the pollution tax is set below marginal domestic damage in the post-agreement equilibrium. This is clearly inefficient: efficiency requires that the tax be set equal to global marginal damage. This inefficiency reflects both the usual transboundary externality, and the strategic distortion of the pollution tax in response to the trade agreement. This strategic distortion arises because the agreement does not eliminate the incentive for each country to influence trade flows; it simply restricts the instruments that can be used for that purpose. When the subsidy is restricted by agreement, the pollution tax must play the role of both a corrective tax and a strategic trade instrument.

It is not possible to make a definitive welfare comparison between the pre- and post-agreement equilibria when there is transboundary pollution because both equilibria are inefficient. However, the welfare cost associated with the distortion of the pollution tax has the potential to more than offset the welfare gains associated with restricting subsidies. Both countries could be made worse-off by the trade liberalization agreement. To highlight the potential gains and losses, it is perhaps useful to summarize the nature of the distortions in each equilibrium relative to the globally efficient solution when pollution is transboundary. In the pre-agreement equilibrium there is a strategic incentive to set too high a subsidy from a global perspective. This is compounded by the transboundary externality which inflates the non-strategic element of the subsidy (that part directed at the under-production problem). There is therefore too much production. Moreover, the pre-agreement equilibrium tax-setting rule ignores the damage done by transboundary pollution and so is negatively distorted. However, the actual equilibrium tax rate could be higher or lower than in the efficient solution because marginal domestic damage (to which the tax rate is equated) will be higher due to the higher level of production if damage is strictly convex. In contrast, the post-agreement equilibrium is characterized by the efficient subsidy-setting rule but a tax-setting rule that is negatively distorted by both strategic trade incentives and the transboundary externality. The actual post-agreement equilibrium tax rate is too low, and the induced technology is too dirty. This in turn generates more production than is efficient.

9.6 CONCLUSION

This chapter has examined the strategic incentives to distort pollution taxes in open economies with imperfect competition. I began by examining a world economy in which a pollution tax is the only instrument. The associated free-trading equilibrium exhibits pollution taxes that are distorted away from the globally efficient taxes by three separate effects: the usual transboundary externality effect, a rent capture effect and a pollution-shifting effect. The rent capture effect tends to reduce equilibrium taxes as each country attempts to gain a competitive advantage over its trading partner and thereby capture foreign rent through net exports. The pollution-shifting effect tends to raise equilibrium taxes as each country attempts to transfer production and its associated pollution to the other country. This effect vanishes if pollution is perfectly transboundary because shifted pollution causes as much damage to the domestic environment as does domestic pollution. The net effect on symmetric equilibrium taxes is negative, except in the special case of perfect competition with no transboundary pollution. In this case the two effects are mutually offsetting and the Nash equilibrium is efficient.

I then turned to a setting in which both a pollution tax and a production subsidy are available to national governments. My results indicate that the properties of the non-cooperative and restricted equilibria in this setting depend crucially on the degree to which pollution is transboundary. The equilibrium in which taxes and subsidies are set non-cooperatively is globally efficient in the absence of transboundary pollution. In particular, there is no distortion of either pollution taxes or production subsidies relative to the globally efficient solution. Conversely, when pollution is at least partly transboundary then the non-cooperative equilibrium is inefficient. The equilibrium subsidy is positively distorted by the strategic incentive to capture foreign rents (or equivalently, to protect domestic rents). The equilibrium tax-setting rule is negatively distorted, reflecting the usual transboundary externality. There is no purely strategic distortion of the pollution tax; the entire strategic trade role falls on the subsidy. I then examined a restricted equilibrium in which countries agree to prohibit the use of subsidies for trade-related goals. The use of a production subsidy is restricted to correcting for the under-production associated with imperfect competition. In the absence of transboundary pollution, this agreement changes nothing. The pre- and post agreement equilibria coincide exactly and both are efficient. In contrast, when pollution is transboundary, the strategic trade role that before the agreement lay with the subsidy is taken on by the pollution tax in the post-agreement equilibrium. In particular, the tax-setting rule is negatively distorted relative to the pre-agreement equilibrium and relative to the globally efficient solution. The agreement therefore has an ambiguous effect on welfare when pollution is transboundary. Both countries could be made worse off by the agreement. Only when there is no transboundary pollution is the welfare effect of the agreement unambiguous. In that case it is welfare neutral.

The pollution tax distortion and the ambiguous welfare effect associated with the trade liberalization agreement are of course due to the less-than-comprehensive nature of the agreement. A comprehensive coordination agreement between symmetric countries covering all policy variables cannot make either country worse off. However, to be assured of welfare gains, the agreement must be truly comprehensive. All policy instruments may need to be coordinated. Any policy instruments that are not explicitly coordinated potentially become subject to strategic distortions when the strategic use of other instruments is restricted. The key point is that the incentives to distort trade flows are not eliminated by a trade liberalization agreement. A trade agreement only restricts the means by which countries can act on those incentives. International agreements must extend beyond the elimination of traditional trade barriers such as tariffs and "unfair" subsides if they are to reliably deliver the welfare gains promised by textbook models of competitive world trade.

REFERENCES

Barnett, A. (1980) The Pigouvian tax rule under monopoly. *American Economic Review*, 70, 1037-1041.

Barrett, S. (1994) Strategic environmental policy and international trade. forthcoming in *Journal of Public Economics*.

Cumberland, H. (1979) *Interregional pollution spillovers and consistency of environmental policy* in H. Siebert et. al. (eds), *Regional Environmental Policy: the Economic Issues*. New York: New York University Press.

Dixit, A. (1986) Comparative statics for oligopoly. *International Economic Review*, 27, 107-122, (1986).

Kennedy, P. (1994), Equilibrium pollution taxes in open economies with imperfect competition, forthcoming in *Journal of Environmental Economics and Management*.

Lee, D. (1975) Efficiency of pollution taxation and market structure. *Journal of Environmental Economics and Management*, 22, 69-72.

Markusen, J., E. Morey and N. Olewiler (1992) *Non-cooperative equilibria in regional environmental policies when plant locations are endogenous*. mimeo, University of British Columbia.

Mintz J. and H. Tulkens (1986) Commodity tax competition between member states of a federation: equilibrium and efficiency. *Journal of Public Economics*, 29, 133-172.

Oates W. and R. Schwab (1988) Economic competition among jurisdictions; efficiency enhancing or distortion inducing?. *Journal of Public Economics*, 35, 333-354.

Ulph, A. (1992) *The choice of environmental policy instruments and strategic international trade*. in Pethig, R. (ed.), Conflicts and Cooperation in Managing Environmental Resources, Berlin: Springer Verlag.

Wildason, D. (1988) Nash equilibria in models of fiscal competition. *Journal of Public Economics*, 35, 229-240.

Chapter 10 ENVIRONMENTAL POLICY IN A SMALL OPEN ECONOMY WITH DISTORTIONARY LABOUR TAXES: A GENERAL EQUILIBRIUM ANALYSIS

A. Lans Bovenberg, Ruud A. de Mooij[1]

ABSTRACT

This chapter illustrates the trade-off between aggregate private income, employment, equity, and the quality of the environment. Raising pollution taxes on households and using the revenues to cut distortionary labour taxes may raise aggregate private income, employment, and environmental quality. However, the income distribution between households becomes less equitable as non-labour incomes fall. Pollution taxes affecting firms harm employment even if the revenues are used to cut taxes on labour. Subsidies for cutting pollution, in contrast, may expand employment but only if they fail to improve environmental quality.

10.1 INTRODUCTION

European countries face serious environmental problems at a time when the aging of the population and rising social spending are putting pressure on the public finances. Accordingly, policy aims at curbing pollution while at the same time boosting employment to strengthen the revenue base of the public sector. In this connection, environmental taxes may seem particularly attractive.[2] As economic instruments, they may achieve environmental objectives with relatively little costs. Moreover, the revenues these taxes generate may be used to cut distortionary taxes on labour, thereby stimulating employment and reducing the distortions imposed by the tax system (see e.g. Oates (1991)).

This chapter adopts a general equilibrium model of a small open economy with distortionary labour taxes to explore the macro-economic effects of various instruments of environmental policy, such as pollution taxes, environmental subsidies, and environmental levies that allow for tax-free allowances. The model is employed

[1] We are grateful to Peter van Bergeijk, Ben Heijdra, Jarig van Sinderen and Ed Westerhout for helpful suggestions.

[2] Smith (1992) surveys various aspects of environmental taxation.

to perform explicit welfare analysis aimed at examining, in a second best framework, the interactions between environmental and labour-market distortions and between environmental policy and the public finances. In addition to the effects on employment, environmental quality, aggregate private income, and welfare, the impact on the income distribution between households is analysed. In connection to this, we examine alternative ways to use revenues from environmental policy instruments that yield revenue. In particular, either the labour tax or the value-added tax may be cut. Alternatively, the government may raise public transfers.

The results reveal that governments in small open economies face important trade-offs between equity, environmental quality, employment, and aggregate private income. To illustrate, policy packages that boost both environmental quality and employment worsen the income distribution by reducing non-labour incomes. If the government protects these latter incomes, it can cut pollution only at the price of lower employment and lower aggregate private income.

The chapter is structured as follows. Section 10.2 presents the model. Section 10.3 examines environmental taxes and subsidies on households. Pollution taxes and subsidies involving firms are dealt with in section 10.4. Section 10.5 presents some numerical simulations of the macro-economic impact of energy taxes. Finally, section 10.6 contains the main conclusions.

10.2 THE MODEL

This section discusses the model contained in table 10.1. Notation is defined at the end of the table. The model describes a small open economy facing exogenous world-market prices for commodities and capital. As labour is immobile internationally, the wage rate is the only price that is determined endogenously. The model distinguishes four sectors, namely the production, consumption, public and foreign sectors.

Production

Firm behavior is described by a representative firm that supplies a single tradable commodity. The firm maximizes profits subject to a neo-classical production function (10.1), which exhibits constant returns to scale. Capital (K) combines with an input that causes pollution when used in production, called the polluting or dirty input (E), to produce a composite factor (N). The latter combines with labour (L) to yield output (Y) (see figure 10.1). The implicit demand functions for the three inputs are given by (10.2), (10.3) and (10.4), while (10.5) represents the non-profit condition. E^* denotes the pollution that firms may deduct before the pollution tax is assessed. We call E^* the tax-free pollution allowance for producers.[3] Two price equations (10.6) and (10.7) link marginal prices faced by the firm to the market wage (P_l) and the market price of the dirty input (P_e).

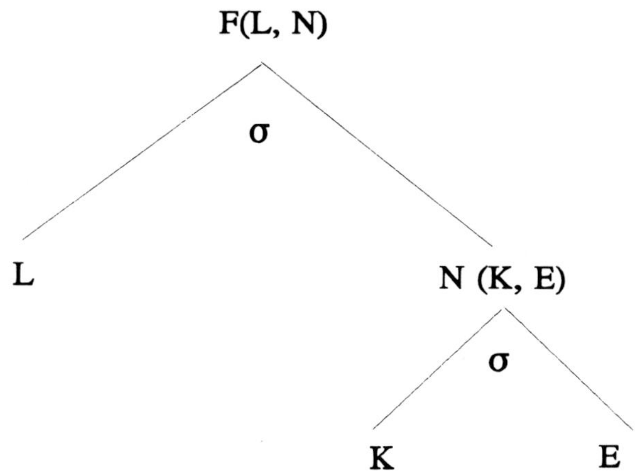

Figure 10.1 The production structure.

Consumption

The economy is populated by two classes of households. Households in different classes feature the same preferences but differ in their sources of income. Households within the same class are identical, both in terms of preferences and income sources. The representative household in class i, which is indexed by i, optimizes a homothetic utility function of the form (10.8). One of the two consumption commodities, the "dirty" commodity, causes environmental damage when consumed (d^i), while the other, the "clean" commodity (c^i), does not. The two commodities are combined to yield a composite consumption bundle, q^i, in the utility function (see figure 10.2). The substitution elasticity between dirty and clean consumption is denoted by σ_q.

[3] The tax-free allowance only refers to existing firms. This prevents the situation that an infinite number of new firms enter the market, that all benefit from the tax-free allowance by producing an infinitely small amount of product.

Table 10.1 The model in levels

Firms

Production function:

$$Y = F[L_d, N(K,E)] \tag{10.1}$$

First order conditions:

$$P_y \frac{\partial Y}{\partial L_d} = P_y \frac{\partial F}{\partial L_d} = P_l^f \tag{10.2}$$

$$P_y \frac{\partial Y}{\partial K} = P_y \frac{\partial F}{\partial N}\frac{\partial N}{\partial K} = r \tag{10.3}$$

$$P_y \frac{\partial Y}{\partial E} = P_y \frac{\partial F}{\partial N}\frac{\partial N}{\partial E} = P_e^f \tag{10.4}$$

Non-profit condition:

$$P_y Y = P_l^f L_d + P_e^f E + rK - t_e E^* \tag{10.5}$$

Price equations:

$$P_l^f = (1+t_l)P_l \tag{10.6}$$

$$P_e^f = P_e + t_e \tag{10.7}$$

Households

Utility function:

$$u^i = U[Q(c^i, d^i), v^i, M] \tag{10.8}$$

Budget restriction:

$$(1+t_c)[P_c c^i + P_d d^i + t_d(d^i - D^*)] = P_l l_s^i + s^i + rw^i \tag{10.9}$$

First order conditions:

$$\frac{\partial U}{\partial v^i} = \lambda P_l \tag{10.10}$$

$$\frac{\partial U}{\partial c^i} = \frac{\partial U}{\partial Q}\frac{\partial Q}{\partial c^i} = \lambda(1+t_c)P_c \tag{10.11}$$

$$\frac{\partial U}{\partial d^i} = \frac{\partial U}{\partial Q}\frac{\partial Q}{\partial d^i} = \lambda(1+t_c)P_d^h \tag{10.12}$$

Price equation:

$$P_d^h = P_d + t_d \tag{10.13}$$

Time constraint:

$$l_s^i = tt^i - v^i \tag{10.14}$$

Aggregates:

$$X = \sum_{i=1}^{2} N^i x^i \quad for \quad x = c,d,l_s,tt,v,w,s \tag{10.15}$$

Government

Balanced budget:

$$P_g G + S = t_e[E - E^*] + (1 + t_c)t_d[D - D^*] + t_l P_l L + t_c[P_c C + P_d D] \tag{10.16}$$

Equilibrium on the labourmarket:

$$L_s = L_d \tag{10.17}$$

Walras law: Balance of payments

$$P_e E + rK + P_d D + P_c C + P_g G = P_y Y + rW \tag{10.18}$$

Environmental quality:

$$M = M(D,E) \tag{10.19}$$

Notation of table 10.1

quantities:

Y = domestic production
K = domestic demand for capital services
E = domestic demand for polluting inputs
L_d = domestic demand for labour
V = aggregate demand for leisure
M = environmental quality

D = aggregate demand for polluting consumption commodities
C = aggregate demand for clean consumption commodities
Q = composite consumption bundle

L_s = aggregate labour supply
tt^i = labour endowment of household i
S = lump-sum transfers to households provided by the government

W = stock of financial2 wealth owned by domestic households
G = demand for public goods
N^i = number of households in class i

prices:

P_y = world market price of domestic output
r = rate of return on capital
P_e = world market price of dirty inputs
P_e^f = price of dirty inputs faced by domestic firms
P_l = market wage
P_l^f = before-tax wage (wage costs faced by domestic producers)
P_c = world market price of clean commodities

P_d = world market price of dirty commodities

P_d^h = price of dirty commodities faced by domestic households
P_g = world market price of public goods

P_q = ideal price index for the composite consumption bundle Q faced by domestic households

λ = marginal utility of income

taxes:

t_e = specific tax on dirty inputs
t_d = specific tax on dirty consumption commodities
t_l = ad-valorem tax on wage income
t_c = ad-valorem tax on consumption

Allowances:

E^* = tax-free pollution allowance for firms
D^* = tax-free pollution allowance for households

In addition, leisure (v^i) and environmental quality (M) enter utility. The quality of the environment is a collective good. Hence, it is exogenous to the household decision problem. σ_u denotes the substitution elasticity between leisure and consumption. The household budget available for consumption consists of three categories: labour income ($P_l l^i$), lump-sum income transfers from the government (s^i), and the part of capital income that is not saved (rw^i). Households differ in their labour endowments, tt^i, and in the non-labour income, $s^i + rw^i$, they receive. D^* represents the tax-free pollution allowance which is the same for both households. The tax-free allowance amounts to a fictive lump-sum transfer because it is independent of labour supply. Accordingly, we'll classify it as part of non-labour income. Equations (10.10), (10.11) and (10.12) represent the implicit expressions for labour supply and the demands for the two consumption commodities, respectively. Equation (10.13) links the consumer price (exclusive of the value-added tax) of the polluting commodity (P_d^h) to its market price (P_d). Moreover, a value-added tax (t_c) that applies to the value of consumption inclusive of the specific environmental tax (t_d), raises the consumer price level of both clean and dirty consumption commodities. Equation (10.15) aggregates a number of variables across households where N^i denotes the number of households in class i.

Government

The government supplies public goods (G) and provides lump-sum income transfers to households. To finance its expenditures the government relies on four taxes: an ad-valorem tax on labour income[4] (t_l), a value-added tax on consumption commodities (t_c), and two specific environmental taxes, one on polluting inputs in production (t_e) and one on polluting consumption commodities (t_d). In addition to environmental levies, tax-free allowances and environmental subsidies are included. In particular, the government may allow a tax-free pollution allowance. This can be interpreted as the free distribution of pollution rights or, alternatively, as lump-sum transfers provided by the government. If households or firms pollute less than the tax-free allowance, they receive a subsidy for each unit of pollution below the allowance.

[4] This tax applies neither to transfer income nor capital income.

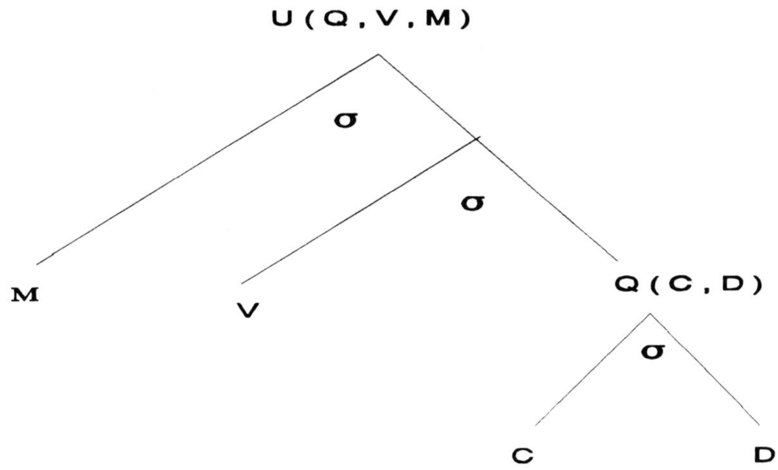

Figure 10.2 The structure of the utility function.

The government budget is assumed to be balanced (see (10.16)). In the rest of this chapter, we examine alternative ways to balance the budget once the government applies a particular instrument of environmental policy. In particular, the government may adjust the tax on labour income, the value-added tax, or (lump-sum) transfers to households.

Foreign sector

All commodities in the economy are tradable. Hence, their market-prices are determined on world-markets and are thus exogenous to the small open economy. Moreover, the rate of return on capital (r) is exogenous because capital is perfectly mobile internationally. Equilibrium on the labour-market is given by (10.17). (10.18) represents the balance of payments equation. It states that the value of the domestic supply of tradables constraints domestic demand for tradables.[5]

[5] Using Walras law, one can derive this equation by combining the non-profit condition (10.5), the household budget constraint (10.9) and the government budget constraint (10.16).

Table 10.2 The model in relative changes

Firms

Domestic production:

$$\tilde{Y} = \alpha_{lf}\tilde{L}_d + \alpha_{ef}\tilde{E} + \alpha_k\tilde{K} \tag{10.20}$$

Labour demand:

$$\tilde{L}_d = \tilde{Y} + \sigma_l[\alpha_{ef}(\tilde{P}_e^f - \tilde{P}_l^f) + \alpha_k(\tilde{r} - \tilde{P}_l^f)] \tag{10.21}$$

Demand for capital:

$$\tilde{K} = \tilde{Y} + \sigma_l\alpha_{lf}[\tilde{P}_l^f - \tilde{r}] + \frac{\alpha_{ef}}{\alpha_{ef} + \alpha_k}[\sigma_n - \sigma_l\alpha_{lf}](\tilde{P}_e^f - \tilde{r}) \tag{10.22}$$

Demand for polluting inputs:

$$\tilde{E} = \tilde{Y} + \sigma_l\alpha_{lf}(\tilde{P}_l^f - \tilde{P}_e^f) + \frac{\alpha_k}{\alpha_{ef} + \alpha_k}[\sigma_n - \sigma_l\alpha_{lf}](\tilde{r} - \tilde{P}_e^f) \tag{10.23}$$

Non-profit condition:

$$\tilde{P}_y = \alpha_{lf}\tilde{P}_l^f + \alpha_{ef}\tilde{P}_e^f + \alpha_k\tilde{r} - \alpha_{lf}\tilde{A}^f \tag{10.24}$$

Real value of the tax free pollution allowance for firms:

$$\alpha_{lf}\tilde{A}^f = \alpha_{ef}^*\tilde{t}_e + \beta_e\tilde{E}^* \tag{10.25}$$

Before-tax wage rate:

$$\tilde{P}_l^f = \tilde{P}_l + \tilde{t}_l \tag{10.26}$$

Producer-price for polluting inputs:

$$\tilde{P}_e^f = (1 - \frac{t_e}{P_e^f})\tilde{P}_e + \tilde{t}_e \tag{10.27}$$

Households

Budget restriction:

$$\tilde{P}_q + \tilde{Q} = (1-\phi_{nl})(\tilde{P}_l+\tilde{L}_s) + \phi_w(\tilde{r} + \tilde{W}) + \tilde{S} + \tilde{A}^h \tag{10.28}$$

Real value of the tax free pollution allowance for households:

$$\tilde{A}^h = \frac{\alpha_{dh}^*}{\alpha_q}\tilde{t}_d + \frac{\beta_d}{\alpha_q}\tilde{D}^* + \frac{\beta_d^*}{\alpha_q}\tilde{t}_c \tag{10.29}$$

Aggregate labour supply:

$$\tilde{L}_s = \theta_{ll}^m \tilde{P}_l^n - \theta_l^m \tilde{S}^* \tag{10.30}$$

Aggregate demand for polluting consumption commodities:

$$\tilde{D} = \tilde{Q} + \phi_c \sigma_q (\tilde{P}_c - \tilde{P}_d^h) \tag{10.31}$$

Aggregate demand for clean consumption commodities:

$$\tilde{C} = \tilde{Q} + (1-\phi_c)\sigma_q(\tilde{P}_d^h - \tilde{P}_c) \tag{10.32}$$

Price level of the consumption bundle:

$$\tilde{P}_q = \phi_c \tilde{P}_c + (1-\phi_c)\tilde{P}_d^h + \tilde{t}_c \tag{10.33}$$

Real after tax wage rate:

$$\tilde{P}_l^n = \tilde{P}_l - \tilde{P}_q \tag{10.34}$$

Real value of aggregate non-labour income:

$$\tilde{S}^* = \tilde{S} + \tilde{A}^h + \phi_w(\tilde{r} + \tilde{W}) - \phi_{nl}\tilde{P}_q \tag{10.35}$$

Household price for polluting consumption commodities:

$$\tilde{P}_d^h = (1 - \frac{t_d}{P_d^h})\tilde{P}_d + \tilde{t}_d \tag{10.36}$$

Government

Balanced budget:

$$\begin{aligned}\alpha_g(\tilde{P}_g + \tilde{G}) + \alpha_q\tilde{S} &= [\alpha_{ef} - \alpha_{ef}^*]\tilde{t}_e + [\alpha_{dh} - \alpha_{dh}^*]\tilde{t}_d \\ &+ \alpha_{lf}\tilde{t}_l + [\alpha_q - \beta_d^*]\tilde{t}_c + \beta_e[\tilde{E} - \tilde{E}^*] \\ &+ \beta_d[\tilde{D} - \tilde{D}^*] + \beta_l(\tilde{P}_l + \tilde{L}) + \beta_c[\phi_c(\tilde{P}_c + \tilde{C}) \\ &+ (1 - \phi_c - \frac{\beta_d}{\alpha_q})(\tilde{P}_d + \tilde{D})]\end{aligned} \tag{10.37}$$

Equilibrium

Labour-market:

$$\tilde{L}_s = \tilde{L}_d \tag{10.38}$$

Walras law

Balance of payments:

$$\begin{aligned}\tilde{Y} + \tilde{P}_y + \alpha_w(\tilde{W} + \tilde{r}) &= \alpha_k(\tilde{K} + \tilde{r}) + \alpha_e(\tilde{E} + \tilde{P}_e) \\ &+ \alpha_d(\tilde{D} + \tilde{P}_d) + \alpha_c(\tilde{C} + \tilde{P}_c) + \alpha_g(\tilde{G} + \tilde{P}_g)\end{aligned} \tag{10.39}$$

Environmental quality:

$$\tilde{M} = -\theta_e \tilde{E} - \theta_d \tilde{D} \tag{10.40}$$

Endogenous variables:

$$\tilde{Y},\ \tilde{L}_d,\ \tilde{K},\ \tilde{E},\ \tilde{P}_l^f,\ \tilde{P}_e^f,\ \tilde{Q},\ \tilde{L}_s,\ \tilde{D},\ \tilde{C},$$

$$\tilde{P}_q,\ \tilde{P}_l^n,\ \tilde{P}_d^h,\ \tilde{P}_l,\ \tilde{M},\ \tilde{S}^*,\ \tilde{A}^f,\ \tilde{A}^h$$

Exogenous variables:

$$\tilde{P}_e = \tilde{P}_d = \tilde{r} = \tilde{P}_y = \tilde{P}_c = \tilde{P}_g = \tilde{G} = \tilde{W} = 0$$

$$\tilde{t}_e,\ \tilde{t}_d,\ \tilde{E}^*,\ \tilde{D}^* \geq 0$$

Either endogenous or exogenous:

$$\tilde{t}_c,\ \tilde{t}_l,\ \tilde{S}$$

Notation table 10.2:

$$\tilde{t}_l = \frac{\Delta t_l}{1+t_l},\quad \tilde{t}_c = \frac{\Delta t_c}{1+t_c},\quad \tilde{t}_e = \frac{\Delta t_e}{P_e^f},\quad \tilde{t}_d = \frac{\Delta t_d}{P_d^h},$$

$$\tilde{S} = \frac{\Delta S}{P_q Q},\quad \tilde{E}^* = \frac{\Delta E^*}{E},\quad \tilde{D}^* = \frac{\Delta D^*}{D}$$

Basic parameters

σ_l = substitution-elasticity between labour and the composite production factor in production
σ_n = substitution-elasticity between capital and polluting inputs in production
σ_q = substitution-elasticity between dirty and clean consumption commodities
σ_u = substitution-elasticity between leisure and consumption
θ_e = elasticity that measures the relative change in M due to a rise in E
θ_d = elasticity that measures the relative change in M due to a rise in D
$\eta = \dfrac{\partial U/\partial M}{\partial U/\partial Q}\dfrac{M}{Q}$ the marginal rate of substitution between M and Q, multiplied by M/Q

Derived parameters

$\theta_l^i = \dfrac{v^i}{tt^i - \phi_{nl}^i v^i}$ income elasticity of labour supply of household i

$\theta_{ll}^i = [\sigma_u - (1-\phi_{nl}^i)]\theta_l^i$ uncompensated wage elasticity of labour supply of household i
$\theta_{ll}^{*i} = \theta_l^i\sigma_u = \theta_{ll}^i + (1-\phi_{nl}^i)\theta_l^i$ compensated wage elasticity of labour supply of household i

$\theta_l^m = \displaystyle\sum_{i=1}^{2}\frac{N^i l_s^i}{L_s}\frac{\phi_{nl}^i}{\phi_{nl}}\theta_l^i$ income elasticity of aggregate labour supply

$\theta_{ll}^{m} = \sum_{i=1}^{2} \frac{N^i l_s^i}{L_s} \theta_{ll}^i$ uncompensated wage elasticity of aggregate labour supply

$\theta_{ll}^{*m} = \quad \theta_l^m \sigma_u = \theta_{ll}^m + (1-\phi_{nl})\theta_l^m$ compensated wage elasticity of aggregate labour supply

$$\sigma_l^* = \frac{\alpha_k}{\alpha_k + \alpha_{ef}} \sigma_n + \frac{\alpha_{ef}}{\alpha_k + \alpha_{ef}} \frac{\sigma_l}{\alpha_{lf}}$$

$$Z = \alpha_d + \alpha_c = \frac{1}{1+t_c}[\alpha_q - \beta_d] > 0$$

$$R = \beta_l + \beta_e + (1-\phi_{nl})[\frac{\beta_d}{1+t_c} + \beta_c] > 0$$

$$\det(1) = (1-\phi_{nl})(Z + \theta_l^m R) - \theta_{ll}^{*m} R > 0$$

$$\det(2) = (1 - \frac{\beta_d^*}{\alpha_q})(Z + \theta_l^m R) - \theta_{ll}^{*m} R > 0$$

$$\det(3) = (Z + \theta_l^m R) - \theta_{ll}^{*m} R > 0$$

Shares in P_yY

α_{lf}	$= P_l^f L/P_y Y$	α_l	$= P_l L/P_y Y$
α_{ef}	$= P_e^f E/P_y Y$	α_e	$= P_e E/P_y Y$
α_k	$= rK/P_y Y$	α_{dh}	$= (1+t_c)P_d^h D/P_y Y$
α_d	$= P_d D/P_y Y$	α_q	$= P_q Q/P_y Y$
α_c	$= P_c C/P_y Y$	α_g	$= P_g G/P_y Y$
α_{nl}	$= (S + rW + (1+t_c)t_d D^*)/P_y Y$	α_{ef}^*	$= P_e^f E^*/P_y Y$
α_{dh}^*	$= (1+t_c)P_d^h D^*/P_y Y$	β_e	$= t_e E/P_y Y$
β_d	$= (1+t_c)t_d D/P_y Y$	β_c	$= t_c P_q Q/(1+t_c)P_y Y$
β_d^*	$= (1+t_c)t_d D^*/P_y Y$	β_l	$= t_l p_l L/P_y Y$

Shares in P_qQ

$\phi_w = rW/P_q Q$
$\phi_{nl} = 1 - (P_l L/P_q Q)$
$\phi_c = (1+t_c)P_c C/P_q Q$

Environment

The quality of the environment depends negatively on the demands for dirty inputs and dirty consumption commodities (10.19).

Linearization and solution

Following Keller (1980), the model is log-linearized around an initial equilibrium. Table 10.2 contains the log-linearized model. Notation is defined at the end of the table. A tilde (~) denotes a relative change, unless indicated otherwise. The prices of tradables, the rate of return on capital, the stock of financial wealth (W) and the supply of public goods (G) are exogenous. To express the endogenous variables in terms of the exogenous policy instruments, we need to specify the way the government balances its budget. The log-linearized government budget constraint can be written as (if we set the changes in the exogenous variables equal to zero):

$$\begin{aligned} &\alpha_q \tilde{S} - (\alpha_q - \beta_d^*)\tilde{t}_c - \alpha_{lf}\tilde{t}_l = \\ &(\alpha_{ef} - \alpha_{ef}^*)\tilde{t}_e + (\alpha_{dh} - \alpha_{dh}^*)\tilde{t}_d + \beta_l \tilde{P}_l \\ &- \beta_e \tilde{E}^* - \beta_d \tilde{D}^* + \alpha_q \tilde{b}^t \end{aligned} \tag{10.41}$$

The right-hand side of equation (10.40) shows the effect of environmental policy instruments on the government budget. The left-hand side reveals that the budget can be balanced by adjusting the labour income tax, t_l, the value-added tax, t_c, transfers, S, or a combination of these variables[6] $\tilde{b}^t$ represents the tax base effect and consists of the effects on the bases of all four taxes:

$$\begin{aligned} \alpha_q \tilde{b}^t &= \beta_l \tilde{L} + \beta_e \tilde{E} + \beta_d (1 - \frac{\beta_c}{\alpha_q})\tilde{D} \\ &+ \beta_c [\phi_c \tilde{C} + (1 - \phi_c)\tilde{D}] \end{aligned} \tag{10.42}$$

Employment

A rise in the real wage rate (P_l^n) boosts aggregate labour supply if the uncompensated wage elasticity of aggregate labour-supply (θ_{ll}^m) is positive. We assume that the labour supply curve indeed slopes upwards.[7] Higher real non-labour income (S^*) reduces labour supply because of a negative income effect. The strength of this effect depends on the income elasticity of aggregate labour supply, θ_l^m:

$$\tilde{L}_s = \theta_{ll}^m \tilde{P}_l^n - \theta_l^m \tilde{S}^* \tag{10.43}$$

[6] Appendix A,B and C solve the log-linearized model contained in table 10.2 under alternative assumptions regarding the way the government balances its budget. The various shares and elasticities are defined at the end of table 10.2.

[7] Empirical evidence reveals that the value of the wage elasticity is rather small but positive, see e.g. Hausman (1985). The uncompensated wage elasticity of aggregate labour supply, θ_{ll}^m, is related to the corresponding compensated elasticity, θ_{ll}^{*m}, by: $\theta_{ll}^m = \theta_{ll}^{*m} - (1-\phi_{nl})\theta_l^m$. Hence, the uncompensated elasticity is positive if the substitution effect, which is represented by the compensated elasticity θ_{ll}^{*m}, dominates the income effect, $(1-\phi_{nl})\theta_l^m$. Furthermore, the compensated elasticity is directly related to the substitution elasticity between leisure and the composite consumption good: $\theta_{ll}^{*m} = \theta_l^m \sigma_u$.

Welfare: environmental quality and aggregate private income

The marginal excess burden measures the welfare effects of environmental policies. It corresponds to the additional transfer that must be provided to households to keep their utility, after a policy shock, at its initial level.[8] A positive excess burden consequently implies a loss in welfare. The marginal excess burden, $\tilde{b}$, consists of a (public) environmental component and a private component (see appendix D):

$$\tilde{b} = -\tilde{b}^{e} - \tilde{b}^{p} \tag{10.44}$$

The first term at the right-hand side of (10.44) represents the welfare effect due to a change in environmental quality. This term, $\tilde{b}^{e}$, will be called the effect on 'environmental income' and can be written as:

$$\tilde{b}^{e} = -\eta\theta_{e}\tilde{E} - \eta\theta_{d}\tilde{D} \tag{10.45}$$

If the demand for polluting inputs or polluting consumption commodities declines, environmental quality and thus welfare improves

(i.e. $\eta \equiv \dfrac{(\partial U/\partial M)}{(\partial U/\partial Q)}\dfrac{M}{Q} > 0, \quad \theta_{j} \equiv -\dfrac{\partial M}{\partial J}\dfrac{J}{M} > 0$ for J=E,D).

The second term at the right-hand side of (10.44) measures changes in aggregate real private household income ($\tilde{b}^{p}$), which can be written as the sum of effects on the real after-tax wage rate and the real value of aggregate non-labour income:

$$\tilde{b}^{p} = (1-\phi_{nl})\tilde{P}_{l}^{n} + \tilde{S}^{*} \tag{10.46}$$

Using the government budget constraint (10.37), we derive that the effect on aggregate real private income is equal to the tax-base effect defined in (10.42) (see appendix E):

$$\tilde{b}^{p} = \tilde{b}^{t} \tag{10.47}$$

Intuitively, $\tilde{b}^{t}$ represents the consequences of policy shocks for the efficiency of the tax system as a device to raise public revenues. In particular, an eroding tax base implies less efficient revenue-raising because it requires higher marginal tax rates to raise the same amount of revenues. Given public spending, this additional burden of revenue raising is reflected in a lower aggregate private income.

To assess the overall welfare-effect, we need to combine the consequences for (public) environmental income and aggregate private income. A particular policy yields a *weak* double dividend if it succeeds in raising both environmental income and employment. If, in addition, aggregate private income increases, the policy produces

[8] The excess burden corresponds to the compensating variation. For infinitely small shocks, the compensating variation equals the equivalent variation, see Keller (1980).

a *strong* double dividend.[9] In sections 10.3 and 10.4 we examine whether particular environmental policies can yield such double dividends.

Income distribution

The impact on the income distribution between the two classes of households is an important aspect of public policy. Households of class 1, which are called 'productive' households, feature a relatively large labour endowment compared to the non-labour income they collect. Households of class 2, the 'less productive' households, rely more on non-labour income (i.e. $\frac{tt^1}{s^1+rw^1} > \frac{tt^2}{s^2+rw^2}$). The effect on private income of household i is given by:[10]

$$\tilde{b}^{pi} = (1-\phi^i_{nl})\tilde{P}^n_l + \frac{\phi^i_{nl}}{\phi_{nl}}\tilde{S}^* \qquad i = 1,2 \tag{10.48}$$

It can easily be verified that the income share of labour income, $1-\phi_{nl}^i$, is largest for the productive household (i.e. $\phi^2_{nl} > \phi^1_{nl}$). Hence, (10.48) reveals that the productive household benefits relatively more from higher real wages. The less productive household, in contrast, profits more from higher non-labour incomes. A change in the distribution of labour and non-labour incomes thus affects the income distribution between the two households. The impacts on the real after-tax wage rate and the real value of non-labour income can be expressed as follows:

$$\tilde{P}^n_l = \tilde{P}_l - [(1-\phi_c)\tilde{t}_d + \tilde{t}_c] \tag{10.49}$$

$$\tilde{S}^* = \tilde{S} + \tilde{A}^h - \phi_{nl}[(1-\phi_c)\tilde{t}_d + \tilde{t}_c] \tag{10.50}$$

The second term at the right hand side of (10.50) represents changes in the real value of the tax-free pollution allowance, which is defined as:

$$\tilde{A}^h = \frac{\alpha^*_{dh}}{\alpha_q}\tilde{t}_d + \frac{\beta^*_d}{\alpha_q}\tilde{t}_c + \frac{\beta_d}{\alpha_q}\tilde{D}^* \tag{10.51}$$

[9] Expressions (10.2) and (10.7) reveal that, if the demand for polluting inputs or -commodities declines (i.e. $\tilde{E}$, $\tilde{D} < 0$), a rise in private income (i.e. $\tilde{b}^t = \tilde{b}^p > 0$) is a stronger condition than a rise in employment (i.e. $\tilde{L} > 0$).

[10] The relative change in the real value of non-labor income is assumed to be the same for both households, i.e. $\tilde{s}^{*i} = \frac{\phi^i_{nl}}{\phi_{nl}}\tilde{S}^*$.

The real value of the tax-free allowance can be increased in three ways. The first term at the right hand side of (10.51) indicates that higher environmental taxes raise the real value of the tax-free allowance because the right to freely pollute becomes more valuable at higher tax rates. The second term shows that a higher value-added tax strengthens this effect. This is because the value-added tax is applied on top of the environmental tax and, therefore, raises the effective tax rate on polluting consumption commodities. The third term corresponds to an increase in the tax-free allowance itself.

How environmental policy affects the income distribution depends on the way the government balances its budget. In order to avoid political tensions, the government may want to prevent large changes in the income distribution. In particular, the government may aim at income neutrality, defined as the same relative changes in incomes of productive and less productive households (i.e. $\tilde{b}^{p1} = \tilde{b}^{p2}$). This requires that the relative changes in labour- and non-labour incomes are the same:

$$\tilde{S}^* = \phi_{nl}\tilde{P}_l^n \tag{10.52}$$

If (10.52) holds, both households face the same relative change in income, irrespective of how their income is distributed over labour and non-labour incomes.

10.3 ENVIRONMENTAL TAXES AND SUBSIDIES ON HOUSEHOLDS

Environmental taxes (subsidies) on households can be imposed (granted) in order to discourage the consumption of dirty commodities. Table 10.3 contains the impact of such policies on employment, labour and non-labour incomes, and aggregate private income. The revenue is used either to cut the tax on labor incomes or the value-added tax (the first and second columns, respectively). The third column contains the impact of environmental taxes if the government ensures income neutrality.[11] Finally, the fourth column examines the effects of raising the tax-free pollution allowance for (or lump-sum income transfers to) households if the labour income tax is used to balance the government budget.

This section is organized as follows. To set the stage for the discussion of more complex and realistic settings, subsection 10.3.1 explores the case in which initial environmental taxes, tax-free allowances and non-labour incomes are all zero. The effects of raising environmental taxes from an initial equilibrium with positive environmental taxes are examined in subsection 10.3.2. Subsection 10.3.3 focuses on the case with positive non-labour incomes and tax-free allowances in a setting with

[11] The government can adopt various instruments to achieve income neutrality. For example, it may employ tax-free allowances or raise public transfers.

zero initial environmental taxes. It shows how environmental policy may impact the economy by changing the distribution between labour and non-labour incomes. Finally, subsection 10.3.4 examines the overall effects of environmental policy with non-zero initial taxes, tax-free allowances and non-labour incomes.

10.3.1 Introducing an environmental tax

We start by examining the case with a tax rate, a tax-free pollution allowance, and non-labour incomes of zero in the initial equilibrium (i.e. $\beta_d = \beta_d^* = \alpha_{dh}^* = \phi_{nl} = 0$). In this case, the introduction of a tax on dirty consumption commodities affects neither employment nor aggregate private income.[12] The reason for this is the following. The tax on polluting commodities does not directly impact firm behavior. Hence, the before-tax wage is unaffected according to the non-profit condition (10.24):

$$\tilde{P}_l^f = 0 \tag{10.53}$$

The impact on the after-tax wage, which determines labour supply, thus depends on the effect on the tax wedge between before-tax and real after-tax wages. The overall wedge is the sum of two components: first, the wedge between the before-tax wage and the market wage and, second, the wedge between the market wage and the real after-tax wage:

$$\begin{aligned} \tilde{P}_l^f - \tilde{P}_l^n &= (\tilde{P}_l^f - \tilde{P}_l) + (\tilde{P}_l - \tilde{P}_l^n) = \\ &\tilde{t}_l + (1-\phi_c)\tilde{t}_d + \tilde{t}_c \end{aligned} \tag{10.54}$$

$$\textit{tax wedge} = \textit{wedge}1 + \textit{wedge}2$$

On the one hand, environmental taxes raise wedge 2. On the other hand, the reduction in the labour income tax reduces wedge 1. In order to determine the effect on the aggregate wedge, we need to find the cut in the labour income tax that is made possible by the higher tax on polluting commodities. The government budget constraint yields the following relationship between these tax changes:

$$\begin{aligned} \alpha_l[\tilde{t}_l + (1-\phi_c)\tilde{t}_d + \tilde{t}_c] &= -\alpha_q\tilde{b}^t + \alpha_q\tilde{A}^h + \alpha_q\tilde{S} - \\ &\alpha_{nl}[(1-\phi_c)\tilde{t}_d + \tilde{t}_c] \end{aligned} \tag{10.55}$$

All terms at the right hand side of (10.55) are zero if employment is unaffected. The tax base effect, $\tilde{b}^t$, vanishes because initial environmental taxes are zero (see

[12] Note that in absence of non-labour income households are identical.

expression (10.42)). The absence of non-labour incomes and tax-free allowances implies that the other terms at the right hand side of (10.55) vanish as well (as $\tilde{A}^h = \tilde{S} = \alpha_{nl} = 0$). Substituting (10.55) into (10.54), we find that the higher wedge between the market and after-tax wages due to a higher environmental tax exactly offsets the lower wedge between the before-tax and market wages on account of a lower tax on labour incomes. The policy experiment affects only the *composition* of the total tax wedge on labour incomes without affecting its overall *magnitude*. While the quality of the environment improves, employment and aggregrate private income remain unaffected.

Employment remains constant also if the revenues from the environmental tax are used to cut the value-added tax. Intuitively, replacing the value-added tax by a tax on polluting commodities does not affect the overall magnitude of the total tax wedge, because both taxes are fully borne by labour incomes in absence of non-labour income.

10.3.2 Increasing an existing environmental tax[13]

If the initial tax on polluting consumption commodities is positive (i.e. $\beta_d > 0$), replacing either the labour income tax or the value-added tax by an environmental tax harms employment and aggregate private income. This is because the environmental tax encourages households to substitute away from dirty towards clean consumption. At positive initial pollution taxes, the resulting erosion of the base of the environmental tax decreases public revenues. Accordingly, the room to cut the endogenous tax is reduced. In terms of expression (10.55), the tax base effect, $\tilde{b}^t$, is negative, which raises the overall tax wedge on labour in (10.54). Hence, real after-tax wages fall and, if the uncompensated wage elasticity of labour supply (θ_{ll}^m) is positive, labour supply declines.

The loss in aggregate private income due to an eroding base of environmental and labour taxes (see expression (10.47)) represents the costs of a cleaner environment. These costs are not reflected in tax revenues collected by the government. Hence, they may be called 'hidden costs' of public intervention and reflect the 'excess burden' of taxation. In order to assess the overall welfare effect of government intervention, this additional excess burden has to be weighed against the 'excess benefit' of a cleaner environment.

[13] This subsection assumes that non-labour incomes and tax-free pollution allowances are zero (i.e. $\phi_{nl} = \alpha_{dh}^* = 0$).

A number of structural parameters affect the magnitude of the decline in employment. A large substitution elasticity between clean and polluting consumption commodities, σ_q, strengthens the erosion of the base of the environmental tax, thereby worsening the negative effects on aggregate private income and employment. At the same time, a large value for this parameter yields a more substantial improvement in the quality of the environment. Another important parameter is the initial environmental tax rate, represented by $ß_d$. A higher initial tax rate strengthens the adverse revenue effect of the erosion of the tax base and, therefore, reduces the room to cut the labour income tax. Another relevant factor is the size of the public sector, represented by the term R in the reduced-form equations (see appendix A). In particular, the determinant in the reduced form coefficients declines if the public sector grows. Therefore, a larger public sector implies larger adverse effects on employment. Intuitively, environmental taxes become a less efficient instrument to raise revenues if high public spending requires high initial tax rates.[14] If initial taxes are high, an erosion of the bases of these taxes exert a large negative effect on the revenues. Hence, the endogenous tax rate on labour incomes can fall less and the overall tax wedge rises more.

10.3.3 Changing the income distribution

In the presence of non-labour incomes and tax-free allowances, environmental policy may impact the income distribution between productive and less productive households by changing the distribution between labour and non-labour incomes. Through this channel, environmental taxes may yield additional economic effects. As illustrated in the fourth column of table 10.3, higher lump-sum income transfers or the tax-free allowances, that are financed by a higher tax on labour income, reduce employment and aggregate private income. Intuitively, the overall tax wedge on labour in (10.55) widens due to a transfer from labour to non-labour incomes.

Positive non-labour incomes

To clearly identify how environmental taxes affect the economy through changing the income distribution, we assume that non-labour income is positive (for at least the

[14] Simulations with a large computable general equilibrium model of the United States confirm this. In particular, Goulder (1993) finds that the welfare costs of carbon taxes are quite sensitive to initial marginal rates of pre-existing income taxes. Indeed, his results show that imposing these taxes in an economy with higher levels of distortionary taxes raises the welfare costs of carbon taxes.

less productive households, i.e. $\phi_{nl}^2 > \phi_{nl}^1 \geq 0$) while both the tax-free allowance and the initial environmental tax allowance are zero. In that case, a shift from a labour income tax towards a pollution levy on households boosts employment and raises aggregate private income. The reason is that, in contrast to the case without non-labour income, the government is able to more than compensate real labour incomes for the real income loss due to a higher environmental tax. In terms of expressions (10.54) and (10.55), the overall wedge between the before-tax wage and the real after-tax wage is reduced. Intuitively, in contrast to the labour tax, the environmental tax is not only borne by labour but also by non-labour incomes. Hence, the tax experiment redistributes income from non-labour income (which pays the higher tax on dirty consumption but is not compensated through lower income taxes) to labour income.

The government can thus reduce the tax *rate* on wage income by more than the increase in the average tax *rate* on consumption. As a consequence, real wages rise, thereby boosting labour supply. Hence, a better environmental quality is accompanied by a higher level of employment. Furthermore, the labour-market distortion is alleviated and aggregate private income rises. Accordingly, the policy reaps both a weak and a strong double dividend. Reducing the value added-tax leaves employment unaffected even if non-labour incomes are positive. Using $\tilde{t}_l = \tilde{A}^h = \tilde{S} = 0$, we can rewrite (10.55) as:

$$(1-\phi_c)\tilde{t}_d + \tilde{t}_c = -\tilde{b}^t \qquad (10.56)$$

The overall wedge is not changed if the tax base effect is zero due to the absence of initial environmental taxes. Intuitively, this tax experiment does not affect the distribution between labour and non-labour incomes as both incomes bear both the tax that is cut (i.e. the value-added tax) and the tax that is raised (i.e. the environmental tax).

Positive tax-free allowances

Environmental taxes exert an additional effect on the income distribution if the initial tax-free allowance is positive (i.e $\alpha_{dh}^* > 0$). In particular, higher environmental taxes raise the real value of the tax-free pollution allowance because they make the right to pollute freely more valuable. A higher value of the tax-free allowance raises non-labour incomes but reduces labour income on account of a higher endogenous labour income tax. As a consequence, the overall tax wedge on labour income increases, thereby reducing employment and aggregate private income.

The overall effect on aggregate non-labour income due to switching from wage taxes towards environmental taxes is determined by the following condition:

$$\tilde{S}^* > 0 \quad iff \quad \tilde{L} < 0 \quad iff \quad \phi_{nl}\alpha_{dh} - \alpha_{dh}^* < 0 \qquad (10.57)$$

Table 10.3 Reduced form coefficients of t_d (and S) when the revenue is used to cut t_l, t_c, or when income neutrality is imposed.

	$\tilde{t}_d \uparrow, \tilde{t}_l \downarrow$	$\tilde{t}_d \uparrow, \tilde{t}_c \downarrow$	$\tilde{t}_d \uparrow$ *income neutrality*	$\tilde{S} \uparrow$ *or* $\frac{\beta_d}{\alpha_q}\tilde{D}^* \uparrow, \tilde{t}_l \uparrow$
		employment		
$\tilde{L}$	$-[\theta_{ll}^{*m}-(1-\phi_{nl})\theta_l^m]\frac{\beta_d}{1+t_c}\frac{\phi_c\sigma_q}{\det(1)} < 0$ $+\phi_{nl}\frac{\alpha_{dh}}{\alpha_q}\frac{Z\theta_{ll}^{*m}}{\det(1)} > 0$ $-\frac{\alpha_{dh}^*}{\alpha_q}\frac{Z\theta_{ll}^{*m}}{\det(1)} < 0$	$-[\theta_{ll}^{*m}-(1-\frac{\beta_d^*}{\alpha_q})\theta_l^m]\frac{\beta_d}{1+t_c}\frac{\phi_c\sigma_q}{\det(2)} < 0$ $-\frac{\alpha_{dh}^*}{\alpha_q}\frac{Z\theta_{ll}^{*m}}{\det(2)} < 0$ $+\frac{\alpha_{dh}}{\alpha_q}\frac{\beta_d^*}{\alpha_q}\frac{Z\theta_{ll}^{*m}}{\det(2)} > 0$	$-[\theta_{ll}^{*m}-\theta_l^m]\frac{\beta_d}{1+t_c}\frac{\phi_c\sigma_q}{\det(3)} < 0$	$-\frac{Z\theta_{ll}^{*m}}{\det(1)} < 0$
		income distribution		
$\tilde{S}^*$	$-\phi_{nl}\frac{\alpha_{dh}}{\alpha_q} < 0$ $+\frac{\alpha_{dh}^*}{\alpha_q} > 0$	$-[\phi_{nl}-\frac{\beta_d^*}{\alpha_q}]\frac{\beta_d}{1+t_c}\frac{\phi_c\sigma_q}{\det(2)}$ $+\frac{\alpha_{dh}^*}{\alpha_q}\frac{\det(1)}{\det(2)} > 0$ $-\frac{\alpha_{dh}}{\alpha_q}\frac{\beta_d^*}{\alpha_q}\frac{\det(1)}{\det(2)} < 0$	$-\phi_{nl}\frac{\beta_d}{1+t_c}\frac{\phi_c\sigma_q}{\det(3)} < 0$	$+1$
$\tilde{P}_l^n$	$-\frac{\beta_d}{1+t_c}\frac{\phi_c\sigma_q}{\det(1)} < 0$ $+\phi_{nl}\frac{\alpha_{dh}}{\alpha_q}\frac{Z+\theta_l^m R}{\det(1)} > 0$ $-\frac{\alpha_{dh}^*}{\alpha_q}\frac{Z+\theta_l^m R}{\det(1)} < 0$	$-\frac{\beta_d}{1+t_c}\frac{\phi_c\sigma_q}{\det(2)} < 0$ $-\frac{\alpha_{dh}^*}{\alpha_q}\frac{Z+\theta_l^m R}{\det(2)} < 0$ $+\frac{\alpha_{dh}}{\alpha_q}\frac{\beta_d^*}{\alpha_q}\frac{Z+\theta_l^m R}{\det(2)} > 0$	$-\frac{\beta_d}{1+t_c}\frac{\phi_c\sigma_q}{\det(3)} < 0$	$-\frac{Z+\theta_l^m R}{\det(1)} < 0$
		private income		
$\tilde{b}^p$	$-(1-\phi_{nl})\frac{\beta_d}{1+t_c}\frac{\phi_c\sigma_q}{\det(1)} < 0$ $+\phi_{nl}\frac{\alpha_{dh}}{\alpha_q}\frac{\theta_{ll}^{*m}R}{\det(1)} > 0$ $-\frac{\alpha_{dh}^*}{\alpha_q}\frac{\theta_{ll}^{*m}R}{\det(1)} < 0$	$-(1-\frac{\beta_d^*}{\alpha_q})\frac{\beta_d}{1+t_c}\frac{\phi_c\sigma_q}{\det(2)} < 0$ $-\frac{\alpha_{dh}^*}{\alpha_q}\frac{\theta_{ll}^{*m}R}{\det(2)} < 0$ $+\frac{\alpha_{dh}}{\alpha_q}\frac{\beta_d^*}{\alpha_q}\frac{\theta_{ll}^{*m}R}{\det(2)} > 0$	$-\frac{\beta_d}{1+t_c}\frac{\phi_c\sigma_q}{\det(3)} < 0$	$-\frac{R\theta_{ll}^{*m}}{\det(1)} < 0$

$$\det(1) = (1-\phi_{nl})(Z + \theta_l^m R) - \theta_{ll}^{*m}R$$

$$\det(2) = (1-\frac{\beta_d^*}{\alpha_q})(Z+\theta_l^m R) - \theta_{ll}^{*m}R$$

$$\det(3) = (Z + \theta_l^m R) - \theta_{ll}^{*m}R$$

Higher environmental taxes raise the real value of non-labour income if the initial tax-free allowance (and hence α_{dh}^{*}) is large, while non-labour income (and hence ϕ_{nl}) is small. The second condition illustrates the trade-off between, on the one hand, a positive effect on employment (and hence the tax base) and, on the other hand, a rise in non-labour incomes. Employment can rise only if non-labour incomes fall. In the presence of a large initial tax-free allowance, higher environmental taxes boost non-labour income and thus can no longer yield a double dividend. In the case of a higher environmental subsidy (i.e. $\alpha_{dh} < \alpha_{dh}^{*}$), a double dividend is excluded.

If the value-added tax is replaced by an environmental tax, the interaction of positive initial environmental taxes and positive tax-free allowances (i.e $\beta_{d}^{*} > 0$) yields an additional effect on the income distribution. This is because the value-added tax is applied on top of the environmental tax. Hence, a lower value-added tax reduces the effective tax rate on polluting consumption commodities, thereby cutting the real value of the tax-free allowance. As a consequence, employment and aggregate private income expand.

10.3.4 The overall effect

Environmental policy affects employment and aggregate private income through two channels. Subsection 10.3.2 examined the first channel, namely the erosion of the base of the environmental tax. This channel raises the marginal tax burden on labour incomes, thereby reducing employment. It reflects the lower pollution level on account of the higher price for pollution and operates only if the initial environmental tax is non-zero. The second channel involves redistribution of income between labour and non-labour incomes. It works only in the presence of positive tax-free pollution allowances or positive non-labour incomes. In particular, instruments that redistribute income toward labour income at the expense of other incomes reduce the marginal tax burden on labour. This makes it more attractive to work, thereby alleviating labour-market distortions.

The various environmental policy experiments can be decomposed into two elements reflecting the two channels described above. The first element is the higher price for pollution. This improves environmental quality but at the same time, if initial environmental taxes are positive, erodes the tax base and employment. The second element involves redistribution of income. A policy can reap a double dividend if it favors labour incomes at the expense of non-labour incomes and at the same time does not produce a serious erosion of the environmental tax base. To illustrate, a higher environmental tax combined with a cut in labour taxes yields a double dividend if the first (tax base) effect is small due to small initial environmental taxes while the second (redistribution) effect is sizable because of a large non-labour share.

Table 10.4 Environmental taxes on households: trade-offs between employment, aggregate private income, non-labour income, and environmental quality.

	$t_d \uparrow$ and $t_l \downarrow$	$t_d \uparrow$ and $t_c \downarrow$ (or income neutrality)
zero initial tax, tax-free allowance, and non-labour income ($ß_d = \alpha_{dh}^{*} = \phi_{nl} = 0$)		
employment	0	0
private income	0	0
environmental income	+	+
positive initial tax ($ß_d > 0$), zero non-labour income and tax-free allowance ($\phi_{nl} = \alpha_{dh} = 0$)		
employment	-	-
private income	-	-
environmental income	+	+
zero initial tax and tax-free allowance ($ß_d = \alpha_{dh}^{*} = 0$), positive non-labour income ($\phi_{nl} > 0$)		
employment	+	0
private income	+	0
environmental income	+	+
non-labour income	-	0
positive initial tax, tax-free allowance, and non-labour income ($ß_d$, ϕ_{nl}, $\alpha_{dh}^{*} > 0$)		
employment	+/-[1]	-
private income	+/-[1]	-
environmental income	+	+
non-labour income	+/-[1]	-

[1] Impacts on employment and private income can be positive only if non-labour incomes decline (see subsection 10.3.4).

The scope for a double dividend diminishes if the government attaches a higher priority to protecting incomes of less productive households who suffer from lower non-labour incomes. Indeed, environmental policy instruments can alleviate labour-market distortions by expanding employment only by reducing the real value of non-labour incomes, thereby changing the income distribution away from less productive households. This illustrates the fundamental trade-off between equity and efficiency (see table 10.4). In particular, environmental policy cannot produce a (weak) double dividend if income neutrality is imposed (see the third column of table 10.3). Substituting (10.51) into (10.55), one finds that, if the government ensures income neutrality, the overall tax wedge on labour income is directly related to the tax-base effect, $\tilde{b}^{t}$:

$$\tilde{t}_l + (1-\phi_c)\tilde{t}_d + \tilde{t}_c = -\tilde{b}^{t} \qquad (10.58)$$

Accordingly, the overall tax wedge remains unaffected only if a zero initial environmental tax produces a zero tax-base effect. If the inital environmental taxes are positive, however, the overall tax wedge expands, and the consumption wage and labour supply thus decline.

10.4 ENVIRONMENTAL TAXES AND -SUBSIDIES ON FIRMS

The government can impose environmental taxes on firms in order to discourage the use of polluting inputs in the production proces. This section investigates the effects of these taxes if the revenues from these levies are used to cut the tax on labour incomes. Subsection 10.4.1 explores the case in which the initial tax rate and the tax-free pollution allowance are zero (i.e. $ß_e = \alpha_{ef}^{*} = 0$). Subsection 10.4.2 investigates the case with a positive initial tax rate while the tax-free pollution allowance is zero. The impact of positive tax-free allowances is examined in subsection 10.4.3. Finally, subsection 10.4.4 explores the case in which environmental taxes are raised from an initial equilibrium with both positive intitial tax rates and positive tax-free allowances.

10.4.1 Introducing an environmental tax

The first column of table 10.5 reveals that introducing a tax on polluting inputs does not affect employment if tax-free allowances are absent (i.e. $ß_e = \alpha_{ef}^{*} = 0$). The reason is that after-tax wages and, therefore, labour-supply are unaffected. In particular, since consumption prices remain constant, the effect on the real after-tax wage is equal to the impact on the market wage. The latter is linked to the before-tax wage by:

$$\tilde{P}_l = \tilde{P}_l^{f} - \tilde{t}_l \qquad (10.59)$$

Table 10.5 The effects of an environmental tax on firms, t_e

	$\tilde{t}_e \uparrow, \tilde{t}_l \downarrow$	$\tilde{E}^* \uparrow, \tilde{t}_l \uparrow$
	employment	
$\tilde{L}$	$-[\theta_{ll}^{*m}-(1-\phi_{nl})\theta_l^m]\dfrac{\beta_e\sigma_l^*}{\det(1)} < 0$ $+[\theta_{ll}^{*m}-(1-\phi_{nl})\theta_l^m]\dfrac{\alpha_{ef}^*}{\alpha_{lf}}\dfrac{\beta_e\sigma_l}{\det(1)} > 0$	$+[\theta_{ll}^{*m}-(1-\phi_{nl})\theta_l^m]\dfrac{\beta_e}{\alpha_{lf}}\dfrac{\beta_e\sigma_l}{\det(1)} > 0$
	real after-tax wage	
	$-\dfrac{\beta_e\sigma_l^*}{\det(1)} < 0$ $+\dfrac{\alpha_{ef}^*}{\alpha_{lf}}\dfrac{\beta_e\sigma_l}{\det(1)} > 0$	$+\dfrac{\beta_e}{\alpha_{lf}}\dfrac{\beta_e\sigma_l}{\det(1)} > 0$
	real private income	
$\tilde{b}^p$	$-(1-\phi_{nl})\dfrac{\beta_e\sigma_l^*}{\det(1)} < 0$ $+(1-\phi_{nl})\dfrac{\alpha_{ef}^*}{\alpha_{lf}}\dfrac{\beta_e\sigma_l}{\det(1)} > 0$	$+(1-\phi_{nl})\dfrac{\beta_e}{\alpha_{lf}}\dfrac{\beta_e\sigma_l}{\det(1)} > 0$
	environmental income	
$\tilde{b}^e$	$+\eta\theta_e(\dfrac{[\theta_{ll}^{*m}-(1-\phi_{nl})\theta_l^m]\beta_e}{\det(1)}+1)\,\sigma_l^* < 0$ $+\eta\theta_d(1-\phi_{nl})(1-[\theta_{ll}^{*m}-(1-\phi_{nl})\theta_l^m]\dfrac{\beta_e\sigma_l^*}{\det(1)} < 0$ $+\eta\theta_e(\dfrac{[\theta_{ll}^{*m}-(1-\phi_{nl})\theta_l^m]\beta_e}{\det(1)}+1)\,\dfrac{\alpha_{ef}^*}{\alpha_{lf}}\sigma_l < 0$ $+\eta\theta_d(1-\phi_{nl})(1-[\theta_{ll}^{*m}-(1-\phi_{nl})\theta_l^m]\dfrac{\alpha_{ef}^*}{\alpha_{lf}}\dfrac{\beta_e\sigma_l}{\det(1)} < 0$	$+\eta\theta_e(\dfrac{[\theta_{ll}^{*m}-(1-\phi_{nl})\theta_l^m]\beta_e}{\det(1)}+1)\,\dfrac{\beta_e}{\alpha_{lf}}\sigma_l < 0$ $+\eta\theta_d(1-\phi_{nl})(1-[\theta_{ll}^{*m}-(1-\phi_{nl})\theta_l^m]\dfrac{\beta_e}{\alpha_{lf}}\dfrac{\beta_e\sigma_l}{\det(1)} < 0$

The environmental tax on firms reduces the demand for polluting inputs, thereby decreasing the marginal productivity of labour. Hence, the before-tax wage falls. In particular, the non-profit condition (10.24) yields:

$$\tilde{P}_l^f = -\frac{\alpha_{ef}}{\alpha_{lf}}\tilde{t}_e + \tilde{A}^f \tag{10.60}$$

Substitution of (10.60) into (10.59) yields for the market wage:

$$\tilde{P}_l = -\frac{\alpha_{ef}}{\alpha_{lf}}\tilde{t}_e - \tilde{t}_l + \tilde{A}^f \tag{10.61}$$

Expression (10.61) reveals that, if the value of the tax-free allowance remains unaffected (i.e. $\tilde{A}^f = 0$), the overall effect on the market wage depends on the balance between, on the one hand, a lower before-tax wage rate due to a higher environmental tax and, on the other hand, the cut in the labour income tax.
Using the government budget constraint (10.37), we derive the following relationship between the changes in environmental and labour taxes:

$$\alpha_l\left[\frac{\alpha_{ef}}{\alpha_{lf}}\tilde{t}_e + \tilde{t}_l - \tilde{A}^f\right] = -\alpha_q\tilde{b}^t \tag{10.62}$$

At zero intial taxes and constant employment, the tax base effect at the right hand side of (10.62) is zero (see expression (10.42)). Accordingly, the market wage (10.61) remains unaffected; the lower labour income tax exactly offsets the negative impact of the lower before-tax wage on the market wage. In fact, the tax on dirty inputs amounts to an indirect tax on labour incomes because output, polluting inputs, and capital are all perfectly mobile internationally and are thus able to shift the entire burden of the pollution levy to the immobile factor labour. The change in the tax structure therefore amounts to change in the *composition* of the (implicit and explicit) tax burden on labour, without affecting the overall *level* of that burden.

10.4.2 Increasing an existing environmental tax

The first column of table 10.5 reveals that increasing an existing tax on polluting inputs (i.e. $\beta_e > 0$) harms employment and aggregate private income if the tax-free allowance is zero (i.e. $\alpha_{ef}^* = 0$). This is because the substitution away from polluting inputs erodes the base of the environmental tax. At positive initial taxes this produces a negative tax base effect $\tilde{b}^t$(see expression (10.42)). According to (10.61) and (10.62), the tax rate on labour incomes cannot fall sufficiently to offset the negative impact of a lower before-tax wage rate on the market wage if the tax base shrinks. Hence, the real after-tax wage declines, thereby harming the incentives to supply labour. A cleaner environment is therefore accompanied by lower aggregate private income on account of eroding bases of both the environmental tax and the labour income tax.

The size of the adverse employment effect depends on the elasticity σ_l^*, which is a composite of the substitution elasticities in production:

$$\sigma_l^* = \frac{\alpha_k}{\alpha_k + \alpha_{ef}} \sigma_n + \frac{\alpha_{ef}}{\alpha_k + \alpha_{ef}} \frac{\sigma_l}{\alpha_{lf}} \tag{10.63}$$

This elasticity reflects the substitution away from the dirty input due to both the rise in P_e^f and the fall in P_l^f. Hence, a large elasticity σ_l^* produces a large erosion of the tax base and, therefore, a substantial fall in the after-tax wage. Moreover, just as in case of a higher environmental tax on households, a higher initial tax rate (β_e) and a larger public sector (R) raise the absolute value of the negative effect on employment and aggregate private income.

10.4.3 Tax-free allowances

If the initial tax-free allowance is positive (i.e. $\alpha_{ef}^* > 0$), a higher environmental tax[15] raises the real value of the initial tax-free allowance. The first column of table 10.5 reveals that this constitutes another channel through which the environmental tax impacts employment and aggregate private income. In fact, a higher tax-free allowance (E^*) financed by a tax on labour induces similar effects (see the second column of table 10.5).

Expression (10.60) reveals that a higher value of the tax-free allowance benefits labour as the before-tax wage rises. The corresponding higher marginal productivity of labour is achieved through higher demands for capital and polluting inputs. The base of the environmental tax therefore widens, thereby boosting the revenues from the environmental tax. Accordingly, the endogenous tax on labour income can decline. Part of the higher tax-free allowance is financed by higher revenues from the environmental tax instead of a higher labour income tax. This allows for a rise in the market wage and thereby the real after-tax wage. On the one hand, this raises the incentives to supply labour, thereby expanding employment and aggregate private income. On the other hand, environmental quality declines because of the increased demand for polluting inputs.

Note that the relevant elasticity is σ_l rather than the composite elasticity σ_l^*. This is because the producer price of polluting inputs (P_e^f) is not affected by a change in the tax-free allowance. Hence, the elasticity of substitution between capital and the

[15] The tax becomes a subsidy if α_{ef}^* becomes sufficiently large (i.e. $\alpha_{ef}^* > \alpha_{ef}$).

dirty input, σ_n, is not relevant. Indeed, the before-tax wage rate is the only producer price that changes and thus only the elasticity of substitution between labour and the other inputs, σ_l, determines factor substitution.

10.4.4 Overall effect of environmental taxes and subsidies

If both the initial tax rate on polluting inputs and the initial tax-free allowance are positive, the overall effect of higher environmental taxes on private and (public) environmental incomes depends on the relative magnitude of two effects. The first effect, due to a rise in the environmental tax, is positive for environmental income but negative for employment (section 10.4.2). In contrast, the second effect, due to a higher value of the tax-free pollution allowance, harms the environment but benefits employment (section 10.4.3).

The first column of table 10.5 reveals that a higher environmental tax raises real wages, employment and aggregate private income if and only if:

$$\frac{\alpha_{ef}^*}{\alpha_{lf}}\sigma_l - \sigma_l^* > 0 \tag{10.64}$$

In contrast, environmental income rises if and only if condition (10.64) is not met. This implies that higher environmental income is not compatible with higher aggregate private income. Hence, policies raising the marginal costs of pollution do not yield a double dividend. A fundamental trade-off exists between aggregate private income and public environmental income. Whether environmental or aggregate private income increases depends on condition (10.64). Using (10.63), we can rewrite the left hand side of (10.64) as:

$$\frac{\alpha_{ef}^*}{\alpha_{lf}}\sigma_l - \sigma_l^* = \frac{1}{\alpha_{ef}+\alpha_k}\Big[(\alpha_{ef}^* - \alpha_{ef})\frac{\sigma_l}{\alpha_{lf}} - (\alpha_k\sigma_n + \alpha_{ef}^*\sigma_l)\Big] \tag{10.65}$$

In case of a tax regime, the initial demand for polluting inputs exceeds the tax-free allowance (i.e. $\alpha_{ef}^* < \alpha_{ef}$). Hence, the right hand side of (10.65) cannot be positive. Higher environmental taxes thus always harm aggregate private income, but improve environmental quality. As far as an environmental subsidy is concerned, expression (10.65) reveals that this instrument may actually worsen environmental quality and expand employment. In particular, in case of a subsidy-regime (i.e. $\alpha_{ef}^* > \alpha_{ef}$), the sign of (10.65) depends, among others, on the magnitudes of σ_n, σ_l, and α_{lf}. If capital and polluting inputs are complementary (i.e. σ_n is small relative to σ_l) and if labour accounts for a small production share (i.e. α_{lf} is small), a higher subsidy may boost aggregate private income. This is because the higher shadow price for dirty inputs does not cause much substitution away from polluting inputs towards capital due to the small elasticity σ_n. At the same time, the small production share of labour produces a large rise in the wage rate. The higher wage rate, in turn, induces a lot of substitution towards the dirty input because of the large elasticity σ_l.

10.5 NUMERICAL SIMULATIONS

This section presents some illustrative numerical results for the consequences of environmental taxes in the Netherlands. We abstract from tax-free allowances and assume that revenues are used to cut the labour income tax whether or not in combination with higher income transfers in order to ensure income neutrality.

Table 10.6 Initial shares and parameter values.

intitial values	shares in P_yY	shares in P_qQ
$P_yY = 385$		
$P_l^fL = 325$	$\alpha_{lf} = 0.8400$	
$rK = 51$	$\alpha_k = 0.1300$	
$P_e^fE = 9$	$\alpha_{ef} = 0.0240$	
$P_qQ = 281$	$\alpha_q = 0.7300$	
$P_cC = 264$	$\alpha_c = 0.6800$	$\phi_c = 0.94$
$P_d^hD = 17$	$\alpha_{dh} = 0.0500$	$1\text{-}\phi_c = 0.06$
$P_lL = 165$	$\alpha_l = 0.4300$	$1\text{-}\phi_{nl} = 0.59$
$S + rW = 116$	$\alpha_{nl} = 0.3000$	$\phi_{nl} = 0.41$
$t_lP_lL = 160$	$\beta_l = 0.4200$	
$t_cP_cC = 35$	$\beta_c = 0.0900$	
$t_eE = 1$	$\beta_e = 0.0023$	
$t_dD = 3$	$\beta_d = 0.0078$	
$P_gG = 83$	$\alpha_g = 0.2200$	
$P_eE = 8$	$\alpha_e = 0.0200$	
$P_dD = 14$	$\alpha_d = 0.0400$	
basic parameters		derived parameters
$\sigma_n = 1.3$		$\sigma_l^* = 1.25$
$\sigma_l = 0.8$		$\theta_l^m = 0.15$
$\sigma_q = 0.3$		$\theta_{ll}^{*m} = 0.28$
$\sigma_u = 1.9$		$\theta_{ll}^m = 0.20$

Table 10.6 contains the numerical values for the shares and parameters. The initial shares are calibrated employing Dutch data for 1989. The various elasticities are 'guestimated' using empirical estimates from the literature.For polluting inputs and polluting consumption commodities, we employ energy data. Hence, the environmental taxes must be interpreted as taxes on energy. The substitution elasticities in production imply that capital and energy are substitutes.[16] Using the estimates contained in Hausman (1985), we calibrate the uncompensated wage elasticity of aggregate labour supply, θ_{ll}^{m}, and the income elasticity of aggregate labour supply, θ_{l}^{m}, at, respectively, 0.2 and 0.15. The environmental parameters η and θ_{j} are not specified because of a lack of information.

The model in relative changes can only investigate small policy shocks. In case of large shocks, the linear approximation may produce large linearization errors. We apply a method presented by Bovenberg and Keller (1984) to alleviate this bias. This method of stepwise linearization in effect splits up large shocks into a number of shocks that are small enough to be approximated by the linear model. Tables 10.7, 10.8 and 10.9 contain the simulation results of tax increases that raise the user price of energy by, respectively, 1%, 50% and 100%.

Energy tax on households

Table 10.7 reveals that an energy tax on households raises employment, aggregate private income and environmental income if revenues are used to cut the tax on labour incomes. Hence, this policy does indeed yield both a weak and a strong double dividend. The reason is that the initial tax rate (β_{d}) is small. Accordingly, the erosion of the base of the energy tax produces only a small adverse revenue effect. In addition, the positive employment effect due to a decline in real non-labour income is relatively large due to the high initial share of transfer income. Indeed, a 100% increase in energy prices raises employment by 1,0%.
At higher tax rates, the eroding base of the energy tax yields a more substantial reduction in tax revenues. At the same time, non-labour income declines, thereby mitigating the positive employment effect associated with the redistribution of income from non-labour to labour incomes. Hence, the scope for double dividends diminishes at larger tax increases. To illustrate, a 50% tax increase yields a larger rise in aggregate private income (+0.11%) than a 100% tax increase (+0.01%).

The income distribution changes in favor of labour incomes. A 100% tax increase raises the real after-tax wage rate by 3.7%, while non-labour income declines by 5.6%.

[16] The condition for substitutability is: $\sigma_{ll} - \alpha_{lf}\sigma_{l} > 0$. The shares and parameter values imply that $\sigma_{ll} - \alpha_{lf}\sigma_{l} = 0.63$, See Hesse and Tarkka (1986).

If the government ensures income neutrality, real wages fall by 0.7% while employment declines by 0.1%.

Energy tax on firms

Table 10.8 reveals that a 100% energy tax increase on firms accompanied by a cut in the labour income tax reduces real after-tax wages by 2.0%, employment by 0.4% and aggregate private income by 1.2%. Energy demand declines more substantially by imposing the energy tax on firms (-24.0%) than by imposing a tax of the same magnitude on households (-10.5%). This is in part because firms are more sensitive to changes in relative prices than households.

Increasing the tax rate raises the adverse employment and income effects more than proportionally. The reason is that the eroding base of the energy tax exerts larger adverse revenue effects at higher tax rates. A linear approximation of large shocks thus underestimates these effects.

Table 10.7 The effects (in % changes) of increases in energy taxes on households (t_d).

	1% increase in energy prices		50% increase in energy prices		100% increase in energy prices	
	$t_l \downarrow$	income neutrality	$t_l \downarrow$	income neutrality	$t_l \downarrow$	income neutrality
employment	+0.014	-0.001	+0.57	-0.06	+1.0	-0.1
domestically employed capital stock	+0.014	-0.001	+0.57	-0.06	+1.0	-0.1
real aftertax wage	+0.054	-0.003	+2.16	-0.29	+3.7	-0.7
real non labour income	-0.062	-0.003	-2.90	-0.29	-5.6	-0.7
aggregate private income	+0.006	-0.003	+0.11	-0.29	+0.01	-0.7
total energy demand	-0.164	-0.180	-6.35	-6.99	-10.5	-11.6

Table 10.8 The effects (in % changes) of increases in energy taxes on firms (t_e).

	1% increase in energy prices	50% increase in energy prices	100% increase in energy prices
employment	-0.002	-0.18	-0.4
domestically employed capital stock	+0.053	+1.96	+3.2
real after-tax wage	-0.009	-0.89	-2.0
aggregate private income	-0.005	-0.52	-1.2
total energy demand	-0.471	-15.85	-24.0

Broad-based energy tax

Table 10.9 indicates that a broad-based energy tax on both firms and households yields a weak, but not a strong, double dividend at large increases. In particular, a 100% increase in energy prices, boosts employment by 0.7%. Intuitively, the positive employment effect due to a decline in real non-labour income dominates the negative employment effects associated with the eroding base of the energy tax. Despite higher employment, aggregate private income declines by 1.1%. If income neutrality is imposed, the employment effect turns negative (-0.3%), while aggregate private income falls more substantially by 1.7%.

Table 10.9 The effects (in % changes) of increases in energy taxes on both households and firms.

	1% increase in energy prices		50% increase in energy prices		100% increase in energy prices	
	t_l ↓	income neutrality	t_l ↓	income neutrality	t_l ↓	income neutrality
employment	+0.012	-0.002	+0.42	-0.14	+0.7	-0.3
domestically employed capital stock	+0.067	+0.053	+2.57	+2.00	+4.3	+3.3
real aftertax wage	+0.045	-0.008	+1.32	-0.73	+1.9	-1.7
real nonlabour income	-0.062	-0.008	-2.90	-0.73	-5.6	-1.7
aggregate private income	+0.001	-0.008	-0.39	-0.73	-1.1	-1.7
total energy demand	-0.635	-0.644	-22.15	-22.14	-34.2	-33.9

10.6 CONCLUSIONS

This chapter illustrates the fundamental trade-off between aggregate private income, employment, equity, and the quality of the natural environment in a small open economy. Raising pollution taxes on consumption and using the revenues to cut distortionary labour taxes may yield a double dividend by raising aggregate private income, employment, and environmental quality. However, environmental taxes on households can alleviate labour-market distortions only by cutting the real value of non-labour incomes. Thus, the double dividend involves costs in terms of a less equitable income distribution.

If the government imposes the restriction that incomes of different households must be affected in the same way (so-called income neutrality), environmental policy always reduces employment and aggregate private income and thus fails to produce a double dividend. The government can employ various instruments to ensure income neutrality. In particular, tax-free allowances and lump-sum transfers change the trade-off between equity and employment in similar ways. An alternative way to protect non-labour incomes is to employ the revenues from environmental taxes to cut the value-added tax instead of the labour income tax.

The analysis in section 10.4 of environmental policy instruments affecting firms illustrates the trade-off between employment and environmental quality. In particular, higher pollution taxes harm employment -- even if the revenues are used to cut taxes on labour. Subsidies for cutting pollution, in contrast, may boost employment. However, this is only possible if these subsidies, somewhat paradoxically, worsen environmental quality. Indeed, higher tax-free allowances for firms change the trade-off between employment and environmental quality in favor of employment. Tax-free allowances for households, in contrast, primarily affect the trade-off between employment and non-labour incomes in favor of non-labour incomes but, in contrast to tax-free allowances in production, at the expense of employment. The reason is that tax-free allowances on consumption favor non-labour incomes. In production, these allowances benefit labour-incomes by raising wages.

APPENDIX A: THE REDUCED FORM EQUATIONS IF THE Labour-INCOME TAX IS ENDOGENOUS

In this appendix, we assume that $\tilde{t}_l$ is the endogenous tax rate that balances the government budget. Hence, the variables $\tilde{t}_c$, $\tilde{t}_e$, $\tilde{t}_d$, $\tilde{E}^*$, $\tilde{D}^*$ *and* $\tilde{S}$ are exogenous policy instruments. The following exogenous variables do not change:

$\tilde{P}_y$, $\tilde{P}_e$, $\tilde{r}$, $\tilde{P}_c$, $\tilde{P}_d$, $\tilde{P}_g$, $\tilde{G}$, $\tilde{W}$. First, we write equation (10.39) in case the above conditions hold:

$$\tilde{Y} = \alpha_k\tilde{K} + \alpha_e\tilde{E} + \alpha_d\tilde{D} + \alpha_c\tilde{C} \tag{A1}$$

Using (A1) and the equation for the supply of goods (10.20), we eliminate Y:

$$\alpha_{lf}\tilde{L}_d = -\beta_e\tilde{E} + \alpha_d\tilde{D} + \alpha_c\tilde{C} \tag{A2}$$

Here we have used the following relationship between the initial shares that can be derived from the definitions at the end of table 10.2:

$$\alpha_{ef} = \alpha_e + \beta_e \tag{A3}$$

We now express the three endogenous variables at the right-hand side of (A2) in terms of $\tilde{L}$ and $\tilde{P}_l$. To find the expression for $\tilde{E}$, we subtract (10.21) from (10.23) and use (10.27) to eliminate $\tilde{P}_e^f$:

$$\tilde{E} = \tilde{L}_d + \sigma_l\tilde{P}_l^f - \left[\frac{\alpha_k}{\alpha_{ef}+\alpha_k}\sigma_n + \frac{\alpha_{ef}}{\alpha_{ef}+\alpha_k}\sigma_l\right]\tilde{t}_e \tag{A4}$$

Using the non-profit condition (10.24), we can eliminate $\tilde{P}_l^f$:

$$\tilde{P}_l^f = -\frac{\alpha_{ef}}{\alpha_{lf}}\tilde{t}_e + \tilde{A}^f \tag{A5}$$

Subtitution of (A5) into (A4) yields:

$$\tilde{E} = \tilde{L}_d - \sigma_l^*\tilde{t}_e + \sigma_l\tilde{A}^f \tag{A6}$$

$$\textit{where}\quad \sigma_l^* = \left[\frac{\alpha_k}{\alpha_k+\alpha_{ef}}\sigma_n + \frac{\alpha_{ef}}{\alpha_k+\alpha_{ef}}\frac{\sigma_l}{\alpha_{lf}}\right] \tag{A7}$$

To find $\tilde{C}$ and $\tilde{D}$ in terms of $\tilde{P}_l$ and $\tilde{L}$, we write $\tilde{Q}$ in (10.28) as the sum of an effect on the real after-tax wage, $\tilde{P}_l^n$, and an effect on real non-labour income $\tilde{S}^*$:

$$\tilde{Q} = (1-\phi_{nl})(\tilde{P}_l^n + \tilde{L}) + \tilde{S}^* \tag{A8}$$

Here we have used (10.34) and (10.35) for $\tilde{P}_l^n$ and $\tilde{S}^*$. Substituting (A8) into (10.31) and (10.32), we find:

$$\tilde{D} = (1-\phi_{nl})(\tilde{P}_l^n + \tilde{L}) + \tilde{S}^* - \phi_c\sigma_q\tilde{t}_d \tag{A9}$$

$$\tilde{C} = (1-\phi_{nl})(\tilde{P}_l^n + \tilde{L}) + \tilde{S}^* + (1-\phi_c)\sigma_q\tilde{t}_d \tag{A10}$$

Here, we have employed price equation (10.36) to eliminate $\tilde{P}_d^h$. Substituting (A6), (A9) and (A10) into (A2) yields:

$$\alpha_{lf}\tilde{L} = [\alpha_d + \alpha_c][(1-\phi_{nl})(\tilde{P}_l^n + \tilde{L}) + \tilde{S}^*] - \beta_e\sigma_l\tilde{A}^f - \beta_e\tilde{L} + \beta_e\sigma_l^*\tilde{t}_e + [(1-\phi_c)\alpha_c - \phi_c\alpha_d]\sigma_q\tilde{t}_d \tag{A11}$$

We can write the last term in square brackets at the right-hand side of (A11) as:

$$(1-\phi_c)\alpha_c - \phi_c\alpha_d = \frac{(1+t_c)P_d^h D}{P_q Q}\frac{P_c C}{P_y Y} - \frac{(1+t_c)P_c C}{P_q Q}\frac{P_d D}{P_y Y} =$$

$$\frac{(1+t_c)P_c C}{P_q Q}\frac{t_d D}{P_y Y} = \frac{\beta_d}{1+t_c}\phi_c \tag{A12}$$

Substituting (A12) into (A11), we find:

$$[\alpha_{lf}+\beta_e-(1-\phi_{nl})(\alpha_d+\alpha_c)]\tilde{L} =$$

$$\beta_e\sigma_l^*\tilde{t}_e - \beta_e\sigma_f\tilde{A}^f + \frac{\beta_d}{1+t_c}\phi_c\sigma_q\tilde{t}_d + [\alpha_d+\alpha_c][(1-\phi_{nl})\tilde{P}_l^n + \tilde{S}^*] \tag{A13}$$

We define:

$$Z = \alpha_d + \alpha_c = \frac{1}{1+t_c}[\alpha_q - \beta_d] > 0 \tag{A14}$$

$$R = \alpha_{lf} + \beta_e - (1-\phi_{nl})Z = \beta_l + \beta_e + (1-\phi_{nl})\frac{\beta_d}{1+t_c} + (1-\phi_{nl})\beta_c > 0 \tag{A15}$$

The term Z in (A14) can be interpreted as the ratio to production of marginal after-tax expenditure of housholds on consumer goods. R in (A15) is rewritten by using (10.5), (10.9), (10.16) and (10.18) from table 10.1. The term R represents the marginal size of the public sector, i.e. the change in tax revenue as a result of one additional unit of employment.[17] Using the new notation, we write (A13) as:

$$R\tilde{L} - (1-\phi_{nl})Z\tilde{P}_l^n = Z\tilde{S}^* + \frac{\beta_d}{1+t_c}\phi_c\sigma_q\tilde{t}_d + \beta_e\sigma_l^*\tilde{t}_e - \beta_e\sigma_f\tilde{A}^f \tag{A16}$$

We write aggregate labour supply (10.30) as:

$$\tilde{L}_s = \theta_{ll}^m\tilde{P}_l^n - \theta_l^m\tilde{S}^* \tag{A17}$$

In matrix notation, (A16) and (A17) can be expressed as:

[17] Note that $(1-\phi_{nl})Z + R = \alpha_{lf} + \beta_e$ can be interpreted as marginal national income, i.e. the rise in national income as a result of one additional unit of employment. The term $(1-\phi_{nl})Z$ stands for marginal disposable income of workers and represents the marginal size of the private sector. The term R represents the increase in public revenues as a result of additional employment and thus can be interpreted as the marginal size of the public sector.

$$\begin{bmatrix} R & -(1-\phi_{nl})Z \\ 1 & -\theta_{ll}^{m} \end{bmatrix} \begin{bmatrix} \tilde{L} \\ \tilde{P}_l^{\,n} \end{bmatrix} = \begin{bmatrix} -\beta_e\sigma_l & \beta_e\sigma_l^{*} & \dfrac{\beta_d}{1+t_c}\phi_c\sigma_q & Z \\ 0 & 0 & 0 & -\theta_l^{m} \end{bmatrix} \begin{bmatrix} \tilde{A}^{f} \\ \tilde{t}_e \\ \tilde{t}_d \\ \tilde{S}^{*} \end{bmatrix} \tag{A18}$$

We can solve for the endogenous variables by inverting the matrix on the left-hand side of (A18).

$$\det(1) \begin{bmatrix} \tilde{L} \\ \tilde{P}_l^{\,n} \end{bmatrix} = \begin{bmatrix} -\theta_{ll}^{m} & (1-\phi_{nl})Z \\ -1 & R \end{bmatrix} \begin{bmatrix} -\beta_e\sigma_l & \beta_e\sigma_l^{*} & \dfrac{\beta_d}{1+t_c}\phi_c\sigma_q & Z \\ 0 & 0 & 0 & -\theta_l^{m} \end{bmatrix} \begin{bmatrix} \tilde{A}^{f} \\ \tilde{t}_e \\ \tilde{t}_d \\ \tilde{S}^{*} \end{bmatrix} \tag{A19}$$

$$\textit{where} \quad \det(1) = (1-\phi_{nl})Z - \theta_{ll}^{m}R > 0 \tag{A20}$$

The term "det(1)" must be positive for the equilibrium to be stable. In other words, the Laffer-curve should be upward sloping. The determinant, det(1), consists of two elements. The first term on the right-hand side of (A20) can be interpreted as the real value of after-tax wage income. This net private income constitutes the base of the public sector. The second term stands for the impact of higher real after-tax wages on labour supply and, therefore, on the tax base. If θ_{ll}^{m} and R are large relative to $(1-\phi_{nl})Z$, the model may be unstable. This is because lower wages reduce the incentives to supply labour, thereby eroding the tax base. This requires higher tax rates on labour in order to meet the revenue constraint. This further reduces net wages and, therefore, the tax base. Indeed, the model is unstable if the Laffer curve is downward slooping. This is the case if the public sector is large relative to the private sector and if private behavior is sensitive to tax incentives. To compare the determinant with the models explored in appendix B and appendix C, we can write (A20) as:

$$\det(1) = (1-\phi_{nl})(Z+\theta_l^{m}R) - \theta_{ll}^{*m}R \tag{A21}$$

The reduced form equations for $\tilde{P}_l^{\,n}$ and $\tilde{L}$ can now be computed from (A19):

$$\det(1)\tilde{L} = -\theta_{ll}^{*m}Z\tilde{S}^{*} - \theta_{ll}^{m}\frac{\beta_d}{1+t_c}\phi_c\sigma_q\tilde{t}_d - \theta_{ll}^{m}\beta_e\sigma_l^{*}\tilde{t}_e + \theta_{ll}^{m}\beta_e\sigma_f\tilde{A}^{f} \tag{A22}$$

$$\det(1)\tilde{P}_l^{n} = -[Z + \theta_l^{m}R]\tilde{S}^{*} - \frac{\beta_d}{1+t_c}\phi_c\sigma_q\tilde{t}_d - \beta_e\sigma_l^{*}\tilde{t}_e + \beta_e\sigma_f\tilde{A}^{f} \tag{A23}$$

Recall from (10.35) that:

$$\tilde{S}^{*} = \tilde{S} - \phi_{nl}[\tilde{t}_c + (1-\phi_c)\tilde{t}_d] + [\frac{\alpha_{dh}^{*}}{\alpha_q}\tilde{t}_d + \frac{\beta_d}{\alpha_q}\tilde{D}^{*} + \frac{\beta_d^{*}}{\alpha_q}\tilde{t}_c] \tag{A24}$$

The value-added tax, $\tilde{t}_c$, as well as the tax-free allowance, $\tilde{D}^{*}$, affect employment and the real wage through their impact on real non-labour income $\tilde{S}^{*}$. The reduced form equation for $\tilde{b}^{p}$ can be derived by substituting the reduced forms of $\tilde{P}_l^{n}$ and $\tilde{S}^{*}$ from (A23) and (A24) into (10.46).

APPENDIX B: REDUCED FORM EQUATIONS IF THE VALUE-ADDED TAX IS ENDOGENOUS

This appendix solves the model in case the government balances its budget by adjusting the value-added tax, $\tilde{t}_c$. We assume that the exogenous tax on wage income, the tax on polluting inputs, and the tax-free allowance for firms do not change (i.e. $\tilde{t}_l = \tilde{t}_e = \tilde{E}^{*} = 0$). First we express the three endogenous variables at the right hand side of (A2) in terms of $\tilde{L}$ and $\tilde{t}_c$. To find an expression for $\tilde{E}$, we use (A5) which implies that the before-tax wage does not change:

$$\tilde{P}_l^{f} = 0 \tag{B1}$$

By substituting (B1) into (A4), we derive:

$$\tilde{E} = \tilde{L} \tag{B2}$$

We now express $\tilde{C}$ and $\tilde{D}$ in terms of $\tilde{L}$ and $\tilde{t}_c$. (B1) and the assumption that $\tilde{t}_l = 0$ imply that the market wage does not change:

$$\tilde{P}_l = \tilde{P}_l^{f} - \tilde{t}_l = 0 \tag{B3}$$

To ease notation, we define:

$$\tilde{H} = \frac{\alpha_{dh}^{*}}{\alpha_q}\tilde{t}_d + \frac{\beta_d}{\alpha_q}\tilde{D}^{*} \tag{B4}$$

Substitution of (B3) and (B4) into (10.31) and (10.32) yields:

$$\tilde{D} = [(1-\phi_{nl})\tilde{L} + \tilde{S} + \tilde{H} - (1-\phi_c)\tilde{t}_d - \tilde{t}_c + \frac{\beta_d^*}{\alpha_q}\tilde{t}_c] - \phi_c\sigma_q\tilde{t}_d \tag{B5}$$

$$\tilde{C} = [(1-\phi_{nl})\tilde{L} + \tilde{S} + \tilde{H} - (1-\phi_c)\tilde{t}_d - \tilde{t}_c + \frac{\beta_d^*}{\alpha_q}\tilde{t}_c] + (1-\phi_c)\sigma_q\tilde{t}_d \tag{B6}$$

where we have used (10.28), (10.29), (10.33) and (10.36) to eliminate $\tilde{Q}$, $\tilde{A}^h$, $\tilde{P}_q$ and $\tilde{P}_d^h$. Substituting (B2), (B5) and (B6) into (A2), we arrive at:

$$R\tilde{L} + Z(1-\frac{\beta_d^*}{\alpha_q})\tilde{t}_c = Z[\tilde{S} + \tilde{H} - (1-\phi_c)\tilde{t}_d] + \frac{\beta_d}{1+t_c}\phi_c\sigma_q\tilde{t}_d \tag{B7}$$

The last term at the right hand side of (B7) is derived by using (A12). Substitution of (B3), (B4), (10.33), (10.34) and (10.35) into (10.30) yields:

$$\tilde{L} + [\theta_{ll}^{*m} - (1-\frac{\beta_d^*}{\alpha_q})\theta_l^m]\tilde{t}_c = -(\theta_{ll}^{*m} - \theta_l^m)(1-\phi_c)\tilde{t}_d - \theta_l^m[\tilde{S} + \tilde{H}] \tag{B8}$$

We write (B7) and (B8) in matrix notation as:

$$\begin{bmatrix} R & (1-\frac{\beta_d^*}{\alpha_q})Z \\ 1 & \theta_{ll}^{*m} - (1-\frac{\beta_d^*}{\alpha_q})\theta_l^m \end{bmatrix} \begin{bmatrix} \tilde{L} \\ \tilde{t}_c \end{bmatrix} = \begin{bmatrix} Z & \frac{\beta_d}{1+t_c}\phi_c\sigma_q - (1-\phi_c)Z \\ -\theta_l^m & -(1-\phi_c)(\theta_{ll}^{*m} - \theta_l^m) \end{bmatrix} \begin{bmatrix} \tilde{H}+\tilde{S} \\ \tilde{t}_d \end{bmatrix} \tag{B9}$$

We can solve for the endogenous variables by inverting the matrix on the left-hand side of (B9).

$$-\det(2)\begin{bmatrix} \tilde{L} \\ \tilde{t}_c \end{bmatrix} = \begin{bmatrix} \theta_{ll}^{*m} - (1-\frac{\beta_d^*}{\alpha_q})\theta_l^m & -(1-\frac{\beta_d^*}{\alpha_q})Z \\ -1 & R \end{bmatrix} \begin{bmatrix} Z & \frac{\beta_d}{1+t_c}\phi_c\sigma_q - (1-\phi_c)Z \\ -\theta_l^m & -(1-\phi_c)(\theta_{ll}^{*m} - \theta_l^m) \end{bmatrix} \begin{bmatrix} \tilde{H}+\tilde{S} \\ \tilde{t}_d \end{bmatrix} \tag{B10}$$

$$\text{where} \quad \det(2) = (1 - \frac{\beta_d^*}{\alpha_q})[Z + \theta_l^m R] - \theta_{ll}^{*m} R > 0 \qquad \text{(B11)}$$

The term "det(2)" must be positive for the equilibrium to be stable. This determinant reflects the slope of the Laffer-curve and represents the overall revenue effect of a higher value-added tax. It consists of three elements. First, a higher value-added tax increases the revenues of the existing tax base (Z) (tax rate effect). Second, the tax base itself is affected through two channels: labour supply increases on account of the income effect ($\theta_l^m R$), but decreases on account of the substitution effect ($-\theta_{ll}^{*m} R$). The third element in the determinant is β_d^*/α_q. Intuitively, a higher value-added tax increases the real value of the tax-free pollution allowance and therefore raises non-labour income. Hence, the government collects less additional revenues by increasing the value-added tax. In general det(2) exeeds det(1) (see (A21)) because the value-added tax constitutes a broader base than the labour income tax as it is also applied to non-labour income.[18]

The reduced form equations for $\tilde{L}$ and $\tilde{t}_c$ can now be computed from (B10).

$$\begin{aligned} \det(2)\tilde{L} &= -[\theta_{ll}^{*m} - (1 - \frac{\beta_d^*}{\alpha_q})\theta_l^m] \frac{\beta_d}{1+t_c} \phi_c \sigma_q \tilde{t}_d \\ &+ \frac{\beta_d^*}{\alpha_q}(1-\phi_c) Z \theta_{ll}^{*m} \tilde{t}_d - Z\theta_{ll}^{*m}(\tilde{H} + \tilde{S}) \end{aligned} \qquad \text{(B12)}$$

$$\begin{aligned} \det(2)\tilde{t}_c &= + \frac{\beta_d}{1+t_c} \phi_c \sigma_q \tilde{t}_d - (1-\phi_c)[Z - (\theta_{ll}^{*m} - \theta_l^m) R] \tilde{t}_d \\ &+ (Z + \theta_l^m R)(\tilde{H} + \tilde{S}) \end{aligned} \qquad \text{(B13)}$$

The reduced forms for $\tilde{S}^*$ and $\tilde{P}_l^n$ can be derived by substituting $\tilde{t}_c$ from (B13) into (10.35) and (10.34) respectively. Real private income, $\tilde{b}^P$, can be found by substituting the reduced form equation for $\tilde{S}^*$ and $\tilde{P}_l^n$ into (10.46).

APPENDIX C: REDUCED FORMS IF INCOME NEUTRALITY IS IMPOSED

This appendix solves the model in case the government maintains income neutrality, which implies:

[18] det(1) = det(2) only if aggregate non-labour income is zero (i.e. $S + rW + t_d D^* = 0$). In that case, the base of the value-added tax corresponds to the base of the labour income tax.

$$\tilde{S}^* = \phi_{nl}\tilde{P}_l^n \tag{C1}$$

Substitution of (C1) into (A9) and (A10) yields:

$$\tilde{D} = \tilde{P}_l^n + (1-\phi_{nl})\tilde{L} - \phi_c\sigma_q\tilde{t}_d \tag{C2}$$

$$\tilde{C} = \tilde{P}_l^n + (1-\phi_{nl})\tilde{L} + (1-\phi_c)\sigma_q\tilde{t}_d \tag{C3}$$

Following the procedure of appendix A, we substitute (A6), (C2), and (C3) into (A2) and arrive at:

$$R\tilde{L} - Z\tilde{P}_l^n = \frac{\beta_d}{1+t_c}\phi_c\sigma_q\tilde{t}_d + \beta_e\sigma_l^*\tilde{t}_e - \beta_e\sigma_f\tilde{A}^f \tag{C4}$$

Substituting condition (C1) into (10.30), we find for labour supply:

$$\tilde{L}_s = (\theta_{ll}^{*m} - \theta_l^m)\tilde{P}_l^n \tag{C5}$$

We write (C4) and (C5) in matrix notation as:

$$\begin{bmatrix} R & -Z \\ 1 & -(\theta_{ll}^{*m} - \theta_l^m) \end{bmatrix} \begin{bmatrix} \tilde{L} \\ \tilde{P}_l^n \end{bmatrix} = \begin{bmatrix} -\beta_e\sigma_l & \beta_e\sigma_l^* & \frac{\beta_d}{1+t_c}\phi_c\sigma_q \\ 0 & 0 & 0 \end{bmatrix} \begin{bmatrix} \tilde{A}^f \\ \tilde{t}_e \\ \tilde{t}_d \end{bmatrix} \tag{C6}$$

We solve for the endogenous variables by inverting the matrix on the left-hand side of (C6).

$$\det(3)\begin{bmatrix} \tilde{L} \\ \tilde{P}_l^n \end{bmatrix} = \begin{bmatrix} -(\theta_{ll}^{*m} - \theta_l^m) & -Z \\ -1 & R \end{bmatrix} \begin{bmatrix} -\beta_e\sigma_l & \beta_e\sigma_l^* & \frac{\beta_d}{1+t_c}\phi_c\sigma_q \\ 0 & 0 & 0 \end{bmatrix} \begin{bmatrix} \tilde{A}^f \\ \tilde{t}_e \\ \tilde{t}_d \end{bmatrix} \tag{C7}$$

$$\textit{where} \quad \det(3) = (Z + \theta_l^m R) - \theta_{ll}^{*m}R > 0 \tag{C8}$$

The term "det(3)" must be positive for the equilibrium to be stable. It stands for the overall revenue effect of a rise in the labour income tax if income neutrality is imposed. If non-labour income is reduced by the same amount as labour income, the tax base is broadest. Hence, det(3) ≥ det(2) ≥ det(1). (Compare (C8) with (A21) and (B11)). The reduced form equations for $\tilde{P}_l^n$ and $\tilde{L}$ can now be computed from (C7).

$$\det(3)\tilde{L} = -(\theta_{ll}^{*m} - \theta_l^m)\frac{\beta_d}{1+t_c}\phi_c\sigma_q\tilde{t}_d - (\theta_{ll}^{*m} - \theta_l^m)\beta_e\sigma_l^*\tilde{t}_e + (\theta_{ll}^{*m} - \theta_l^m)\beta_e\sigma_f\tilde{A}^f \quad \text{(C9)}$$

$$\det(3)\tilde{P}_l^n = -\frac{\beta_d}{1+t_c}\phi_c\sigma_q\tilde{t}_d - \beta_e\sigma_l^*\tilde{t}_e + \beta_e\sigma_f\tilde{A}^f \quad \text{(C10)}$$

APPENDIX D: THE MARGINAL EXCESS BURDEN

The compensating variation is defined as the transfer necessary to maintain, after a policy shock, household utility at its initial level:

$$0 = \frac{\partial U}{\partial v^i}dv^i + \frac{\partial U}{\partial c^i}dc^i + \frac{\partial U}{\partial d^i}dd^i + \frac{\partial U}{\partial M}dM \quad \text{(D1)}$$

Using (10.10), (10.11) and (10.12), we can write (D1) as:

$$-\lambda P_l dl_s^i + \lambda(1+t_c)P_c dc^i + \lambda(1+t_c)P_d^h dd^i + \frac{\partial U}{\partial M}dM = 0 \quad \text{(D2)}$$

where we have used that $dl_s^i = -dv^i$. The household budgetconstraint (10.9) yields:

$$l_s^i dP_l + P_l dl_s^i + ds^i + db^i = [P_c c^i + P_d d^i + t_d(d^i - d^*)]dt_c + (1+t_c)[P_c dc^i + P_d dd^i + (d^i - d^*)dt_d + t_d(dd^i - dd^*)] \quad i = 1,2 \quad \text{(D3)}$$

where db^i is the compensating transfer for household i. Substituting (D2) into (D3) and summing up over two households, we arrive at the following expression:

$$db = -dS - LdP_l + [P_q Q - t_d D^*]dt_c + (1+t_c)[(D-D^*)dt_d - t_d dD^*] - \frac{\partial U}{\partial M}\frac{dM}{\lambda} \quad \text{(D4)}$$

where db represents the aggregate compensating variation. Here we have used the fact that environmental quality is a public good. The marginal excess burden, $\tilde{b}$, is defined as the change in the compensating variation relative to aggregate household income, $P_q Q$. Hence, the marginal excess burden can be written as:

$$\tilde{b} = -\tilde{S} - \tilde{A}^h - (1-\phi_{nl})\tilde{P}_l + (1-\phi_c)\tilde{t}_d + \tilde{t}_c - \frac{(\partial U/\partial M)}{(\partial U/\partial Q)}\frac{M}{Q}\tilde{M} \quad \text{(D5)}$$

where we used $\frac{\partial U}{\partial Q} = \lambda P_q$. Alternatively, using (10.34) and (10.35), we can write the marginal excess burden as:

$$\tilde{b} = -\tilde{S}^* - (1-\phi_{nl})\tilde{P}_l^n + \eta(\theta_e\tilde{E} + \theta_d\tilde{D}) \quad \text{(D6)}$$

Here, we employed (10.40) and the definition of η.

APPENDIX E: PRIVATE INCOME AND THE TAX BASE

Substituting the real after-tax wage rate from (10.34) and real non-labour income from (10.35) into (10.46), we find:

$$\tilde{b}^{p} = \tilde{S} + (1-\phi_{nl})\tilde{P}_{l} + \frac{\alpha_{dh}^{*}}{\alpha_{q}}\tilde{t}_{d} + \frac{\beta_{d}}{\alpha_{q}}\tilde{D}^{*} + \frac{\beta_{d}^{*}}{\alpha_{q}}\tilde{t}_{c} - (1-\phi_{c})\tilde{t}_{d} - \tilde{t}_{c}$$

$$= \frac{1}{\alpha_{q}}[\alpha_{q}\tilde{S} + \alpha_{l}\tilde{P}_{l} - (\alpha_{dh}-\alpha_{dh}^{*})\tilde{t}_{d} - (\alpha_{q}-\beta_{d}^{*})\tilde{t}_{c} + \beta_{d}\tilde{D}^{*}] \tag{E1}$$

where we have used $\frac{\alpha_{dh}}{\alpha_{q}} = 1-\phi_{c}$ and $\frac{\alpha_{l}}{\alpha_{q}} = 1-\phi_{nl}$. Asuming that $\tilde{P}_{g} = \tilde{P}_{d} = \tilde{P}_{c} = \tilde{G} = 0$, the government budget constraint (10.37) amounts to:

$$\alpha_{q}\tilde{S} = [\alpha_{ef}-\alpha_{ef}^{*}]\tilde{t}_{e} + [\alpha_{dh}-\alpha_{dh}^{*}]\tilde{t}_{d} + \alpha_{lf}\tilde{t}_{l} + [\alpha_{q}-\beta_{d}^{*}]\tilde{t}_{c} + \beta_{e}[\tilde{E}-\tilde{E}^{*}]$$

$$+ \beta_{d}[\tilde{D}-\tilde{D}^{*}] + \beta_{l}(\tilde{P}_{l} + \tilde{L}) + \beta_{c}[\phi_{c}\tilde{C} + (1-\phi_{c} - \frac{\beta_{d}}{\alpha_{q}})\tilde{D}] \tag{E2}$$

Using the non-profit condition (10.24) and assuming that $\tilde{P}_{y} = \tilde{P}_{e} = \tilde{r} = 0$, we write $\alpha_{lf}\tilde{t}_{l}$ as:

$$\alpha_{lf}\tilde{t}_{l} = -\alpha_{lf}\tilde{P}_{l} - (\alpha_{ef}-\alpha_{ef}^{*})\tilde{t}_{e} + \beta_{e}\tilde{E}^{*} \tag{E3}$$

Substituting (E3) into (E2) yields:

$$\alpha_{q}\tilde{S} + \alpha_{l}\tilde{P}_{l} - [\alpha_{dh}-\alpha_{dh}^{*}]\tilde{t}_{d} - [\alpha_{q}-\beta_{d}^{*}]\tilde{t}_{c} + \beta_{d}\tilde{D}^{*}$$

$$= \beta_{e}\tilde{E} + \beta_{d}\tilde{D} + \beta_{l}\tilde{L} + \beta_{c}[\phi_{c}\tilde{C} + ((1-\phi_{c}) - \frac{\beta_{d}}{\alpha_{q}})\tilde{D}] \tag{E4}$$

where we have used $\alpha_{lf} - \beta_{l} = \alpha_{l}$. The term at the left hand side of (E4) is equal to the term between square brackets at the (second) right hand side of (E1). Substituting (E6) into (E1), we find for real private income:

$$\tilde{b}^{p} = \frac{1}{\alpha_{q}}\{\beta_{e}\tilde{E} + \beta_{d}\tilde{D} + \beta_{l}\tilde{L} + \beta_{c}[\phi_{c}\tilde{C} + ((1-\phi_{c}) - \frac{\beta_{d}}{\alpha_{q}})\tilde{D}]\} \tag{E5}$$

Equation (E5) states that the effect on real private income is equal to the tax-base effect as defined in (10.42).

REFERENCES

Bovenberg, A.L. and W.J. Keller (1984) Non-linearities in applied general equilibrium models. *Economic Letters*, 14, PP. 53-59.

L.H. Goulder (1993) Effects of carbon taxes in an economy with prior tax distortions: An intertemporal general equilibrium analysis. Working paper, Stanford University, January.

Hausman, J.A. (1985) Taxes and labor supply. in A.J. Auerbach and M.S. Feldstein (eds.) *Handbook of public economics*. Vol. 1 North Holland, Amsterdam .

Hesse, D.M. and H. Tarkka (1986) The demand for capital, labor and energy in European manufacturing industry before and after oil price shocks, *Scandinavian Journal of Economics*, Vol. 88, No.3, pp. 529-546.

Keller, W.J. (1980) *Tax incidence: A general equilibrium approach*. North-Holland, Amsterdam .

Oates, W.E. (1991) Pollution charges as a source of public revenues. University of Maryland, Department of Economics, Working Papers no. 91-22.

Smith, S. (1992) Taxation and the Environment: A Suvey. *Fiscal Studies*, vol.13, no.4, pp. 21-57.

PART III

ACIDIFICATION AND TRADEABLE PERMITS

Chapter 11 REDUCTION OF SULFUR EMISSIONS IN EUROPE: A COMBINATION OF THE 'BEST AVAILABLE TECHNOLOGY'-CONCEPT AND THE CRITICAL LOADS-APPROACH

Markus Amann, Leen Hordijk, Ger Klaassen, Wolfgang Schöpp

ABSTRACT

This chapter explores the two major principles which are currently discussed in the negotiations on international reductions of SO_2 emissions in Europe. According to the critical loads concept, emissions are to be reduced until exposure levels are attained that are safe for the natural environment. The 'best available technology' (BAT)-approach calls for uniform implementation of certain emissions standards throughout the region. The chapter develops a critical load-based scenario, in which the gap between current sulfur deposition and critical loads should be reduced everywhere by at least 50 percent. The resulting national emission reductions are then compared to five different BAT scenarios, requesting certain emission standards for different groups of emitters. It is shown that a uniform application of 'best available technology' does not result in a cost-effective allocation of abatement measures. However, a regional zoning of BAT standards could produce results which are very close to the cost-minimal solution. Finally, the robustness of BAT strategies against variations in energy demand is analyzed.

11.1 INTRODUCTION

Now that the current protocol of the UN/ECE Convention on Long-range Transboundary Air Pollution on the reduction of emissions of SO_2 (UN, 1985) is being renegotiated, the principles underlying a protocol are also being discussed. In the current discussions at least three different principles can be noted:

- The present protocols for the reduction of SO_2 and NO_x are based on a principle of *equal reduction percentages* for all signatories to the protocol (the so-called flat-rate or blanket approach).

- Another principle is the use of *Best Available Technology* (BAT) for controlling emissions, sometimes under the condition of 'Not Entailing Excessive Cost' (BATNEEC). This principle is frequently used in Directives of the Commission of the European Communities.

- A third principle is called the *Critical Loads* (CL)-approach, developed under the Convention on Long-Range Transboundary Air Pollution since the preparation of the NO_x Protocol (UN, 1988; Nilsson & Grennfelt, 1988).

It is clear that each principle has its advantages and disadvantages; it is also clear that each principle has strong advocates and strong adversaries among the countries negotiating a protocol. In this chapter we try to combine two principles as a potential basis for a new SO_2 Protocol. We leave out the flat-rate reduction approach, because currently no party to the Convention seems to be in favor of that principle anymore. That leaves us with a source-oriented principle (BAT) and a receptor-oriented principle (Critical Loads).

In this chapter we will first analyze the sulfur deposition in 1990 and compare it with the established critical loads. Based on this, alternative principles for emission reductions will be discussed (Section 11.1). In Section 11.2 an optimized strategy to reduce the difference between sulfur deposition and critical loads by 50 percent will be developed. Section 11.3 explores five different BAT strategies and assesses their environmental effectiveness. In Section 11.4 a CL/BAT strategy combining features of both principles is introduced. The robustness of such a strategy is investigated in Section 11.5 and in Section 11.6 the major conclusions of the analysis are presented.

11.1.1 General approaches to reduce emissions

It has been a generally accepted principle that the signatories of the Convention shall control and reduce their emissions of sulfur in such a way as to minimize the damage to the environment caused by the deposition of sulfur compounds from the atmosphere. Since it is still very difficult to provide a reliable quantitative estimate of monetary costs of the damage by air pollutants, international obligations to reduce emissions can not be derived directly from this principle.

Recently, however, progress towards the physical quantification of damage caused by air pollution has been made. Ecological sciences succeeded in determining the maximum levels of air pollutants to which various ecosystems can be exposed without damage in the long-term (Nilsson & Grennfelt, 1988). In an international effort such 'critical loads' for sulfur deposition have been estimated for the European ecosystems endangered by acid deposition (Hettelingh *et al.*, 1991).

Consequently, basic obligations for a protocol could use the difference between critical loads and sulfur deposition as a measure for damage in physical terms. The larger the difference between deposition and critical loads, the larger the damage is assumed to be. Current knowledge about critical loads does not (yet) allow conclusions about the nature of the relationship (linearity, etc.).

In an ideal situation countries should reduce their emissions until the critical loads of sensitive ecosystems are reached. However, our current understanding of the critical loads suggests rather low exposure levels to be acceptable for natural ecosystems. Analysis has shown that such low sulfur deposition will not be achievable without drastic reductions in energy consumptions and changes in life styles (Amann *et al.*, 1992). Since such modifications are not on the political agenda for the near future, critical loads are not directly usable as short-term policy targets, although they can provide extremely important guidance for long-term strategies.

In the past, the establishment of 'national target loads' for sulfur deposition was suggested, which should reflect politically and socially accepted levels of environmental damage (Persson, 1991). There was never consensus on the question of whether such target loads should require only national acceptance of the environmental damage incurred (taking into account economic and other considerations) or whether they should be internationally accepted and agreed.

An alternative approach to determine emission abatement schedules is based on the precautionary principle. This principle requires the application of all technically and economically feasible measures to reduce emissions when the effects of air pollution are uncertain. Thereby, no direct quantitative relation to environmental sensitivities is established; the selection of measures to be taken is mainly determined by what is considered 'technically and economically feasible'.

11.1.2 The situation in 1990

Between the reference year of the first sulfur protocol 1980 and the year 1990, SO_2 emissions in Europe have been reduced by 26 percent (Table 11.1).

A comparison of sulfur deposition with the critical loads reveals that this decline is not sufficient to bring sulfur deposition below critical loads everywhere in Europe and, consequently, natural ecosystems in significant parts of Europe are not yet protected against damage from acidification.

Critical loads have been estimated for ecosystems within grid cells of 150*150 km (in Europe). For each of these cells the critical loads have been combined into cumulative distribution functions. The '5-percentile' of this function defines the deposition level below which 95 percent of the ecosystems is protected.

Figure 11.1 displays the difference between computed sulfur deposition in 1990 and the 5-percentile of critical loads taking into account the neutralizing effect of deposition of base cations.

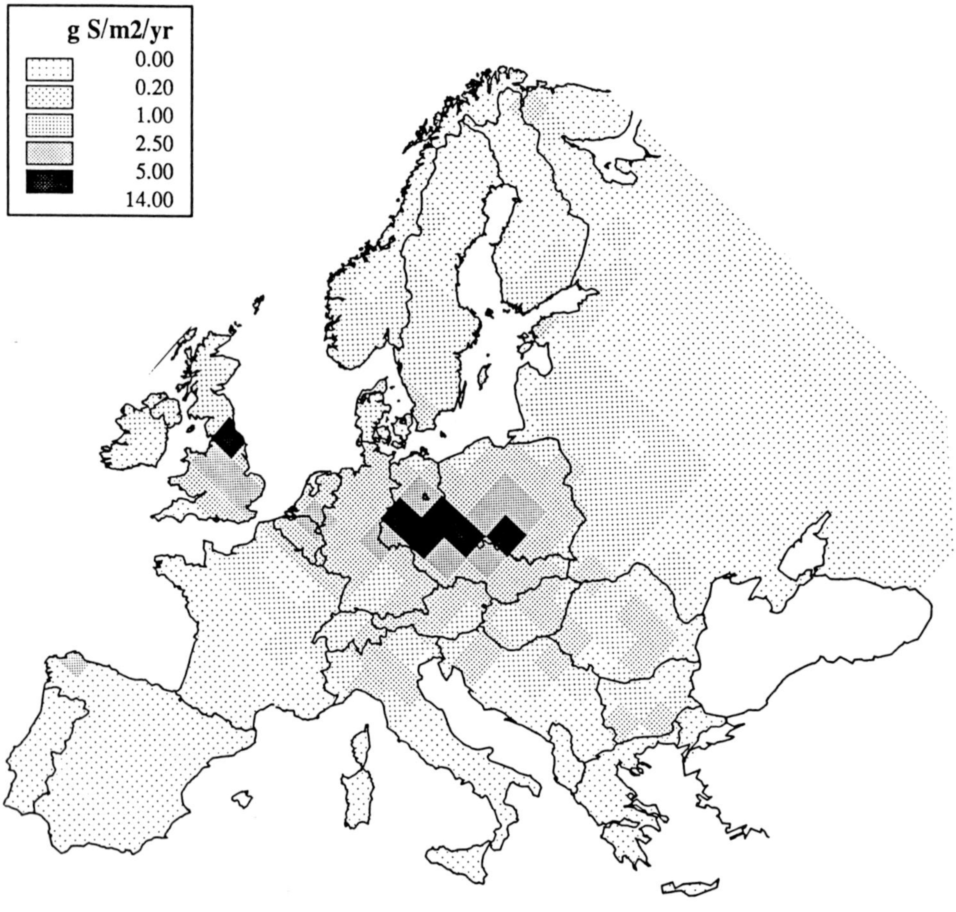

Figure 11.1 Gap between sulfur deposition in 1990 and critical loads (in grams sulfur/m^2/year).

Table 11.1 Emissions of SO_2 and the protection of ecosystems in the years 1980 and 1990. Unless otherwise noted emission data are extracted from UN 1992.

	1980		1990		
Country	Emissions (kton SO_2)	Ecosystems protected	Emissions (kton SO_2)	Change compared to 1980	Ecosystems protected
Albania	101[1)]	100 %	140[1)]	+39 %	96 %
Austria	390	12 %	98	-75 %	44 %
Belgium	828	4 %	414[3)]	-50 %	5 %
Bulgaria	1034	33 %	1266[3)]	+22 %	44 %
CSFR	3100	3 %	2443	-21 %	7 %
Denmark	448	46 %	184	-59 %	53 %
Finland	584	51 %	260	-55 %	77 %
France	3338	39 %	1260	-62 %	75 %
FRG	3194	10 %	940	-71 %	22 %
GDR	4300	10 %	4800	+12 %	22 %
Greece	400	99 %	740	+85 %	97 %
Hungary	1632	25 %	1010	-38 %	31 %
Ireland	222	74 %	168	-24 %	82 %
Italy	3800	55 %	2180[3)]	-43 %	75 %
Luxembourg	24	0 %	16[3)]	-33 %	10 %
Netherlands	466	16 %	207	-56 %	19 %
Norway	140	51 %	54	-61 %	60 %
Poland	4100	3 %	3210	-22 %	10 %
Portugal	266	100 %	204[3)]	-23 %	100 %
Romania	1800	29 %	1800	0 %	70 %
Spain	3319	97 %	2316	-30 %	98 %
Sweden	519	49 %	169	-67 %	70 %
Switzerland	126	29 %	62	-51 %	56 %
Turkey	860[1)]	98 %	1459[1)]	+70 %	97 %
UK	4898	42 %	3774	-23 %	54 %
Yugoslavia	1300	72 %	1480	+14 %	81 %
Baltic Sea	73[1)]	.	73[1)]	0 %	.
North Sea	173[1)]	.	173[1)]	0 %	.
Atlantic Ocean	316[1)]	.	316[1)]	0 %	.
Kola-Karelia	640[1)]	.	701[1)]	+10 %	.
St. Petersburg	496[1)]	.	409[1)]	-18 %	.
Baltic Region	621[1)]	.	653[1)]	+5 %	.
Belarus	740	.	596	-19 %	.
Ukraine	3850	.	2782	-28 %	.
Moldavia	330[1)]	.	313[1)]	-5 %	.
Russia[2)]	5926	.	3350	-43 %	.
EUROPE	54354	59 %	40020	-26 %	72 %

Notes: 1) IIASA estimate

2) Russian Federation within the EMEP area, Kola-Karelia and St. Petersburg regions excluded

3) Data apply to the year 1989

The regional differences in environmental sensitivities and in sulfur deposition levels result in a wide range of excess deposition: whereas in Scandinavia and Southern Europe deposition is generally less than 1 gram $S/m^2/yr$ above critical loads, in some places in Central Europe the excess amounts to more than 13 grams $S/m^2/yr$. It is also important to mention that the extent to which ecosystems were protected does not always follow the national development in emissions due to transboundary pollution (e.g. in the case of Bulgaria, Yugoslavia, etc.).

11.2 CLOSING THE GAP BETWEEN CURRENT DEPOSITION AND CRITICAL LOADS

As discussed in the previous section, the reduction of sulfur deposition below critical loads throughout Europe requires substantial emission reductions, which are far beyond the technical feasibility for current levels of energy consumption.

An alternative, effect-based approach aiming at international equality could be to reduce the difference between current deposition and critical loads everywhere by an equal percentage. Such a 'gap closure'-strategy (of e.g., the 'gap' displayed in Figure 11.1) would then aim at bringing all ecosystems in Europe equally closer to full achievement of the critical loads. It implicitely considers both regional differences in the environmental sensitivities (the critical loads) and some notion of economic efforts for achieving these targets (introduced by consideration of the current gap). Countries with high emissions and large exceedances are supposed to do more than countries in which ecosystems show only small excess of critical loads.

11.2.1 The RAINS model

Whereas the selection of the environmental targets (e.g., the uniform reduction of excess deposition throughout Europe) is a subject for political decisions, the determination of the resulting schedule for national abatement measures is a rather complex process. Integrated assessment models combining the various aspects of emission reduction strategies can help to structure the problem systematically and, even more important, derive the cost-optimal international allocation of reduction measures to achieve such deposition targets.

The analysis carried out in this chapter makes use of the 'Regional Acidification Information and Simulation' (RAINS)-model developed at the International Institute for Applied Systems Analysis (IIASA) in Laxenburg, Austria. The RAINS model focuses on acidification of Europe's natural environment and on the deposition of sulfur and nitrogen compounds that leads to acidification. The model consists of a set of submodules that cover the cause-effect chain: pollutant generation (energy scenarios, emission abatement options, cost of control), atmospheric transport and deposition and environmental effects (forest soils, Scandinavian lakes and

Table 11.5 Comparison of the current national legislations with the various BAT/BATNEEC scenario standards.

	SCEN E	SCEN D	SCEN C	SCEN B	SCEN A
Less than SCEN E	Minimum fuel standards	Min. fuel standards +BAT new	Stricter fuel standards +BAT new	Min. fuel standards +BAT new +BAT old	Stricter fuel standards +BAT new +BAT old[2]
Albania, Bulgaria, F.CSFR[1], Hungary[1], Poland[1], Romania, Turkey, F.Yugoslavia, Kola-Karelia, S.Petersburg,	Norway	Belgium, France, Italy, Luxembourg, UK		Switzerland	Austria, Finland, FRG-W, FRG-E, Netherlands, Sweden
Baltic Region, Belarus, Ukraine, Moldavia, Russia, all sea regions	**D-E: partial BAT new:** Greece, Ireland, Portugal, Spain		**B-C: partial retrofit:** Denmark		

Notes: [1] Recent changes in national legislation not taken into account
[2] Similar or lower emission levels can also be achieved by stricter control of industrial emissions instead of the tight fuel standards.

Table 11.6 European SO_2 emissions, control costs and the protection of ecosystems of the BAT/BATNEEC scenarios.

	SCEN E: Minimum fuel standards	SCEN D: Minimum fuel standards + BAT new	SCEN C: Stricter fuel standards + BAT new	SCEN B: Min. fuel standards + BAT new + BAT retro	SCEN A: Stricter fuel standards + BAT new + BAT retro
Emissions (Mton SO_2)	44.3	36.6	24.6	15.6	10.5
Control costs (Billion DM/yr)	12.0	13.7	26.9	37.8	45.1
Ecosystems protected	68 %	75 %	86 %	94 %	96 %

Based on these examples it can be concluded that as long as a quantification of environmental targets is available, the BAT principle alone does not necessarily promote a cost-effective use of resources.

11.4 A 'CL/BAT'-SCENARIO ACHIEVING A 50 % GAP CLOSURE

Although in recent years substantial progress has been made in quantifying environmental sensitivities, atmospheric dispersion processes and other indicators, significant uncertainties in many fields necessary to determine cost-optimal emission reduction strategies remain yet unresolved. Major uncertainties, which have direct impact on an optimal allocation of resources, are associated with, e.g.,

- the predictions of future economic development,
- the accuracy of future emission estimates,
- the technical development of emission control technologies,
- limited international experience on emission control costs,
- the long-term and the inter-annual variability of the meteorology,
- imperfections of atmospheric models,
- the understanding of processes determining the critical loads,
- lack of regional data on soil characteristics, vegetation, base cation deposition.

Acknowledging these uncertainties, it may appear as rather ambitious to derive from model calculations 'hard facts' on necessary emission reductions. In some cases, minor changes in some input parameters (or in mere assumptions) can change the cost-minimal allocation considerably, leaving a previous cost-optimal solution behind as rather inefficient.

In such a situation a more robust principle to determine the actual extent of emission reductions might add considerable stability to optimization results. In the preceding section it has been demonstrated that the BAT/BATNEEC approach is not able to derive cost-effective targets for emission reductions necessary to achieve certain environmental effects. However, properly designed, the BAT/BATNEEC concept is able to promote the cost-effective selection of measures for achieving emission targets within a region: a major principle underlying the BAT/BATNEEC concept is the selection of measures or application areas according to increasing marginal costs. Thereby, BAT/BATNEEC does not define the overall level of emission reductions, but determines the set of cost-effective measures. The actual amount of emissions reduced through application of these measures will depend crucially on the activity level (e.g., the volume of coal consumption in power plants), determined, e.g., by the economic development.

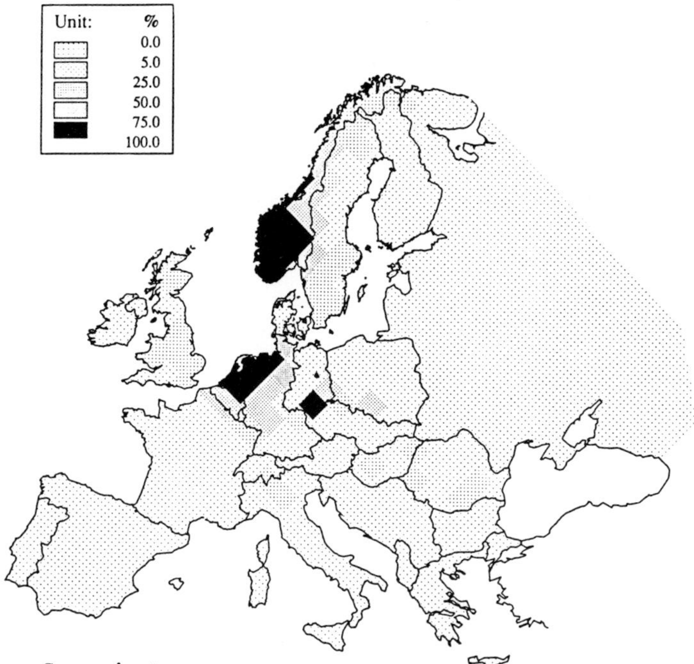

Scenario A

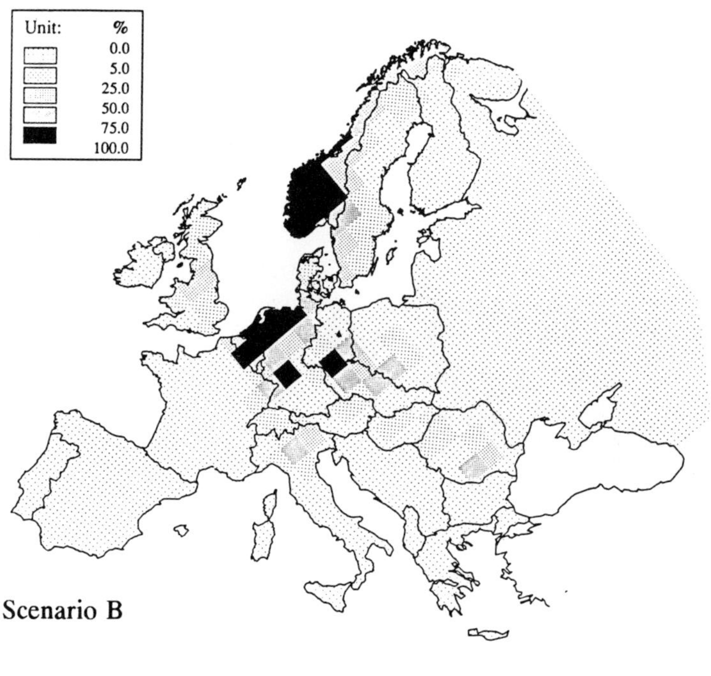

Scenario B

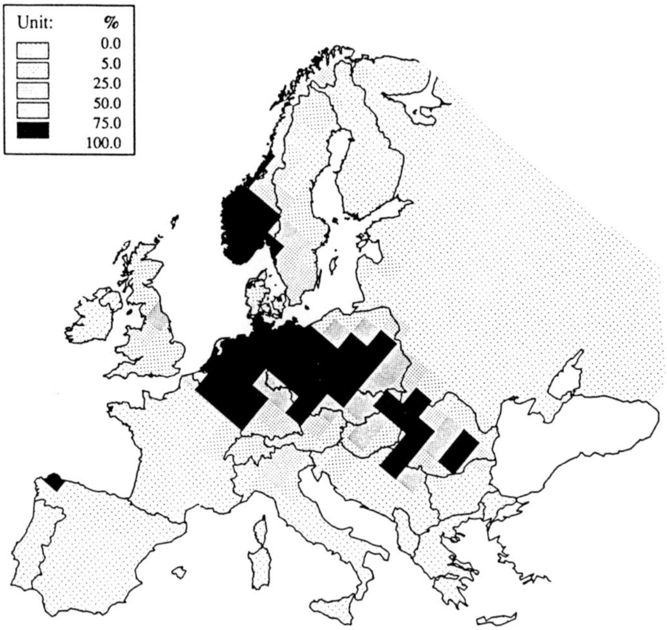

Scenario C

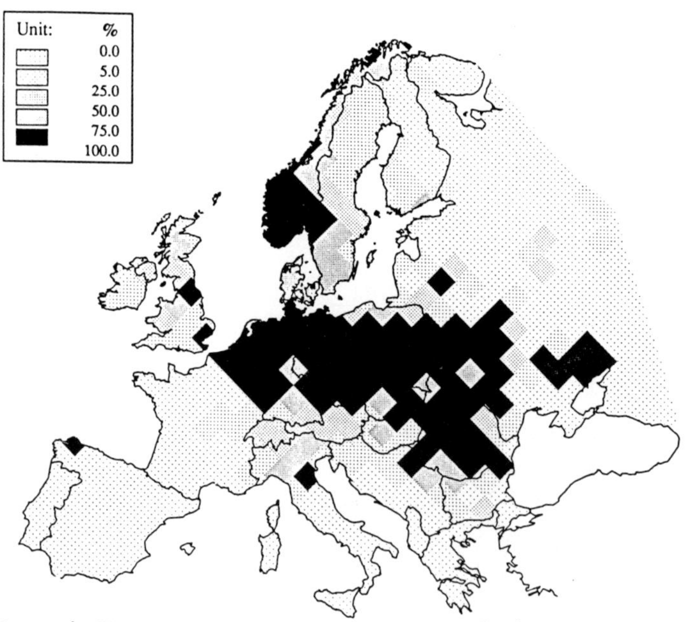

Scenario D

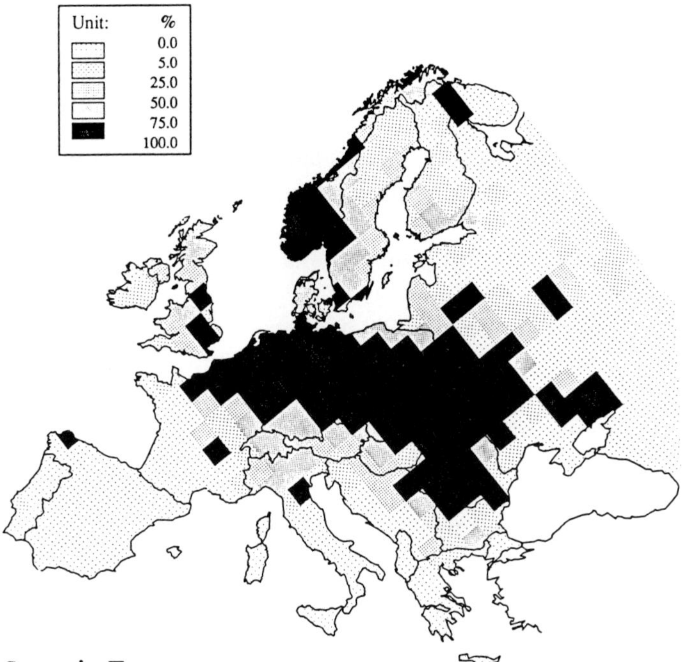

Scenario E

Figure 11.3 The regional distribution of ecosystems having sulfur deposition above critical loads.

The marginal costs of reductions, however, will be similar for the different economic scenarios.

11.4.1 Combining the critical loads and the BAT approaches

In this section an attempt is made to merge the effect-based international optimization inherent to the critical loads approach with the robustness of the source-oriented BAT concept. The 50 percent gap closure scenario described in Section 11.2 (both in terms of costs and environmental impacts) served as the reference case to be reconstructed by proper selection of different BAT standards. For this purpose the national emission levels resulting from the optimization are compared with the emissions from the BAT scenarios A to E developed in Section 11.3 and, for each country, the BAT standard with emissions closest to the optimized levels is identified.

As demonstrated in Table 11.7, it is possible to group all European countries into only three classes of BAT standards to mimic the optimized 50 percent gap closure scenario:

BAT-Class 1:
The most stringent class of emission standards is equivalent to the requirements outlined in BAT-Scenario A:

- Flue gas desulfurization for all new power stations and large industrial boilers;
- Retrofitting of all existing large boilers in power generation and the industrial sector with flue gas desulfurization
- Stringent limits on sulfur content in heavy fuel oil and hard coal (0.6 %), if no desulfurization is applied;
- Reduction of industrial process emissions by 80 percent.

To reach results close to the optimized 50 percent gap closure scenario, such standards would be required in Austria, Belgium, Denmark, Finland, France, Germany (eastern and western part), Ireland, the Baltic countries, Luxembourg, Netherlands, Norway, Sweden, Switzerland, the United Kingdom and the Kola-Karelia and St. Petersburg regions of the Russian federation.

BAT-Class 2:
For the second BAT-class the retrofit requirement, which is in Class 1 obligatory for all large boilers, is relaxed and applicable only to a part of the existing plant stock operating in the year 2000.

- Flue gas desulfurization for all new power stations and large boilers;
- Partial retrofitting of existing large boilers;
- Stringent limits on sulfur content in heavy fuel oil (0.6 %), if no FGD is applied;
- Reduction of industrial process emissions by 80 percent.

The actual extent for the retrofit program could be subject to negotiations, depending, e.g., on the specific economic situation in a country. For reasons of simplicity, the following analysis assumes the application of flue gas desulfurization to 50 percent of the existing power stations.

These standards would apply mainly to Central and Eastern European countries, i.e., Belarus, Bulgaria, Czechoslovakia, Hungary, Italy, Poland, Romania, Spain, Ukraine, former Yugoslavia and the areas of the Russian Federation not contained in Class 1.

BAT-Class 3:
To satisfy environmental constraints imposed by acidification, stringent emission standards apply in countries of Class 3 only for new installations. Existing boilers may be operated without advanced emission control:

- Flue gas desulfurization for all new power stations and large industrial boilers;
- Limits on the sulfur content in heavy fuel oil (1 percent) if no FGD is applied.

This class would contain Albania, Greece, Portugal and Turkey. In addition, low sulfur fuel oil would be required for all marine vessels in the North Sea, the Baltic Sea and the Atlantic Ocean. It has to be mentioned, however, that safe levels of ambient concentrations of SO_2 (e.g., to protect human health) might still require more stringent measures than would be necessary to prevent acidification.

Figure 11.4 displays the regional zoning of the three BAT classes. It shows that the effect-based determination of reduction requirements introduces a regional grouping of countries: regions with the tightest standards of Class 1 (full retrofit) are all located in the northwest and the center of Europe. Around this area, to the east and the south, countries have to follow Class 2 (partial retrofit), and in the rest of Europe BAT is only prescribed for new plants. In cases where the current reduction plans are result in stricter emission ceilings than the relevant BAT scenario, the current legislation is used.

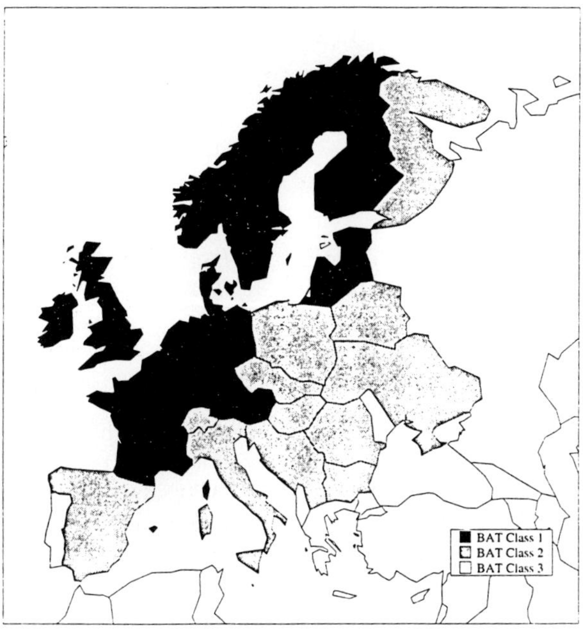

Figure 11.4 BAT standards in Europe assumed in the CL/BAT Scenario to achieve a 50 percent gap-closure.

Table 11.7 Country emissions, control costs and protection of ecosystems for the CL/BAT Scenario.

	The CL/BAT Scenario				
	Emissions			Deposition	
Country	kton SO_2	rel. to 1980	BAT class	Costs (10^6 DM/yr)	Ecosystems protected
Albania	122	+21 %	3	40	100 %
Austria	78	-80 %	CRP (1)	494	90 %
Belgium	179	-78 %	1	946	79 %
Bulgaria	520	-50 %	CRP (2)	315	74 %
CSFR	854	-72 %	2	1300	80 %
Denmark	64	-86 %	1	627	93 %
Finland	116	-80 %	CRP (1)	853	96 %
France	401	-88 %	1	1105	99 %
FRG, West	520	-84 %	CRP (1)	3864	72 %
FRG, East	230	-95 %	CRP (1)	2153	72 %
Greece	557	+39 %	3	282	100 %
Hungary	522	-68 %	2	360	98 %
Ireland	45	-80 %	1	241	99 %
Italy	1290	-66 %	2	2270	89 %
Luxembourg	8	-67 %	1	7	89 %
Netherlands	106	-77 %	CRP (1)	688	47 %
Norway	59	-58 %	1	122	79 %
Poland	1400	-66 %	2	2780	72 %
Portugal	307	+15 %	3	58	98 %
Romania	1280	+29 %	2	1125	100 %
Spain	1645	-50 %	2	1850	98 %
Sweden	100	-81 %	CRP (1)	560	92 %
Switzerland	60	-52 %	2	15	82 %
Turkey	2887	+235 %	CRP	0	100 %
UK	638	-87 %	1	3443	97 %
Yugoslavia	1140	-12 %	2	500	92 %
Baltic Sea	12	-84 %	1	100	.
North Sea	41	-76 %	1	238	.
Atlantic Ocean	76	-76 %	1	434	.
Kola-Karelia	135	-79 %	1	611	.
St. Petersburg	83	-83 %	1	353	.
Baltic Region	142	-77 %	1	702	.
Belarus	156	-79 %	2	300	.
Ukraine	773	-80 %	2	3023	.
Moldavia	231	-30 %	3	90	.
Russia[1)]	3723	-38 %	3	1334	.
EUROPE	19539	-62 %		33138	94 %

Note: 1) Russian Federation within the EMEP area. The Kola-Karelia and St. Petersburg regions are excluded.

The CL/BAT Scenario reduces SO_2 slightly more than the optimized 50 percent gap-closure scenario (down to 19.5 million tons instead of 20.2 million tons) at slightly higher cost (33.1 instead of 30.0 billion DM/year). The extent to which ecosystems are protected against sulfur deposition is almost equal in both cases (94 %). However, these deviations from the theoretical optimum could be regarded as minor, given all the uncertainties inherent in the calculations.

A comparison of the CL/BAT Scenario with current legislation (Table 11.5) reveals that, in order to achieve the gap closure target, most countries have to tighten their emission standards. The only exceptions are countries with already stringent regulations in force (Austria, Finland, West and East Germany, Netherlands, Sweden, Switzerland) and countries with lower sensitivity towards acid deposition (Turkey and Belarus).

11.5 ROBUSTNESS OF THE CL/BAT SCENARIO

In this section the performance of the CL/BAT Scenario under different exogenous conditions is analyzed. Out of the many elements which introduce uncertainties into the design of optimal strategies the analysis is restricted to the impacts of alternative future economic developments and, in particular, to the uncertainties in the energy scenario which was used.

To explore the environmental effectiveness of the CL/BAT Scenario the effect of the various BAT guidelines specified in the scenario on remaining SO_2 emissions are simulated for various energy systems. Three different cases are investigated:

- The 'reference-case' assumes the validity of the official governmental energy projections which have been used in the calculations above.
- As an alternative, a case is analyzed in which stronger economic growth results in higher energy consumption. The assessment is based on the 'high growth'-energy scenario developed by the EFOM-ENV energy model for eight European countries (Rentz *et al.*, 1992).
- Reactions to an 'energy-efficiency' future are studied by assuming energy consumption patterns satisfying strict constraints on national CO_2 emissions as calculated by the EFOM-ENV model.

Table 11.8 lists the resulting emission reductions after implementation of the BAT guidelines of the CL/BAT Scenario for the eight countries, for which the EFOM-ENV analysis has been performed,

The table shows that, although differences among energy scenarios are considerable (resulting in variations of ± 20 percent of unabated emissions), the ranges of remaining emissions after application of BAT are rather small. Furthermore, these variations are smallest in areas with the most sensitive ecosystems, in which the most stringent BAT standards are also required (typically ± 1 percent in Class 1). With less demanding reduction requirements, differences in energy scenarios will have larger influences, although even in the worst case analyzed in this chapter (Italy) variations are within $\pm$ 5 percent. The current lack of equivalent energy analysis for additional countries prohibits a full exploration of the environmental effects of such variations in reduction levels. However, it can be assumed that exceedances of critical loads will not differ dramatically from each other.

Table 11.8 Remaining emissions and emission reduction percentages (related to the emissions of 1980) after applying the emission standards of the CL/BAT Scenarios to different energy scenarios.

		Reference case Official Energy Pathway		EFOM - High energy demand scenario		EFOM - CO_2 reduction scenario	
	BAT class	kton SO_2	change[1)]	kton SO_2	change[1)]	kton SO_2	change[1)]
Finland	1	116	-85 %	120	-84 %	108	-86 %
FRG-W	1	520	-84 %	550	-83 %	483	-85 %
Netherlands	1	106	-77 %	116	-75 %	103	-78 %
UK	1	638	-87 %	669	-86 %	668	-86 %
France	2	500	-80 %	474	-82 %	482	-84 %
Hungary	2	661	-59 %	634	-61 %	602	-63 %
Spain	2	1027	-68 %	965	-70 %	870	-73 %
Italy	2	1450	-62 %	1129	-70 %	1360	-64 %

Note: [1)] related to 1980

11.6 CONCLUSIONS

Different principles for the allocation of international emission reduction requirements exist. This chapter shows that in reality actual abatement schedules derived from the different principles can be rather similar:

An *effect-oriented approach* could set an equal closure of the gap between the current situation and a general long-term target (e.g., the achievement of critical loads) as the objective for all countries. For the reduction of SO_2 emissions in

Europe, an optimized strategy aiming at a 50 percent gap closure would require the control of SO_2 emissions in Europe by 63 percent (compared to 40 percent of the current reduction plans) at 93 percent higher costs. As a consequence, the share of ecosystems protected against sulfur deposition would increase from 82 to 94 percent. Such a strategy would require drastic emission reductions in northwestern and central Europe. Although countries in eastern and western Europe would have to undertake considerable additional efforts, their control levels would be lower than in the northwest. Southern Europe would have to reduce emissions moderately.

Solely *source-oriented approaches*, such as the 'Best Available Control Technology' concept, do not take environmental sensitivity into account for the determination of required emission control levels. The analysis demonstrates that 'uniform' BAT strategies, requiring equal measures from all countries in Europe, do not generally promote a cost-effective use of resources.

However, the BAT concept can be combined with effect-oriented approaches to derive more cost-effective and more robust strategies. A specification of country-specific BAT standards, guided by effect-based reduction requirements, could create three different BAT standards throughout Europe. In a deterministic case, such a strategy would only be slightly less cost-effective than an optimized strategy.

As a basic characteristic of such a strategy, application of 'best available control technology' to new plants is generally necessary in almost all European countries. The high environmental sensitivity in northwestern Europe requires a full retrofit program for existing sources in this region, whereas in the southwest and the east of Europe only a partial retrofit of existing emitters is demanded. In southern Europe, the necessity for retrofit measures will mainly depend on the local air quality situation. Emission levels to be achieved by the implementation of BAT control strategies are rather robust against changes in energy consumption.

REFERENCES

Alcamo J., Shaw R. W., Hordijk L. (eds.) (1990) *The RAINS model of Acidification. Science and Strategies in Europe.* Kluwer Academic Publishers, Dordrecht, Netherlands

Amann M., Hordijk L., Klaassen G., Schöpp W., Sørensen L. (1992a) Economic restructuring in Eastern Europe and acid rain abatement strategies. *Energy Policy*, 12/1992, pp.1186-1197.

Amann M., Bertok I., Cofala J., Klaassen G., Schöpp W. (1992b) *Strategies for reducing sulfur dioxide emissions in Europe*. SR-92-08, International Institute for Applied Systems Analysis, Laxenburg, Austria.

Batterman S., Amann M. (1991) Targeted Acid Rain Strategies Including Uncertainty. *Journal of Environmental Management*, 32,57-72.

Hettelingh J.-P., Downing R.J., de Smet P. (1991) *Mapping Critical Loads for Europe.* CCE Technical Report No. 1, National Institute of Public Health and Environmental Protection, Bilthoven, The Netherlands.

IEA (1991) *Coal information 1991.* International Energy Agency, Paris, France.

Nilsson J., Grennfelt P. (1988) *Critical Loads for Sulphur and Nitrogen.* Miljorapport 1988:16, Nordic Council of Ministers, Copenhagen, Denmark.

Persson G. (1991) *Critical Loads Approach as a Policy Option.* In: Chadwick M.J., Hutton M. (eds.) *Acid Deposition in Europe.* Stockholm Environment Institute, Stockholm, Sweden.

Rentz O., Haasis H.-D., Jattke A., Ruß P., Wietschel M., Amann M. (1992) *Impacts of Energy Supply Structure on National Emissions Reduction Potentials and Emission Reduction Costs.* Final Report. University of Karlsruhe, Germany.

Sandnes H., Styve H. (1992) *Calculated Budgets for Airborne Acidifying Components in Europe, 1985, 1987, 1988, 1989, 1990 and 1991.* Meteorological Synthesizing Centre - West, Norwegian Meteorological Institute, Oslo, Norway.

Schärer B., Haug N. (1987) The Cost of Flue Gas Desulfurization and Denitrification in the Federal Republic of Germany. *Economic Bulletin for Europe*, 39(1), Pergamon Press, Oxford, UK.

UN/ECE (1991) *Energy data base* (on tape). United Nations Economic Commission for Europe, Geneva, Switzerland.

UN/ECE (1992) *Annual Review of Strategies and Policies for Air Pollution Abatement.* United Nations, Economic Commission for Europe, Geneva, Switzerland.

UN (1985) *Protocol to the 1979 Convention on Long-Range Transboundary Air Pollution on the Reduction of Sulphur Emissions or their Transboundary Fluxes by at least 30 per cent.* United Nations Economic Commission for Europe, New York.

UN (1988) *Protocol to the 1979 Convention on Long-Range Transboundary Pollution on the Reduction of Emissions of Nitrogen Oxides.* United Nations, Economic Commission for Europe, New York.

Chapter 12 IMPROVING THE ENVIRONMENT THROUGH PERMIT TRADING: THE LIMITS TO A MARKET APPROACH

Tomasz Żylicz[1]

ABSTRACT

The aim of the chapter is to extend the standard tradeable pollution permit approach to the case where there are several sources of environmental impact affecting several environmental quality indicators in a non-necessarily linear way. Particular emphasis is on how these cause-effect relationships can be reflected in designing market-oriented policy instruments.

The author argues that there is a potential for cost savings if various sources of environmental impact are regulated in a coordinated way. However, permit markets (as well as taxes) could achieve a least cost solution only under very restrictive assumptions. In particular, it is pointed out that some of the known synergistic interactions - implying non-convexity of the constraint function delineating the permit market (or determining a tax) - clearly violate these conditions. This basically limits the usefulness of the tradeable permit and tax approach to simple linear cases.

12.1 INTRODUCTION

There have been attempts to improve the effectiveness and efficiency of environmental policies by achieving a better integration and coordination across various pollutants, media, and exposure characteristics. While some simple measures directly aimed at policy integration have been already implemented in several countries (Opschoor & Vos 1989), the problem still largely remains at the theoretical level. A number of interesting concepts are, however, discussed in the literature. Portney (1988) introduced the notion of a "risk bubble" or "risk trading" to analyze alternative policies to reduce environmental risk in a cost effective way. Along these lines Roumasset & Smith (1990) discussed alternative policies to improve efficiency of reducing health risks linked to air pollutants, and found a large cost-saving potential in addressing outdoor and indoor exposures simultaneously. Hahn (1989)

[1] An earlier draft of this chapter was presented at the Fourth Annual Conference of the European Association of Environmental and Resource Economists in Fontainebleau (June 30 -July 3, 1993). Comments from Ger Klaassen, who agreed to be the discussant of the paper, are gratefully acknowledged.

analyzed the problem of multiple objectives in policy design. He suggested an algorithm for trading environmental permits originally issued for different activities (types of pollution) yet contributing to the same final effect such as human health, safety or welfare. These permits could be traded for each other at the rate corresponding to their relative contributions to the final outcome in order to equate marginal costs of protection not only within a given range of economic activities, but also across these ranges. This intuitively obvious algorithm works well only if the relationship between various activities is linear. Unfortunately, it may misallocate permits if this relationship is a synergistic one. This complicates matters considerably since there is evidence that many important interactions environmental policy makers have to account for are of this kind.

Synergism in environmental interactions occurs whenever the joint effect of two or more factors is different from the sum of outcomes that would have happened had each of these factors acted alone. Thus any non-linear impact function of two variables can be thought of as a manifestation of some synergistic relationship. Once discharged into the environment, many pollutants act in this way. Lower atmospheric ozone is formed as a result of interacting nitrogen oxides and hydrocarbons; sharp increases of morbidity and mortality have been linked to episodes of simultaneously increased ambient concentrations of sulphur dioxide and suspended particulates; discharges of cyanides have much more severe effects on aquatic life if they are accompanied by metals, and so on.

Even though synergistic effects are quite often referred to by environmental economists, there is little theoretical and modelling work devoted to this problem. Beavis & Walker (1979) derived and discussed first order conditions for minimising abatement costs in a watershed under the assumption of synergism. Endres (1986) observed that a cost-effective solution to multiple pollutant abatement problems can be achieved through taxes or tradeable permits if the pollutants interact linearly. At the same time, he indicated that both taxes and permits may fail in this role if the pollutants interact non-linearly. Apart from that, there has been no systematic inquiry into extending market-oriented approaches to the synergistic case and identifying specific conditions under which these might be expected to work. Since it was first mentioned by Crocker (1966) the tradeable permit approach, both in theory and practical implementations, has been confined to single pollutants linearly contributing to ambient concentrations.

The theoretical literature on tradeable permits distinguishes between two basic systems: emission permits and ambient permits. Under the former polluters receive rights to emit certain amount of a harmful substance, and it is the reduction in the emission of this substance that gives the polluter a credit and eventually allows a transfer of the right. Under the latter polluters are endowed with rights to certain shares in ambient concentrations at one or more locations; it is then these shares that can be priced and traded. Most of the implementations of the U.S. "bubble policy" are emission oriented with relevant ambient concentrations acting only as constraints on permit transfers (Liroff 1986, Tietenberg 1994). However, some state jurisdictions did require that transfers resulted in net decrease of emissions, the longer the distance between trading sources the larger the decrease (Tietenberg 1985). This can be seen as a very rough approximation of the ambient approach, whose fuller implementation was considered too cumbersome. Here the analysis is confined to ambient permit systems, which are directly aimed at dealing with complex relationships between sources of environmental stress and their effects.

To focus attention, throughout the chapter we will use the conventional terminology of pollution control. Thus we will refer to *emissions* contributing to *ambient concentrations*, but in fact the whole analytical framework developed here can be applied to any factors affecting welfare conditions of ultimate policy concern. In contrast to the emission rates which are usually expressed in tons per a given unit of time, ambient quality data are typically represented by average annual or daily concentrations in, for example, $\mu g/m^3$, or parts per million. They can also be measured in logarithmically (or otherwise) transformed units such as pH. In addition, if direct welfare effects are of ultimate interest, these effects - e.g. increase in mortality and morbidity, crop damage, additional days with poor visibility, decline in sport fishing in a water body - can be naturally substituted for "physical" concentrations in the analysis. It is also possible to construct aggregate measures of environmental quality for policy purposes. In this chapter we do not assume any particular measurement unit for an ambient concentration. We use the latter term for simplicity and in order to follow the tradition established in the literature even though perhaps "aggregate ambient quality", "exposure factor" and the like might have been more adequate. Likewise, for brevity, we will use the term *synergism* in cases where any non-linear functional relationship could be thought of.

The chapter consists of 8 sections. In the first introductory one the question of interacting pollutants and multiple policy objectives is discussed briefly. The next two sections review some standard results for permit markets designed for pollutants which linearly contribute to ambient environmental quality. The framework to deal with two interacting pollutants and the proof of the efficiency of a permit market combined with a special tax are established in section (12.4). The problem is solved by using a second order Taylor series approximation for an unknown relationship between emissions and ambient quality indicators. Further extensions of the model - to account for multiple pollutants and multiple monitoring locations - are outlined in section (12.5). Sections (12.6)-(12.7) are devoted to a discussion of information and institutional requirements of the model, as well as of possible violations of the assumptions that were used to prove the efficiency of the market. Some tentative conclusions are offered in the last section.

12.2 SIMPLE AMBIENT PERMIT MODEL FOR A SINGLE POLLUTANT

Let us assume that there are n plants emitting respectively e_i ($i = 1,...,n$) of the same pollutant per given unit of time. The emission rate can be reduced if the plant chooses to abate some of its potential discharges. This however is associated with abatement costs $C_i(e_i)$. Let us suppose further that the actual emissions uniformly mix in the homogenous environment and contribute to the ambient concentrations A_0 according to the following formula (in this chapter Σ denotes summation over $i = 1,...,n$):

$$A_0 = a + \Sigma b \cdot e_i = a + b \cdot \Sigma e_i, \quad a \geq 0, \quad b > 0 \qquad (12.2.F)$$

Parameters a and b are pollutant and region specific. In particular, a can be interpreted as "background" concentrations which are independent of the emissions from the plants, and b as an inverse measure of the environment's assimilative capacity. Discharges of phosphorus into a lake may serve as an example for this model.

From the point of view of social welfare an efficient allocation requires (as a necessary but not sufficient condition) that abatement costs be reduced to minimum provided ambient concentrations do not exceed certain predetermined level A:

$$\min_{e1,...,en}\{\Sigma C_i(e_i): \; a + b \cdot \Sigma e_i \leq A, \; e \geq 0, \; i = 1,...,n\}.$$

The Lagrange function of this mathematical programming problem is:

$$L(e_1,...,e_n;\lambda) = \Sigma C_i(e_i) + \lambda \cdot (a + b \cdot \Sigma e_i - A)$$

which yields the following Kuhn-Tucker conditions:

$$\mathrm{d}C_i(e_i)/\mathrm{d}e_i + \lambda b \geq 0, \quad \text{for } i = 1,...,n, \qquad (12.2.1^{\circ})$$
$$\Sigma(\mathrm{d}C_i(e_i)/\mathrm{d}e_i + \lambda b) \cdot e_i = 0, \qquad (12.2.2^{\circ})$$
$$a + b \cdot \Sigma e_i \leq A, \qquad (12.2.3^{\circ})$$
$$\lambda \cdot (a + b \cdot \Sigma e_i - A) = 0, \qquad (12.2.4^{\circ})$$
$$e_1,...,e_n \geq 0, \quad \lambda \geq 0. \qquad (12.2.5^{\circ})$$

Under the standard assumptions (convexity of the functions $C_1,...,C_n$) these imply an optimum allocation of $e_1,...,e_n$.

Whenever possible, we will be using a concise matrix notation in the text:

$$\mathbf{e} \in \mathrm{R}^n, \ \mathbf{e} = [e_1,...,e_n]',$$
$$\mathbf{C}:\mathrm{R}^n \to \mathrm{R}^n, \ \mathbf{C}(\mathbf{e}) = [C(e_1),...,C(e_n)]',$$
$$\partial\mathbf{C}/\partial\mathbf{e} = [\mathrm{d}C(e_1)/\mathrm{d}e_1,...,\mathrm{d}C(e_n)/\mathrm{d}e_n]',$$
$$\mathbf{1} = [1,...,1]',$$
$$\mathbf{0} = [0,...,0]'$$

(where $'$ denotes transposition). From the context it will always be clear what space **1** or **0** belongs to. In this section $\mathbf{1},\mathbf{0} \in \mathrm{R}^n$. Please also note that for any $\mathbf{x} = [x_1,...,x_n]' \in \mathrm{R}^n$: $\mathbf{1}' \cdot \mathbf{x} = \Sigma x_i$.

Using the matrix notation the model can be rewritten as:

$$\min_{\mathbf{e}}\{\mathbf{1}' \cdot \mathbf{C}(\mathbf{e}): \ a + b \cdot \mathbf{1}' \cdot \mathbf{e} \leq A, \ \mathbf{e} \geq \mathbf{0}\}.$$

Thus the Lagrange function and the Kuhn-Tucker conditions become:

$$L(\mathbf{e},\lambda) = \mathbf{1}' \cdot \mathbf{C}(\mathbf{e}) + \lambda \cdot (a + b \cdot \mathbf{1}' \cdot \mathbf{e} - A),$$
$$\partial\mathbf{C}/\partial\mathbf{e} + \lambda b \cdot \mathbf{1}' \geq \mathbf{0}, \quad (12.2.1)$$
$$(\partial\mathbf{C}/\partial\mathbf{e} + \lambda b \cdot \mathbf{1}') \cdot \mathbf{e} = 0, \quad (12.2.2)$$
$$a + b \cdot \mathbf{1}' \cdot \mathbf{e} \leq A, \quad (12.2.3)$$
$$\lambda \cdot (a + b \cdot \mathbf{1}' \cdot \mathbf{e} - A) = 0, \quad (12.2.4)$$
$$\mathbf{e} \geq \mathbf{0}, \ \lambda \geq 0. \quad (12.2.5)$$

One of the ways of achieving an allocation ($\mathbf{e}^*$) given by (12.2.1)-(12.2.5) is to establish a tradeable discharge permit system. If the ambient concentration limit A is to be met and the relationship between emissions and concentrations is given by the formula (12.2.F), then the allowable emissions are $q^e = (A - a)/b$. The regulatory agency may then issue discharge permits for q^e units of the pollutant and initially allocate them among the plants. These initial endowments $\mathbf{q}^e = [q_1^e,...,q_n^e]'$ sum up to q^e: $\Sigma q_i^e = q^e$. The agency also establishes a market for the permits so that their initial owners are free to trade them, and a uniform price P^e per unit permit will emerge, except in unusual (corner solution) cases.

From an individual (ith) plant's point of view the optimisation problem is seen as:

$$\min_{ei,qi}\{C_i(e_i) + P^e \cdot (q_i - q_i^e): \ e_i \leq q_i, \ e_i,q_i \geq 0\}.$$

This implies the Lagrange function

$$L_i(e_i,q_i,\mu_i) = C_i(e_i) + P^e \cdot (q_i - q_i^e) + \mu_i \cdot (e_i - q_i)$$

and the Kuhn-Tucker conditions

$$dC_i(e_i)/de_i + \mu_i \geq 0, \quad (12.2.1e')$$
$$(dC_i(e_i)/de_i + \mu_i) \cdot e_i = 0, \quad (12.2.2e')$$
$$P^e - \mu_i \geq 0, \quad (12.2.1q')$$
$$\mu_i \cdot (P^e - \mu_i) = 0, \quad (12.2.2q')$$
$$e_i \leq q_i, \quad (12.2.3')$$
$$\mu_i \cdot (e_i - q_i) = 0, \quad (12.2.4')$$
$$e_i, q_i, \mu_i \geq 0. \quad (12.2.5')$$

These conditions - if met by all plants simultaneously - are equivalent to (12.2.1)-(12.2.5) with $P^e = \lambda b$. This is established in section 12.4 where indeed a more general version of the equivalence is proven.

It would be easier to analyze extensions of this basic model if we interpret the permit market in terms of the right to certain portion of ambient concentrations rather than the right to a portion in the allowable sum of emissions. To this end let us observe that for any plant i the formula (12.2.F) establishes a one-to-one correspondence between the plant's actual emissions (e_i) and its contribution to the ambient concentration ($b \cdot e_i$). Thus the vector of initial endowments $\mathbf{q}^e = [q_1^e,...,q_n^e]'$ is equivalent to the vector $\mathbf{q}^o = [q_1^o,...,q_n^o]'$, where $q_i^o = bq_i^e$ (for $i=1,...,n$) and consequently $\Sigma q_i^o = A - a$. The ith plant's optimisation problem can be restated as

$$\min_{ei,qi}\{C_i(e_i) + P \cdot (q_i - q_i^o):\ be_i \leq q_i,\ e_i, q_i \geq 0\}$$

where P is the price charged for a unit increase in ambient concentrations rather than the overall emissions. Its Lagrange function is

$$L_i(e_i, q_i, \mu_i) = C_i(e_i) + P \cdot (q_i - q_i^o) + \mu_i \cdot (be_i - q_i)$$

and the Kuhn-Tucker conditions are

$$dC_i(e_i)/de_i + \mu_i b \geq 0, \quad (12.2.1e")$$
$$(dC_i(e_i)/de_i + \mu_i b) \cdot e_i = 0, \quad (12.2.2e")$$
$$P - \mu_i \geq 0, \quad (12.2.1q")$$
$$(P - \mu_i) \cdot q_i = 0, \quad (12.2.2q")$$
$$be_i \leq q_i, \quad (12.2.3")$$
$$\mu_i \cdot (be_i - q_i) = 0, \quad (12.2.4")$$
$$e_i, q_i, \mu_i \geq 0. \quad (12.2.5")$$

The conditions (12.2.1")-(12.2.2") are equivalent to (12.2.1)-(12.2.2) if $P = \lambda$. But the formula (12.2.F) implies that $P = P^e/b$ and it has already been observed that

$P^e = \lambda b$; hence, indeed, the market will set the price P equal to λ thus stimulating the correct allocation.

12.3 EXTENSION I: TWO NON-SYNERGISTIC POLLUTANTS

Suppose now that there are two pollutants whose emissions result in the same detrimental effects. Let $\mathbf{e}_1=[e_{11},...,e_{n1}]'$ denote emissions of the first pollutant from plants 1,...,n and $\mathbf{e}_2=[e_{12},...,e_{n2}]'$ - of the second one. And let the costs of keeping these emissions at the level (e_{i1},e_{i2}) in respective plants be denoted by $C_i(e_{i1},e_{i2})$. It is also assumed that actual emissions $\mathbf{e}_k$ linearly contribute to the ambient concentrations A_k:

$$A_k = \mathbf{b}_k' \cdot \mathbf{e}_k, \quad \mathbf{b}_k = [b_{1k},...,b_{nk}]', \quad \mathbf{b}_k > \mathbf{0}, \quad k = 1,2$$

and these contributions simply add up. In other words, we assume that the joint ambient concentration is:

$$A_o = a + \mathbf{b}_1' \cdot \mathbf{e}_1 + \mathbf{b}_2' \cdot \mathbf{e}_2. \tag{12.3.F}$$

The optimisation problem thus becomes:

$$\min_{e1,e2}\{\mathbf{1}' \cdot \mathbf{C}(\mathbf{e}_1,\mathbf{e}_2): \ a + \mathbf{b}_1' \cdot \mathbf{e}_1 + \mathbf{b}_2' \cdot \mathbf{e}_2 \le A, \ \ \mathbf{e}_1,\mathbf{e}_2 \ge \mathbf{0}\},$$

where, again, A is an allowable level for joint ambient concentration of both pollutants, a stands for a background concentration, and $\mathbf{C}(\mathbf{e}_1,\mathbf{e}_2) = [C_1(e_{11},e_{12}),...,C_n(e_{n1},e_{n2})]'$. Sulphur dioxide and nitrogen oxides emissions adding up to ambient acidity levels provide a good example of this model.

We may introduce an additional notation in order to simplify our algebraic formula. Let

$$\mathbf{B} = [\mathbf{b}_1',\mathbf{b}_2'] = [b_{11},...,b_{n1},b_{12},...,b_{n2}]', \text{ and}$$
$$\mathbf{E} = [\mathbf{e}_1',\mathbf{e}_2'] = [e_{11},...,e_{n1},e_{12},...,e_{n2}]'.$$

Then the mathematical programming problem can be rewritten as:

$$\min_{\mathbf{E}}\{\mathbf{1}' \cdot \mathbf{C}(\mathbf{E}): \ a + \mathbf{B}' \cdot \mathbf{E} \le A, \ \ \mathbf{E} \ge \mathbf{0}\}$$

(note that in the first product above $\mathbf{1}' \in \mathrm{R}^n$, while in the second one $\mathbf{1}' \in \mathrm{R}^{2n}$). Its Lagrange function is

$$L(\mathbf{E};\lambda) = \mathbf{1}' \cdot \mathbf{C}(\mathbf{E}) + \lambda \cdot (a + \mathbf{B}' \cdot \mathbf{E} - A)$$

and the resulting Kuhn-Tucker conditions are:

$$\partial \mathbf{C}/\partial \mathbf{E} + \lambda \mathbf{B}' \geq \mathbf{0}, \tag{12.3.1}$$
$$(\partial \mathbf{C}/\partial \mathbf{E} + \lambda \mathbf{B}') \cdot \mathbf{E} = 0, \tag{12.3.2}$$
$$a + \mathbf{B}' \cdot \mathbf{E} \leq A, \tag{12.3.3}$$
$$\lambda \cdot (a + \mathbf{B}' \cdot \mathbf{E} - A) = 0, \tag{12.3.4}$$
$$\mathbf{E} \geq \mathbf{0}, \lambda \geq 0. \tag{12.3.5}$$

As in the case of the basic model we may ask under what conditions the permit market will stimulate an optimal allocation of abatement effort. Let us assume that the regulatory agency initially allocated rights to the ambient concentrations (for both pollutants jointly) in such a way that plant i received q_i^o, and that the standard A is attained, i.e. $a + \Sigma q_i^o = A$. As in the previous model, let P be the equilibrium price of a unit permit. The ith plant's optimisation problem is then:

$$\min_{ei1,ei2,qi}\{C_i(e_{i1},e_{i2}) + P \cdot (q_i - q_i^o):\ b_{i1}e_{i1} + b_{i2}e_{i2} \leq q_i,\ e_{i1}, e_{i2}, q_i \geq 0\}.$$

The associated Lagrange function is

$$L_i(e_{i1},e_{i2},q_i;\mu_i) = C_i(e_{i1},e_{i2}) + P \cdot (q_i - q_i^o) + \mu_i \cdot (b_{i1}e_{i1} + b_{i2}e_{i2} - q_i).$$

This yields the following Kuhn-Tucker conditions:

$$\partial C_i(e_{i1},e_{i2})/\partial e_{i1} + \mu_i b_{i1} \geq 0,\ \partial C_i(e_{i1},e_{i2})/\partial e_{i2} + \mu_i b_{i2} \geq 0, \tag{12.3.1e$'$}$$
$$(\partial C_i(e_{i1},e_{i2})/\partial e_{i1} + \mu_i b_{i1}) \cdot e_{i1} + (\partial C_i(e_{i1},e_{i2})/\partial e_{i2} + \mu_i b_{i2}) \cdot e_{i2} = 0, \tag{12.3.2e$'$}$$
$$P - \mu_i \geq 0, \tag{12.3.1q$'$}$$
$$(P - \mu_i) \cdot q_i = 0, \tag{12.3.2q$'$}$$
$$b_{i1}e_{i1} + b_{i2}e_{i2} \leq q_i, \tag{12.3.3$'$}$$
$$\mu_i \cdot (b_{i1}e_{i1} + b_{i2}e_{i2} - q_i) = 0, \tag{12.3.4$'$}$$
$$e_{i1},e_{i2},q_i,\mu_i \geq 0. \tag{12.3.5$'$}$$

Again, one can show that conditions (12.3.1´)-(12.3.5´) (if met by all plants simultaneously) are equivalent to (12.3.1)-(12.3.5). The equivalence of the first order conditions again indicates that the permit market will yield an efficient allocation of abatement effort regardless of the initial endowments.

12.4 EXTENSION II: TWO SYNERGISTIC POLLUTANTS

Now we turn to a more interesting case, where two pollutants jointly contribute to the ambient concentrations and their contributions do not simply add up, but "reinforce" each other's effect. In its simplest manifestation this can be modelled by adding to the formula (12.3.F) a term containing products of individual contributions. Thus in this section it will be assumed that the emissions contribute to the ambient concentrations according to the following formula:

$$A_o = a + \mathbf{B}' \cdot \mathbf{E} + \mathbf{E}' \cdot \mathbf{F} \cdot \mathbf{E}, \qquad (12.4.F)$$

where $\mathbf{F}$ is a symmetric, positive semidefinite $2n \times 2n$ matrix. The rest of the notation is the same as in previous sections.

The formula (12.4.F) derives from a more general assumption about the functional relationship between emissions and ambient concentrations. It could be assumed, for instance, that this relationship takes the form of a sufficiently smooth function S,

$$A_o = S(\mathbf{E}) = S(\mathbf{e}_1,\mathbf{e}_2). \qquad (12.4.F^o)$$

If its first, second, and third derivatives exist, the second order Taylor formula can be applied and an approximate relationship reads

$$\begin{aligned} A_o &\approx S(0,0) + S'(0,0) \cdot \mathbf{E} + (1/2) \cdot \mathbf{E}' \cdot S''(0,0) \cdot \mathbf{E} = \qquad (12.4.F') \\ &= S(0,0) + S'(0,0) \cdot [\mathbf{e}_1',\mathbf{e}_2']' + (1/2) \cdot [\mathbf{e}_1',\mathbf{e}_2'] \cdot S''(0,0) \cdot [\mathbf{e}_1',\mathbf{e}_2']', \end{aligned}$$

where

$$S'(x_1,\ldots,x_n,y_1,\ldots,y_n) = [\partial S/\partial x_1,\ldots,\partial S/\partial x_n,\partial S/\partial y_1,\ldots,\partial S/\partial y_n],$$

$$S''(x_1,\ldots,x_n,y_1,\ldots,y_n) =$$

$$= \begin{bmatrix} \partial^2 S/\partial x_1^2,\ldots,\partial^2 S/\partial x_1\partial x_n,\partial^2 S/\partial x_1\partial y_1,\ldots,\partial^2 S/\partial x_1\partial y_n \\ \cdot \quad \cdot \quad \cdot \quad \cdot \\ \cdot \quad \cdot \quad \cdot \quad \cdot \\ \partial^2 S/\partial x_n\partial x_1,\ldots,\partial^2 S/\partial x_n^2,\partial^2 S/\partial x_n\partial y_1,\ldots,\partial^2 S/\partial x_n\partial y_n \\ \partial^2 S/\partial y_1\partial x_1,\ldots,\partial^2 S/\partial y_1\partial x_n,\partial^2 S/\partial y_1^2,\ldots,\partial^2 S/\partial y_1\partial y_n \\ \cdot \quad \cdot \quad \cdot \quad \cdot \\ \cdot \quad \cdot \quad \cdot \quad \cdot \\ \partial^2 S/\partial y_n\partial x_1,\ldots,\partial^2 S/\partial y_n\partial x_n,\partial^2 S/\partial y_n\partial y_1,\ldots,\partial^2 S/\partial y_n^2 \end{bmatrix}$$

The first term in formula (12.4.F′) can be easily recognised as background concentration a independent of the emission level. Then, the second one can be partitioned into two linear forms: $[\partial S(\mathbf{0,0})/\partial x_1,...,\partial S(\mathbf{0,0})/\partial x_n]\cdot\mathbf{e}_1$ and $[\partial S(\mathbf{0,0})/\partial y_1,...,\partial S(\mathbf{0,0})/\partial y_n]\cdot\mathbf{e}_2$. After obvious substitutions these are equivalent to the terms $\mathbf{b}_1'\cdot\mathbf{e}_1$ and $\mathbf{b}_2'\cdot\mathbf{e}_2$ in formula (12.3.F), as well as to the term $\mathbf{B}'\cdot\mathbf{E}$ in formula (12.4.F). Thus we may interpret them as the "non-synergistic" part of the relationship between emissions and ambient concentrations. And lastly, the second order term is equivalent to $\mathbf{E}'\cdot\mathbf{F}\cdot\mathbf{E}$ in formula (12.4.F) - the "synergistic" part of the relationship - if we substitute $\mathbf{F}$ for $(1/2)\cdot S''(\mathbf{0,0})$ (hence the assumption of symmetry of the matrix $\mathbf{F}$). Formula (4.F) can then be justified as a second order approximation for a general case of synergistic relationships between pollution discharges and ambient environment quality indicators.

The cost minimisation problem becomes:

$$\min_{\mathbf{E}}\{\mathbf{1}'\cdot\mathbf{C}(\mathbf{E}):\ a + \mathbf{B}'\cdot\mathbf{E} + \mathbf{E}'\cdot\mathbf{F}\cdot\mathbf{E} \le A,\ \mathbf{E} \ge \mathbf{0}\}.$$

Its Lagrange function and the Kuhn-Tucker conditions are:

$$L(\mathbf{E};\lambda) = \mathbf{1}'\cdot\mathbf{C}(\mathbf{E}) + \lambda\cdot(a + \mathbf{B}'\cdot\mathbf{E} + \mathbf{E}'\cdot\mathbf{F}\cdot\mathbf{E} - A),$$

$$\partial\mathbf{C}/\partial\mathbf{E} + \lambda\cdot(\mathbf{B}' + 2\cdot\mathbf{E}'\cdot\mathbf{F}) \ge \mathbf{0}, \tag{12.4.1}$$

$$(\partial\mathbf{C}/\partial\mathbf{E} + \lambda\cdot(\mathbf{B}' + 2\cdot\mathbf{E}'\cdot\mathbf{F}))\cdot\mathbf{E} = 0, \tag{12.4.2}$$

$$a + \mathbf{B}'\cdot\mathbf{E} + \mathbf{E}'\cdot\mathbf{F}\cdot\mathbf{E} \le A, \tag{12.4.3}$$

$$\lambda\cdot(a + \mathbf{B}'\cdot\mathbf{E} + \mathbf{E}'\cdot\mathbf{F}\cdot\mathbf{E} - A) = 0, \tag{12.4.4}$$

$$\mathbf{E} \ge \mathbf{0},\ \lambda \ge 0. \tag{12.4.5}$$

Convexity of $\mathbf{C}$ coupled with positive semidefiniteness of $\mathbf{F}$ guarantee that this is a convex programming problem, and (12.4.1)-(12.4.5) define its optimum.

Before checking under what conditions the plants can be expected to achieve a cost effective allocation of abatement effort in this case, we need some additional assumptions about the design of the permit market.

In the case of non-synergistic pollutants individual contributions add up to the ambient concentrations independently, so that there are no doubts about who uses what portion of the allowable level. In the synergistic case, however, if the contributions add up according to the formula (12.4.F) the relationship becomes more complex. Allowable emissions do not solely depend on the amount of permits received or bought, but on the actual emissions from all the other sources too. Let us denote

$$\mathbf{E}_i = [0,...,0,e_{i1},0,...0,e_{i2},0,...,0]'.$$

Thus the formula $\Sigma\mathbf{E}_i = \mathbf{E}$ defines a partition of the vector **E** such that each $\mathbf{E}_i$ is the vector representing actual emissions of the *i*th plant. Let us observe that

$$\mathbf{B}'\cdot\mathbf{E} + \mathbf{E}'\cdot\mathbf{F}\cdot\mathbf{E} = \mathbf{B}'\cdot\Sigma\mathbf{E}_i + \mathbf{E}'\cdot\mathbf{F}\cdot\Sigma\mathbf{E}_i = \Sigma(\mathbf{B}'\cdot\mathbf{E}_i + \mathbf{E}'\cdot\mathbf{F}\cdot\mathbf{E}_i).$$

It will be assumed from now on that the above formula establishes ambient permit requirements for plants 1,...,*n*. In other words, in order to calculate the real meaning (in terms of emissions $\mathbf{E}_i$) of its right to the ambient concentrations, the *i*th plant needs to know individual emissions of all the plants (**E**), and has to apply the formula

$$q_i = \mathbf{B}'\cdot\mathbf{E}_i + \mathbf{E}'\cdot\mathbf{F}\cdot\mathbf{E}_i. \qquad (12.4.\mathrm{R}_i)$$

q_i is the amount of permits the *i*th plant must own to discharge $\mathbf{E}_i$ provided that the other emissions are **E**.

It is easy to see that in general the market would fail to give incentives for the optimal allocation here. Equation (12.4.1) implies for the *i*th plant (among other things):

$$[\partial C_i/\partial e_{i1}, \partial C_i/\partial e_{i2}] + \lambda\cdot([b_{i1}, b_{i2}] + 2\cdot\mathbf{E}'\cdot[\mathbf{F}_i, \mathbf{F}_{n+i}]) \geq [0,0], \qquad (12.4.1_i)$$

where $\mathbf{F}_i$ and $\mathbf{F}_{n+i}$ are respectively the *i*th and the $(n+i)$th columns of the matrix **F**. However, if the individual cost minimisation problem of the *i*th plant were simply

$$\min_{ei1,ei2,qi}\{C_i(e_{i1}, e_{i2}) + P\cdot(q_i - q_i^o)\text{: } \mathbf{B}'\cdot\mathbf{E}_i + \mathbf{E}'\cdot\mathbf{F}\cdot\mathbf{E}_i \leq q_i,\ e_{i1}, e_{i2}, q_i \geq 0\},$$

then the corresponding first order conditions would read:

$$[\partial C_i/\partial e_{i1}, \partial C_i/\partial e_{i2}] + \mu_i\cdot([b_{i1}, b_{i2}] + (\mathbf{E} + \mathbf{E}_i)'\cdot[\mathbf{F}_i, \mathbf{F}_{n+i}]) \geq [0,0]. \qquad (4.\mathrm{M}^i)$$

Intuition suggests that μ_i should equal *P* and hence λ (unless the plant abates 100% of its potential emissions), but since in general $2\cdot\mathbf{E} \neq \mathbf{E} + \mathbf{E}_i$, one cannot expect that the market alone would achieve the optimum given by (12.4.1)-(12.4.5).

An obvious way to correct this market failure would be to impose a tax in order to adjust plants' first order conditions to the levels defined by ($12.4.1_i$). By comparing conditions ($12.4.1_i$) and ($12.4.\mathrm{M}_i$) one can conclude that the additional term at the right-hand side of ($12.4.\mathrm{M}_i$) should be $P\cdot(\mathbf{E} - \mathbf{E}_i)'\cdot[\mathbf{F}_i, \mathbf{F}_{n+i}]$. This suggests a tax

$$T_i = P\cdot(\mathbf{E} - \mathbf{E}_i)'\cdot\mathbf{F}\cdot\mathbf{E}_i. \qquad (12.4.\mathrm{T}_i)$$

With the tax the ith plant's cost minimisation problem becomes:

$$\min_{ei1,ei2,qi}\{C_i(e_{i1},e_{i2}) + P\cdot(q_i - q_i^o) + T_i\text{: } \mathbf{B}'\cdot\mathbf{E}_i + \mathbf{E}'\cdot\mathbf{F}\cdot\mathbf{E}_i \le q_i, e_{i1},e_{i2},q_i \ge 0\} =$$
$$= \min_{ei1,ei2,qi}\{C_i(e_{i1},e_{i2}) + P\cdot(q_i - q_i^o) + P\cdot(\mathbf{E} - \mathbf{E}_i)'\cdot\mathbf{F}\cdot\mathbf{E}_i\text{:}$$
$$\mathbf{B}'\cdot\mathbf{E}_i + \mathbf{E}'\cdot\mathbf{F}\cdot\mathbf{E}_i \le q_i,\ e_{i1},e_{i2},q_i \ge 0\}.$$

Consequently, the plant's first order conditions become the same as in (12.4.1$_i$). Thus in the synergistic case the permit market - when supplemented with an additional tax - can efficiently allocate abatement activities too. The tax, however, while eliminating inefficiency, has an equity effect which may be of concern.

The formal proof of the efficiency of the permit market in the synergistic case was given in Żylicz (1989) and, with slight modifications, is included in the Appendix in this chapter. It extends the proof established for the linear case by Montgomery (1972) to account for the non-linear relationships between emissions and ambient concentrations. The following assumptions were used to obtain the result:

$$0 \le a \le A; \qquad (12A.1)$$
C_i are differentiable convex functions (for $i = 1,...,n$); (12A.2)
$\mathbf{F}$ is a symmetric positively semidefinite $2n\times 2n$ matrix; (12A.3)
$\mathbf{B}'+\mathbf{E}'\cdot\mathbf{F} > \mathbf{0}$ for every $\mathbf{E} \ge \mathbf{0}$ such that $\mathbf{E} \ne \mathbf{0}$. (12A.4)

Condition (A1) merely states that the target level A can not be set below the background concentrations. (A2)-(A3) are the usual convexity assumptions required in partial equilibrium models. (A4) would be violated, if there was a plant whose emissions negatively contributed to the ambient concentrations. Condition (A4) is clearly satisfied if e.g. all elements of **B** and **F** are positive, although this is not a requirement.

12.5 FURTHER EXTENSIONS OF THE MODEL

The model developed in section (12.4) easily extends to three or more synergistic pollutants. The only changes required are those in the definitions of $\mathbf{e}_k$, $\mathbf{E}$, $\mathbf{C(E)}$, $\mathbf{B}$, $\mathbf{F}$, and $\mathbf{E}_i$. Namely, let m be the number of pollutants. Then for $k = 1,...,m$, $\mathbf{e}_k = [e_{1k},...,e_{nk}]'$ is the vector of emissions of the kth pollutant from plants $1,...,n$ respectively. Vector $\mathbf{E}\in\mathbf{R}^{nm}$ is defined as $\mathbf{E} = [\mathbf{e}_1',...,\mathbf{e}_m']$. Abatement cost required to keep emissions at the rate $e_{i1},...,e_{im}$ is $C(e_{i1},...,e_{im})$, and $\mathbf{C}:\mathbf{R}^{nm}\rightarrow\mathbf{R}^n$, $\mathbf{C(E)} = \mathbf{C}(\mathbf{e}_1,...,\mathbf{e}_m) = [C_1(e_{11},...,e_{1m}),...,C_n(e_{n1},...,e_{nm})]'$. Vector $\mathbf{B}'\in\mathbf{R}^{nm}$ and positively semidefinite $nm\times nm$ matrix $\mathbf{F}$ can again be interpreted as $\partial S(\mathbf{0})/\partial\mathbf{E}$ and $(1/2)\partial^2 S(\mathbf{0})/\partial\mathbf{E}^2$ - the first and the second order terms from the Taylor expansion of a function S relating emissions to ambient concentrations. Finally $\mathbf{E}_i \in R^{nm}$ is a column vector having e_{ik} on its $((k\text{-}1)n+i)$th coordinate for $k = 1,...,m$ and zeroes elsewhere. Thus, as previously, $\Sigma\mathbf{E}_i = \mathbf{E}$ defines a partition of $\mathbf{E}$ used in calculating ith plant's ambient permit demand and tax T_i.

It should be noted that in particular m - the number of pollutants - can be 1. This corresponds to a single pollutant problem with a non-linear relationship between emissions and ambient concentrations.

Another, perhaps even more important extension of the model takes into account the spatial distribution of ambient concentrations. Thus, we may have v receptors (monitoring locations) with background concentrations $a(r)$, where standards $A(r)$ are to be attained for $r = 1,...,v$. Consequently there should be v permit markets and v vectors of initial endowments $\mathbf{q}^o(r)$. For every receptor r there is the associated vector $\mathbf{B}(r)$ and matrix $\mathbf{F}(r)$ representing the relationship between emissions and ambient concentrations. Solving its cost minimisation problem the ith plant has to decide upon the quantity of permits $q_i(r)$ it needs for any receptor r so that the constraints $(\mathbf{B}(r))' \cdot \mathbf{E}_i + \mathbf{E}' \cdot \mathbf{F}(r) \cdot \mathbf{E}_i \leq q_i(r)$ are satisfied for all $r = 1,...,v$ with related Lagrange multipliers $\mu_i(r)$. The corrective tax for the ith plant would be $T_i(1)+...+T_i(v)$, where $T_i(r) = P(r) \cdot (\mathbf{E} - \mathbf{E}_i)' \cdot \mathbf{F}(r) \cdot \mathbf{E}_i$, and $P(1),...,P(v)$ are the prices of permits at markets $1,...,v$.

The notation becomes much more complex, but the proofs referred to in section 12.4 remain valid in this multi-receptor setting too. In fact, the original proof by Montgomery (1972) explicitly deals with the multi-market case. However, it seemed more appropriate for this chapter to avoid excessive notational complexity by suppressing receptor indices and focusing attention on multiple pollutants and non-linearities (not covered in the Montgomery's chapter) which constitute the main object of the analysis.

12.6 INFORMATION ASPECTS

Tradeable permits for synergistic pollutants differ from the ordinary ones in that their real meaning (in terms of emissions) may not be known at the time of purchase. The emissions $\mathbf{E}_i$ allowed by a given quantity of permits q_i depend on the distribution of emissions from all the plants $\mathbf{E}$. This kind of uncertainty is not, however, totally new in the tradeable permit approach. A one-to-one relationship between permits and allowable discharges does not hold in the case of the pioneering design for the Fox River, Wisconsin (O'Neil *et al.* 1983). What the permit allows to discharge depends on current hydrological conditions and decreases with temperature while increases with the flow in the river. Similarly, priority water pollution rights discussed by Howe

& Lee (1983) assume that the permits allow certain discharges in a probabilistic sense only.

In the case of synergistic pollutants the uncertainty results from the market itself rather than from exogenous factors such as meteorological conditions. It is possible to allow variation of the parameters a, A, **B** and **F** according to current environmental conditions on a daily or an even shorter-term basis thus adding more complexity to the analysis. This would direct our analysis to the problems of an optimal intertemporal allocation and to the recognition of the fact that some pollutants may be storable at least for a short period of time. We will not address these issues here. Instead we will focus the attention on a convenient characteristic of the static permit market discussed in sections (12.4)-(12.5).

It should be pointed out that information requirements of the permit market in the synergistic case are greater than in the non-synergistic one but still conceivable. First, of all we obviously assume that the environmental parameters **B** and **F** are known. Estimating these may be a hard task, but any system of regulating emissions so as to meet ambient standards requires them. What is new, however, is that plant i in order to calculate its allowable emissions has to apply formula (12.4.R_i) which involves - in addition to **B** and **F** - also **E**, i.e. the emissions from all the plants. This is a requirement far from trivial. Nevertheless - what is crucial here - no direct information on abatement costs is needed in order to calculate corrective taxes T_i. Intuitively this should come as no surprise as such information is included in the market price P of the permit.

Quite commonly in environmental policy debates information requirements of a tradeable ambient permit model are contrasted with those of an emission tax one to conclude that the latter are lower. Even though this is a fair judgment, it is then typically extended to any tradeable permit and any tax systems, which is certainly not justified. By its very nature, any ambient quality control requires more information than emission control. If a tax system were designed to protect ambient quality, its physical data input, administrative burden, and monitoring requirements would have been exactly the same, as in the case of a tradeable permit system established for the same purpose. With the essential static symmetry of tax and permit systems under perfect information conditions (Pezzey 1992), the real difference between the two is in what may go wrong if economy responds not as expected by the regulator. As explained in the seminal paper by Weitzman (1974), the main static difference between the two systems is that of priority of regulator's concerns. In the permit (i.e. quantity based) approach, the environmental effect of regulation is certain while the regulator may be surprised with the costs it implies. On the contrary, in the tax (i.e. price based) approach, the regulator knows with certainty the marginal abatement costs she imposes on the economy whilst the environmental outcome is uncertain.

Proving mere existence or even uniqueness of a market equilibrium gives no clues as to its stability and other dynamic properties. These depend on behavioral assumptions and may dramatically change with the way expectations are formed and transactions are organised. Potentially high transaction costs (especially in the multi-receptor case) call for very careful design of this market; and the design itself makes an interesting exercise in experimental economics (Lyon (1986)). The main goal here is to create an auction that would discourage strategic behaviour while simultaneously adjusting the allocation of permits and emissions in as few iterations as possible.

Since the tax option does not require any trading, it may be perceived as free from strategic behaviour problem and transaction costs. In fact, it is not. Any tax which makes firms' payments dependent not only on individual performance, but on the performance of all the others as well - which is inevitable under the assumption of synergism - must be cumbersome. Indeed, not as cumbersome as an auction mechanism required for dealing with permits according to the formula (12.4.R_i). However, the auction provides an information on marginal cost of the desired environmental outcome. In the tax model this information is assumed to be *a priori* known or irrelevant in the sense that environmental effects will be *a posteriori* determined by the level of marginal cost (implied by the tax).

12.7 NON-CONVEXITIES AND RELATED ISSUES

We have assumed throughout this chapter that the optimisation problem to be solved is a convex programming one. This requires assuming convexity of the function **C** and positive semidefiniteness of the matrix **F**. There are important reasons to believe that neither of these assumptions hold in many important cases.

An example of a departure from the assumed functional form of **C** (as a composition of separate cost functions C_i for plants $1,...,n$) is the case of wastewater treatment. Here establishing a joint treatment plant serving several sources of pollution may prove to be cost effective.

Potential non-convexity of **C** questions the efficiency of a permit market - as well as a tax system - in the synergistic case as well as in the linear one. In the latter it is the only point of analytical concern. In the former, however, there is an additional source of non-convexity linked to the non-linearity of constraints. If **F** is not positively semidefinite then the whole problem is not amenable to convex programming solution. This kind of non-convexity will be encountered whenever the synergistic effect depends on the product of the sums of emissions of two (or more) pollutants, but not on their squares. In other words, if the second order term in formula (4.F) takes the form of $\mathbf{E}' \cdot \mathbf{F} \cdot \mathbf{E} = (\Sigma e_{i1}) \cdot (\Sigma e_{i2})$ then the matrix

$$\mathbf{F} = \begin{bmatrix} 0,\ldots,0,1,\ldots,1 \\ \cdot \quad \cdot\cdot \quad \cdot \\ \cdot \quad \cdot\cdot \quad \cdot \\ \cdot \quad \cdot\cdot \quad \cdot \\ 0,\ldots,0,1,\ldots,1 \\ 1,\ldots,1,0,\ldots,0 \\ \cdot \quad \cdot\cdot \quad \cdot \\ \cdot \quad \cdot\cdot \quad \cdot \\ \cdot \quad \cdot\cdot \quad \cdot \\ 1,\ldots,1,0,\ldots,0 \end{bmatrix}$$

fails to be positively semidefinite. Here this is not just a "theoretical curio" as there is some evidence that quadratic terms may not be significant in equations regressing ambient damage indicators on pollution variables (Krumm & Graves 1982).

All in all the assumptions of the model should be empirically validated in each particular case, and it is apparent that many instances of synergism cannot be efficiently solved through the tradeable permit approach.

It should be stressed that the non-convexity problem introduced by the matrix **F** does not vanish under the tax system. Unless the model is "linearised" and thus made tractable for the traditional linear permit/tax approach, there exists no linear tax to correct firms' first order conditions to be consistent with the global optimisation requirements ($12.4.1_i$). To check this, let us observe that if a Pigouvian tax T^P_i was to be introduced, the ith firm's optimisation problem would read:

$$\min_{ei1,ei2,qi}\{C_i(e_{i1},e_{i2}) + T^P_i: e_{i1},e_{i2} \geq 0\}.$$

Its first order conditions would then read:

$$[\partial C_i/\partial e_{i1},\partial C_i/\partial e_{i2}] + [\partial T^P_i/\partial e_{i1},\partial T^P_i/\partial e_{i2}] \geq [0,0].$$

Their consistency with ($12.4.1_i$) requires that

$$[\partial T^P_i/\partial e_{i1},\partial T^P_i/\partial e_{i2}] = \lambda \cdot ([b_{i1},b_{i2}] + 2 \cdot \mathbf{E}' \cdot [\mathbf{F}_i,\mathbf{F}_{n+i}]).$$

It is obvious that the only taxes T^P_i which satisfy the latter condition have the form:

$$T^P_i = \lambda \cdot (\mathbf{B}' \cdot \mathbf{E}_i + \mathbf{E}' \cdot \mathbf{F} \cdot \mathbf{E}_i) + \text{const}_i.$$

The tax is thus non-linear, depends on the distribution of emissions of all pollutants from all sources, and fails to lead to a convex programming problem under exactly the same circumstances as in the tradeable permit system. Additionally, to compute such a tax the regulator has to estimate the parameter λ whose value depends on individual costs functions.

Hence the results of this chapter do not point at the tradeable permit approach as inferior to an alternative option, but rather they indicate difficulties *inherent* in any policy instrument designed to achieve cost effectiveness taking into account the non-linearity of the relationship between an environmental impact and its cause.

12.8 SUMMARY AND CONCLUSIONS

The aim of this chapter was to extend the notion of the tradeable permit system to the case where there are several factors non-linearly contributing to an ambient quality indicator of policy concern. The ambient permit market was chosen as the basis for further extensions and was generalised first to two linearly interacting pollutants and then to the non-linear case. A second order Taylor series approximation was used to analyze the latter. First order conditions for a cost effective allocation of permits were derived for the model. Then, using standard assumptions, it was proved that provided an appropriate corrective measure (tax) was imposed, any market equilibrium allocates the permits in a cost effective way. The assumptions were basically the same (with the exception of those related to the second order term, which was not present in Montgomery's (1972) model) as in the proof of efficiency in the standard linear case.

It has been shown that unlike the linear case the permit market alone will not achieve an efficient solution. Tradeable permits have to be supplemented with a congestion-like tax which depends on the market price of the permit and the emissions distribution. However, calculating the tax does not then depend on any additional information beyond what has to be known anyway in order to relate emissions to ambient concentrations. Nevertheless, both the rigid convexity assumptions and the complexity of trading mechanisms in the non-linear case suggest that, in many instances, synergistic pollutants have to be handled by means of direct regulations. Alternatively, a linearised version of the model (e.g. the first order Taylor series approximation of the problem) should be applied.

Two further extensions of the model have been considered. The number of pollutants may be greater than two (it can be one as well) and the number of receptors - i.e. ambient quality monitoring locations - may be greater than one. In the latter case the plants would have to carry out simultaneous transactions in several ambient markets. This adds to the complexity of an institutional design for the appropriate trading mechanism, but does not preclude its viability.

Even certain desirable static properties of the tradeable permit system that were outlined in the chapter do not guarantee its stability which happens to be crucial for all practical purposes. The ability of the permit market to converge towards an equilibrium position depends on the design of the auction mechanism applied but also on the expectations formation in consecutive rounds of trading. Prospects for a successful collaborative behaviour on the part of the agents acting in the market are difficult to assess and should be experimentally tested before final conclusions and/or policy recommendations are made.

The results reported in this chapter extend to tax systems in an obvious way. Thus the non-linearity of relationships discussed does not imply the failure of tradeable permits specifically. Rather it implies the failure of attacking the problem of cost effectiveness through any market instruments like permits or taxes under certain circumstances. In particular, the results of the chapter do not interfere with recent analyses of specific "niches" where either tradeable permits or taxes may find their most appropriate place (Howe 1992). They do not interfere with the preference for tradeable permits in implementation areas where the policy should not allow to exceed a certain physically determined scale of environmental impact letting the market to determine a cost effective allocation of protection activities (Daly 1992). Finally as an attempt to study the limits of the tradeable permit approach, they are consistent with the potentially prominent role for this instrument in economies in transition (Dudek *et al.* 1992, Żylicz 1994). Tradeable permits provide exactly the tool to address in a politically viable way enormous problems of sectoral and regional economic restructuring without compromising environmental recovery schedules or without resorting to command-and-control.

MATHEMATICAL APPENDIX

We will formally prove the efficiency of the permit market supplemented with the tax given by formula (12.4.$T_{i)}$). Our proof is established in 5 propositions. *Proposition (12.1)* guarantees that (under assumptions (12A.1)-(12A.4)) the abatement cost minimisation problem has a solution. *Propositions (12.2)-(12.4)* reveal some desirable properties of the permit market. In particular, *Proposition (12.2)* characterises a market equilibrium in terms of the Kuhn-Tucker conditions. A proof of existence of such an equilibrium is given in *Proposition (12.3)*. *Proposition (12.4)* - the core of the proof - states that any market equilibrium minimises the overall abatement costs. Finally, in *Proposition (12.5)* it is observed that market equilibria do not depend on initial allocation of permits.

Proposition 12.1
There exist $\mathbf{E}^*$ and λ^* solving Kuhn-Tucker conditions (12.4.1)-(12.4.5). Moreover, if C_i are strictly convex (for $i=1,...,n$) the solution is unique.

In what follows the asterisk (*) will be used to denote a solution to the Kuhn-Tucker conditions (12.4.1)-(12.4.5). e^*_{ik} denotes the coordinate of this solution corresponding to the ith plant and kth pollutant, with $\mathbf{E}^*_i$ defined consistently with previous notation.

Let us state once again the ith plant's optimisation problem:

$$\min_{ei1,ei2,qi}\{C_i(e_{i1},e_{i2}) + P\bullet(q_i - q^o_i) + T_i\text{: } \mathbf{B}'\bullet\mathbf{E}_i + \mathbf{E}'\bullet\mathbf{F}\bullet\mathbf{E}_i \le q_i, e_{i1},e_{i2},q_i \ge 0\} =$$
$$= \min_{ei1,ei2,qi}\{C_i(e_{i1},e_{i2}) + P\bullet(q_i - q^o_i) + P\bullet(\mathbf{E} - \mathbf{E}_i)'\bullet\mathbf{F}\bullet\mathbf{E}_i\text{:} \qquad (12.1)$$
$$\mathbf{B}'\bullet\mathbf{E}_i + \mathbf{E}'\bullet\mathbf{F}\bullet\mathbf{E}_i \le q_i,\ e_{i1},e_{i2},q_i \ge 0\}.$$

Any solution $(e^M_{i1},e^M_{i2},q^M_i)$ to this problem will be marked with the superscript M (^{M}). Also the corresponding $\mathbf{E}^M$ as well as $\mathbf{E}^M_i$ and $\mathbf{q}^M$ will be consistent with our previous notation.

Definition 12.1
By the market equilibrium we will understand in this context a state of the economy where e^M_{i1}, e^M_{i2}, and q^M_i solve (1) for $i=1,...,n$, and the following market-clearing conditions hold:

$$\Sigma(q^M_i - q^o_i) \le 0, \qquad (12.2)$$
$$P\bullet\Sigma(q^M_i - q^o_i) = 0. \qquad (12.3)$$

As previously, P is the market price and $\Sigma q^o_i = q$ is any initial allocation of permits satisfying $q = A - a$, i.e. the ambient concentration requirement.

Definition 12.2
A market equilibrium is efficient if and only if $\mathbf{1}'\bullet\mathbf{C}(\mathbf{E}^*) = \mathbf{1}'\bullet\mathbf{C}(\mathbf{E}^M)$.

Proposition 12.2
Market equilibrium exists if and only if there exist $\mu^M_i,...,\mu^M_i \ge 0$ and $P \ge 0$ such that for $i=1,...,n$:

$$[\partial C_i(e^M_{i1},e^M_{i2})/\partial e_{i1},\partial C_i(e^M_{i1},e^M_{i2})/\partial e_{i2}] + P\bullet(\mathbf{E}^M\text{-}\mathbf{E}^M_i)'\bullet[\mathbf{F}_i,\mathbf{F}_{n+i}] + \qquad (12.4)$$
$$+ \mu^M_i\bullet([b_{i1},b_{i2}] + (\mathbf{E}^M+\mathbf{E}^M_i)'\bullet[\mathbf{F}_i,\mathbf{F}_{n+i}]) \ge [0,0],$$
$$([\partial C_i(e^M_{i1},e^M_{i2})/\partial e_{i1},\partial C_i(e^M_{i1},e^M_{i2})/\partial e_{i2}] + P\bullet(\mathbf{E}^M\text{-}\mathbf{E}^M_i)'\bullet[\mathbf{F}_i,\mathbf{F}_{n+i}] + \qquad (12.5)$$
$$+ \mu^M_i\bullet([b_{i1},b_{i2}] + (\mathbf{E}^M+\mathbf{E}^M_i)'\bullet[\mathbf{F}_i,\mathbf{F}_{n+i}])\bullet[e^M_{i1},e^M_{i2}]' = 0,$$
$$P - \mu^M_i \ge 0, \qquad (12.6)$$

$$q^M_i \cdot (P - \mu^M_i) = 0, \tag{12.7}$$
$$\mathbf{B}' \cdot \mathbf{E}^M_i + (\mathbf{E}^M)' \cdot \mathbf{F} \cdot \mathbf{E}^M_i \leq q^M_i, \tag{12.8}$$
$$\mu^M_i \cdot (\mathbf{B}' \cdot \mathbf{E}^M_i + (\mathbf{E}^M)' \cdot \mathbf{F} \cdot \mathbf{E}^M_i - q^M_i) = 0, \tag{12.9}$$

and the market-clearing conditions (12.2)-(12.3) are satisfied.

Proposition 12.3
Under assumptions (12A.1)-(12A.4) a market equilibrium exists.

Proposition 12.4
Any market equilibrium is efficient.

Proposition 12.5
In Propositions (12.2)-(12.4) P, e^M_{i1}, e^M_{i2}, and q^M_i, for all $i=1,...,n$, are independent of $q^o_1,...,q^o_n$ as long as $q^o_1,...,q^o_n \geq 0$ and $\Sigma q^o_i = q$.

Proposition 12.1 - Proof:
By the Kuhn-Tucker theorem (12.4.1)-(12.4.5) define a solution of

$$\min_E\{\mathbf{1}' \cdot \mathbf{C}(\mathbf{E}): a + \mathbf{B}' \cdot \mathbf{E} + \mathbf{E}' \cdot \mathbf{F} \cdot \mathbf{E} \leq A, \mathbf{E} \geq \mathbf{0}\}$$

if it exists. It is thus sufficient to prove that in fact the function $\mathbf{1}' \cdot \mathbf{C}(\mathbf{E})$ attains its minimum value over the set $\Psi = \{\mathbf{E} \in R^{2n}: a + \mathbf{B}' \cdot \mathbf{E} + \mathbf{E}' \cdot \mathbf{F} \cdot \mathbf{E} \leq A, \mathbf{E} \geq \mathbf{0}\}$. By (12A.1) Ψ is non-empty since at least $\mathbf{0} \in \Psi$. It is also closed and (12A.4) implies its boundedness. Therefore, by the Bolzano-Weierstass theorem, the continuous function $\mathbf{1}' \cdot \mathbf{C}$ achieves its minimum value on the compact set Ψ. If all the functions C_i are strictly convex, then so is their sum and the minimum is unique.

Proposition 12.2 - Proof:
The Lagrange function for the *i*th plant's optimisation problem (12.1) is:

$$L_i(e_{i1},e_{i2},q_i,\mu_i) = C_i(e_{i1},e_{i2}) + P \cdot (q_i - q^o_i) + P \cdot (\mathbf{E} - \mathbf{E}_i)' \cdot \mathbf{F} \cdot \mathbf{E}_i + \mu_i \cdot (\mathbf{B}' \cdot \mathbf{E}_i + \mathbf{E}' \cdot \mathbf{F} \cdot \mathbf{E}_i - q_i).$$

By the Kuhn-Tucker theorem a solution to (12.1) is given by the following differential conditions:

$$\partial L_i/\partial e_{i1} \geq 0, \text{ and } \partial L_i/\partial e_{i2} \geq 0, \tag{12.10}$$
$$[\partial L_i/\partial e_{i1}, \partial L_i/\partial e_{i2}] \cdot [e_{i1},e_{i2}]' = 0, \tag{12.11}$$
$$\partial L_i/\partial q_i \geq 0, \tag{12.12}$$

$$(\partial L_i/\partial q_i) \cdot q_i = 0, \tag{12.13}$$
$$\partial L_i/\partial \mu_i \leq 0, \tag{12.14}$$
$$(\partial L_i/\partial \mu_i) \cdot \mu_i = 0, \tag{12.15}$$
$$e_{i1}, e_{i2}, q_i, \mu_i \geq 0. \tag{12.16}$$

By differentiating the Lagrange function and substituting $e^M{}_{i1}$, $e^M{}_{i2}$, $q^M{}_i$, $\mu^M{}_i$ for e_{i1}, e_{i2}, q_i, μ_i it is seen that (12.10)-(12.15) are identical to (12.4)-(12.9). (12.16) together with the market-clearing conditions complete the definition of the market equilibrium and the proof.

Proposition 12.3 - Proof:
By Proposition (12.1) there exist $\mathbf{E}^*$ and λ^* solving Kuhn-Tucker conditions (12.4.1)-(12.4.5). We will show that $\mathbf{E}^M = \mathbf{E}^*$, $q^M{}_i = \mathbf{B}' \cdot \mathbf{E}^{Mi} + (\mathbf{E}^M)' \cdot \mathbf{F} \cdot \mathbf{E}^M{}_i$, $P = \lambda^*$, and $\mu^M{}_i = \lambda^*$ for $i=1,...,n$ satisfy the conditions derived in Proposition (12.2). Taking the ith and $(i+n)$th coordinates of the vector inequality in (12.4.1) - which by Proposition (12.1) must be satisfied by $\mathbf{E}^*$ and λ^* - we end up with (12.10). Then adding and subtracting $\lambda \cdot \mathbf{E}_i' \cdot [\mathbf{F}_i, \mathbf{F}_{n+i}]$ at the left hand side yields:

$$[\partial C_i/\partial e_{i1}, \partial C_i/\partial e_{i2}] + \lambda \cdot (\mathbf{E} - \mathbf{E}_i)' \cdot [\mathbf{F}_i, \mathbf{F}_{n+i}]$$
$$+ \lambda \cdot ([b_{i1}, b_{i2}] + (\mathbf{E} + \mathbf{E}_i)' \cdot [\mathbf{F}_i, \mathbf{F}_{n+i}]) \geq [0,0].$$

By substituting $\mathbf{E}^M$ for $\mathbf{E}$, P for the first λ, and $\mu^M{}_i$ for the second one we finally get (12.4). Similarly (12.5) follows from (12.4.2). (12.6)-(12.7) and (12.4.3)-(12.4.4) are clearly satisfied (as equalities) because of the choice of $\mu^M{}_i$ and $q^M{}_i$. To check the market-clearing conditions (12.2)-(12.3) let us observe that

$$\Sigma(q^M{}_i - q^o{}_i) = \Sigma q^M{}_i - \Sigma q^o{}_i = \Sigma(\mathbf{B}' \cdot \mathbf{E}^{Mi} + (\mathbf{E}^M)' \cdot \mathbf{F} \cdot \mathbf{E}^M{}_i) - \Sigma q^o{}_i =$$
$$= \mathbf{B}' \cdot \Sigma \mathbf{E}^M{}_i + (\mathbf{E}^M)' \cdot \mathbf{F} \cdot \Sigma \mathbf{E}^M{}_i - q = \mathbf{B}' \cdot \mathbf{E}^M + (\mathbf{E}^M)' \cdot \mathbf{F} \cdot \mathbf{E}^M - q =$$
$$= a + \mathbf{B}' \cdot \mathbf{E}^M + (\mathbf{E}^M)' \cdot \mathbf{F} \cdot \mathbf{E}^M - A = a + \mathbf{B}' \cdot \mathbf{E}^* + (\mathbf{E}^*)' \cdot \mathbf{F} \cdot \mathbf{E}^* - A$$

(by Definition (12.1) $\Sigma q^o{}_i = q = A - a$). Applying now (12.4.3)-(12.4.4) and substituting P for λ we prove the market-clearing conditions are satisfied.

Proposition 12.4 - Proof:
To prove this we start with any $e^M{}_{i1}$, $e^M{}_{i2}$, $q^M{}_i$, $\mu^M{}_i$ $(i=1,...,n)$ and P satisfying conditions stated in Proposition (12.2). We will show that $\mathbf{E}^* = \mathbf{E}^M$ and $\lambda^* = P$ satisfy (12.4.1)-(12.4.5). We will prove (12.4.1)-(12.4.2) first. By (12.6) $P - \mu^M{}_i \geq 0$, for $i=1,...,n$. Take any i. If equality holds, i.e. $P = \mu^*{}_i$, then we are done with the ith and $(n+i)$th coordinates of (12.4.1). It proves that the ith and $(n+i)$th terms in the sum at the left hand side of (12.4.2) are 0.

Suppose now that $P - \mu^M_i > 0$. By (12.7) it follows that $q^M_i = 0$. This in turn implies $\mathbf{E}^M_i = 0$ (if $\mathbf{E}^M_i \neq 0$ then by (12A.4) $(\mathbf{B}' + (\mathbf{E}^M)' \cdot \mathbf{F} \cdot)\mathbf{E}^M_i > 0$ - which contradicts (12.8)). Hence the ith and $(n+i)$th terms in the sum at the left hand side of (12.4.2) are 0. We will proceed now to show that the ith and $(n+i)$th inequalities in (12.4.1) also hold. Indeed, these terms at the left hand side of (12.4) which depend on P and μ^M_i are:

$$
\begin{aligned}
&P \cdot (\mathbf{E}^M - \mathbf{E}^M_i)' \cdot [\mathbf{F}_i, \mathbf{F}_{n+i}] + \mu^M_i \cdot ([b_{i1}, b_{i2}] + (\mathbf{E}^M + \mathbf{E}^M_i)' \cdot [\mathbf{F}_i, \mathbf{F}_{n+i}]) \leq \\
&\leq P \cdot (\mathbf{E}^M)' \cdot [\mathbf{F}_i, \mathbf{F}_{n+i}] + P \cdot ([b_{i1}, b_{i2}] + (\mathbf{E}^M)' \cdot [\mathbf{F}_i, \mathbf{F}_{n+i}]) = \\
&= P \cdot ([b_{i1}, b_{i2}] + 2(\mathbf{E}^M)' \cdot [\mathbf{F}_i, \mathbf{F}_{n+i}]).
\end{aligned}
$$

The inequality was obtained by substituting P for μ^M_i (a greater number for a smaller one) and using the ith and $(n+i)$th coordinates of (12A.4). Thus if (12.4) holds then also holds $[\partial C_i(e^M_{i1}, e^M_{i2})/\partial e_{i1}, \partial C_i(e^M_{i1}, e^M_{i2})/\partial e_{i2}] + P \cdot ([b_{i1}, b_{i2}] + 2(\mathbf{E}^M)' \cdot [\mathbf{F}_i, \mathbf{F}_{n+i}]) \geq [0,0]$ for all $i = 1,...,n$. This completes the proof of (4.1).

By summing up inequalities (12.8) for all $i = 1,...,n$ we get:

$$
\begin{aligned}
\Sigma(\mathbf{B}' \cdot \mathbf{E}^M_i + (\mathbf{E}^M)' \cdot \mathbf{F} \cdot \mathbf{E}^M_i) &= \mathbf{B}' \cdot \Sigma\mathbf{E}^M_i + (\mathbf{E}^M)' \cdot \mathbf{F} \cdot \Sigma\mathbf{E}^M_i = \\
&\mathbf{B}' \cdot \mathbf{E}^M + (\mathbf{E}^M)' \cdot \mathbf{F} \cdot \mathbf{E}^M \leq \Sigma q^M_i.
\end{aligned}
$$

But $\Sigma q^M_i \leq \Sigma q^o_i$ by the market-clearing condition (12.2). By Definition (12.1) the last sum is equal to $A - a$. Thus $\mathbf{B}' \cdot \mathbf{E}^M + (\mathbf{E}^M)' \cdot \mathbf{F} \cdot \mathbf{E}^M \leq A - a$, and (12.4.3) is proved. If $P = 0$ we are done with (12.4.4) too. To complete the proof let us suppose that $P > 0$. Then, by the market-clearing condition (12.3), $\Sigma q^M_i = \Sigma q^o_i$. If $\mathbf{B}' \cdot \mathbf{E}^M_i + (\mathbf{E}^M)' \cdot \mathbf{F} \cdot \mathbf{E}^M_i = q^M_i$ for all $i = 1,...,n$ the proof is done. So let us suppose that for some $i = i'$: $\mathbf{B}' \cdot \mathbf{E}^M_{i'} + (\mathbf{E}^M)' \cdot \mathbf{F} \cdot \mathbf{E}^M_{i'} < q^M_{i'}$. (12.9) implies then that $\mu^M_{i'} = 0$. Hence $P > \mu^M_{i'}$ (as $P > 0$) and - by (12.7) - we end up with $q^M_{i'} = 0$. But $\mathbf{B}' \cdot \mathbf{E}^M_{i'} + (\mathbf{E}^M)' \cdot \mathbf{F} \cdot \mathbf{E}^M_{i'} < 0$ contradicts (12A.4) and the proof is complete.

Proposition 12.5 - Proof:
P and e^M_{i1}, e^M_{i2}, q^M_i are defined by (12.4)-(12.9), the non-negativity constraints and the market-clearing conditions (12.2)-(12.3). Initial endowments $q^o_1,...,q^o_n$ appear in (12.2)-(12.3) but only in the form of their sum. No other conditions depend on them.

REFERENCES

Beavis, B. and M. Walker (1979) Interactive Pollutants and Joint Abatement Costs: Achieving Water Quality Standards with Effluent Charges. *Journal of Environmental Economics and Management*, **6**, 275-286.

Crocker, T.D. (1966) Structuring of Atmospheric Pollution Control Systems. In: H. Wolozin (ed.) *The Economics of Air Pollution*. W.W. Norton and Co., New York.

Daly, H.E. (1992) Allocation, distribution, and scale: towards an economics that is efficient, just, and sustainable. *Ecological Economics* **6**, 185-193.

Dudek, D.J., Z. Kulczyński and T. Żylicz (1992), *Implementing Tradable Rights in Poland. A Case Study of Chorzów*. paper presented at the Third Annual Conference of the European Association of Environmental and Resource Economists, Cracow.

Endres, A. (1986), Charges, Permits and Pollutant Interactions. *Eastern Economic Journal* **XII**, 327-336.

Hahn, R.W. (1989) A New Approach to the Design of Regulation in the Presence of Multiple Objectives. *Journal of Environmental Economics and Management*, **17**, 195-211.

Howe, C.W. (1992) Taxes *versus* Tradable Discharge Permits: A Review in the Light of the U.S. and European Experience. Mimeo, Institute of Behavioral Science, University of Colorado, Boulder.

Howe, C.W. and D.R. Lee (1983) Priority Pollution Rights: Adapting pollution control to a variable environment. *Land Economics*, **59**, 141-149.

Krumm, R.J. and P.E. Graves (1982) Morbidity and Pollution: Model Specification Analysis for Time Series Data on Hospital Admissions. *Journal of Environmental Economics and Management*, **9**, 311-327.

Liroff, R.A. (1986) *Reforming Air Pollution Regulation: The Toil and Trouble of EPA's Bubble*. The Conservation Foundation, Washington, D.C.

Lyon, R.M. (1986) Equilibrium Properties of Auctions and Alternative Procedures for Allocating Transferable Permits. *Journal of Environmental Economics and Management*, **13**, 129-152.

Montgomery, W.D. (1972) Markets in Licenses and Efficient Pollution Control Programs. *Journal of Economic Theory*, **5**, 395-418.

O'Neil, W.M. David, C. Moore and E. Joeres (1983) Transferable Discharge Permits and Economic Efficiency: The Fox River. *Journal of Environmental Economics and Management*, **10**, 346-355.

Opschoor, J.B., and H.B. Vos (1989) *Economic Instruments for Environmental Protection*. OECD, Paris.

Pezzey, J. (1992) The symmetry between controlling pollution by price and controlling it by quantity. *The Canadian Journal of Economics* **25**.

Portney, P.R. (1988) Reforming environmental regulation: Three modest proposals. *Issues in Science and Technology*, **4**(2), 74-81.

Roumasset, J.A. and K.R. Smith, (1990) Exposure Trading: An Approach to More Efficient Air Pollution Control. *Journal of Environmental Economics and Management*, **18**, 276-291.

Tietenberg, T.H. (1985) *Emissions Trading, and Exercise in Reforming Pollution Policy*. Resources for the Future, Washington, D.C.

Tietenberg, T.H. (1994) Market-Based Mechanisms for Controlling Pollution: Lessons from the U.S. In: T. Sterner (ed.), *Economic Policies for Sustainable Development*. Kluwer, Dordrecht (forthcoming).

Weitzman, M. (1974) Prices vs. Quantities. *Review of Economic Studies* **XLI**, 447-491.

Żylicz, T. (1989) *Marketable Permits for Synergistic Pollutants* Mimeo, Institute of Behavioral Science, University of Colorado, Boulder, CO.

Żylicz, T. (1994) Implementing Environmental Policies in Central and Eastern Europe. In: A-M. Jansson, M. Hammer, C. Folke, and R. Costanza (eds.), *Investing in Natural Capital: the Ecological Economics Approach to Sustainability*, Island Press, Washington, DC, pp.408-430 (in print).

Chapter 13 TRADING SULFUR EMISSION PERMITS IN EUROPE USING AN EXCHANGE RATE

Ger Klaassen[1]

ABSTRACT

Emission permit trading offers an alternative approach in achieving cost-effectiveness for emission reduction but is complex when considering non-uniformly dispersed pollutants where the location of the emission source is important. One simple system is to allow countries to trade emission reduction ceilings according to a fixed exchange rate. From a number of formal conditions for the cost-minimum solution, the ratios of the marginal costs of abatement in the optimal solution are introduced as exchange rates for emission trading. Thereby the transfer coefficients of the binding receptors are taken into account. Generally, because other necessary conditions for an optimum solution are neglected, this system will not achieve the optimal solution and such a trading system does not guarantee that environmental deposition constraints are not violated. Based on the IIASA RAINS model a routine has been developed to simulate bilateral, sequential emission trading for SO_2 emissions in Europe. For a set of example target loads for the deposition, based on closing the gap between the sulfur deposition in 1990 and the 5 percentile critical loads by 60%, calculations identified the cost-minimal allocations of emission reductions and the cost minimum, given that countries at least carry out their Current Reduction Plans. Taking the latter as starting point, exchange rate emission trading simulations achieved cost savings but could not attain the cost minimum. However, cost savings are attained at the expense of an increase in emissions, less ecosystem protection and the violation of deposition targets. Furthermore, although some countries experience less costs and more ecosystem protection after trading, other countries have no cost savings and their level of ecosystem protection decreases. In view of the above conclusions, it seems necessary to examine those possibilities for emission trading that ensure that third parties have the same levels of deposition and ecosystem protection before as after trading.

[1] Computations of the atmospheric transport of pollutants in this chapter are based on the results of the EMEP/MSC-W project at the Norwegian Meteorological Institute, Oslo. The maps of critical loads and target loads used were established by the Coordination Center for Effects (CCE) at the National Institute of Public Health and Environmental Protection (RIVM), Bilthoven, the Netherlands. Comments from the editor, reviewer and Scott Atkinson are appreciated.

13.1 INTRODUCTION

Exchange, for Aristotle, amounted to a mutual sharing. Aristotle made the rate of exchange depend on reciprocity. Rather than the interest of the individual, the interest of the community was his governing principle (Ekelund and Hébert, 1983, p 13-17). Current discussions in Europe on a new sulfur protocol also turned to the possibilities of including emission trading provisions. Remarkably, one of the schedules discussed is to allow trade according to an (fixed) exchange rate (Klaassen and Amann, 1992). An exchange rate or offset rate being defined as: 'the volume of emissions that one source has to decrease if another source increases its emissions with one unit'. This concept is similar to the pollution offset trading provisions in the USA in the eighties (Borowski and Ellis, 1986). As Aristotle, also the UN/ECE debates the correct or just exchange rate. There are questions on whether exchange trading can achieve the cost-minimum solution, i.e. when the environmental objective is the attainment of a set of (region specific) targets for the deposition of sulfur in Europe. There are also discussions on the environmental effects and possible violations of deposition objectives that may result from trading. The objective of this chapter is threefold:

1. to analyze the cost-efficiency and environmental impacts of exchange rate trading of emissions,
2. to describe the method and data used to simulate exchange rate trading of emissions as a bilateral, sequential process,
3. to give an example result based on reducing the difference (the gap) between the deposition in 1990 and the 5 percentile critical sulfur deposition values by 60%.

The structure of the chapter is as follows. Section 13.2 elucidates the concept and provides an in-depth analysis of exchange rate trading. The trading simulation method and costs as well as the used atmospheric transport data are the subject of section 13.3. Section 13.4 describes some example results and section 13.5 draws conclusions.

13.2 ANALYSIS OF EXCHANGE RATE TRADING OF EMISSIONS

13.2.1 The concept of exchange rate trading of emissions

An alternative trading system, which is relatively simple and applied in practice, is emission trading subject to an exchange rate. An exchange or offset rate is defined as: 'the volume of emissions that one source has to decrease if another source increases its emissions with one unit'.

In other words, the exchange rate states that if one source increases emissions by one unit (it buys permits) another source has to decrease its emissions by the same amount multiplied with the exchange rate (it will then sell permits).

A small model will assist in structuring the problem. Text Box 1 defines all the variables that will be used.

E^o_x, E^0_y, E^0_i	pre-trade emissions of respectively source x, y and i (i = 1,2,....I and x,y ϵ i)
E_x, E_y, E_i	post-trade emissions of source x, y and i
W_{xy}	the exchange rate (the rate at which y has to decrease emissions if x increases emissions with one unit)
C_x,C_y,C_i	functions representing the costs to source x, y and i of reducing emissions
C'_x,C'_y,C'_i	marginal costs to source x, y and i of reducing emissions
$E_{i,max}$, $E_{i,min}$	is the maximum c.q minimum level of emissions of source i without control (x,y ϵ i)
μ_j	the Lagrange multiplier or shadow price of receptor j
D_j	the level of deposition at receptor j
D_j^*	the desired level of deposition (the target load) at receptor j
a_{ji}	a (linear) transfer coefficient which translates emissions of source i in deposition at receptor j (j =1,2,...,J)

Text box 1 List of variables

The cost functions C_x and C_y, are represented as functions of the remaining emissions after control. Assuming that both sources involved in trading want to minimize costs the problem of emission trading subject to a fixed exchange rate can be stated as follows:

$$\text{Minimize:} \quad C_x + C_y \tag{13.1}$$

$$\text{Subject to:} \quad E_y - E^0_y = -W_{xy}\,(E_x - E^o_x) \tag{13.2}$$

All volumes are non-negative. This shows that the change in emission of y as a result of trading (E_y - E^0_y) depends on the exchange rate (W_{xy}) and the volume traded by x (E_x - E^o_x). If the exchange rate is one, total emissions have to be constant during trading. Equations 13.1 and 13.2 form a classical problem of programming subject to an equality constraint. The solution can be found in formulating the so called Lagrangian function Z:

$$Z = C_x + C_y + \mu \{(E^0_y + W_{xy} * E^o_x) - (E_y + W_{xy} * E_x)\} \tag{13.3}$$

where μ is the Lagrange Multiplier. The first order or necessary conditions for a cost-minimum solution are found by taking the partial derivatives of Z with respect to E_x, E_y and μ:

$$C'_x - \mu * W_{xy} = 0 \tag{13.4a}$$
$$C'_y - \mu = 0 \tag{13.4b}$$
$$E_y + W_{xy} * E_x = E^0_y + W_{xy} * E^o_x \tag{13.4c}$$

Here C' indicates marginal costs. Conditions (13.4a and 13.4b) can be combined by eliminating μ:

$$C'_x = W_{xy} * C'_y \text{ (or } C'_x/C'_y = W_{xy} \text{ if } C'_y \text{ is not 0)} \tag{13.5}$$

In this case the following situations are possible:

- If in the pre-trade situation $C'_x > W_{xy} * C'_y$, then country x could profit by increasing its emissions and paying country y to purify.
- If in the pre-trade situation $C'_x < W_{xy} * C'_y$, then it pays for country x to reduce emissions further (hence increase marginal costs) and allow country y to increase emissions.

Hence, as long as C'_x is unequal to $W_{xy} * C'_y$ there is an incentive to trade since cost savings are possible.

13.2.2 An exchange rate equal to the ratio of the marginal costs in the optimum

Background and rationale

One possibility for selecting the exchange rate is to base the exchange rate on the ratio of the marginal costs in the cost minimum solution. As will be explained, this rule can be based on the understanding that the ratio of the marginal costs in the optimum depends on the shadow prices (relative difficulties) of attaining the binding deposition constraints. It reflects one of the conditions for a cost-minimal solution if the problem is the minimization of costs subject to constraints on the deposition.

In the remainder of this section we will first give the conditions for a cost minimum solution for attaining a set of targets for the deposition for a general case. Then this section will analyze the conditions under which exchange rate trading can achieve this cost minimum. The following additional elements (see also Text Box 1) are defined: D_j (the level of deposition at receptor j), D_j^* (the desired level of deposition (the target load) at receptor j) and a_{ji} (a linear transfer coefficient which translates emissions of source i in deposition at receptor j). The deposition at a specific location is a function of the emissions (E_i) multiplied by their transfer coefficients (we ignore background depositions since inclusion does not affect the analytic solutions). The cost-effective solution requires that the total costs (C_i) of emission reductions are minimized, subject to the constraint that the desired deposition levels are met at each receptor point.

$$Min \ \sum_{i=1}^{I} C_i \qquad (13.6)$$

subject to

$$\sum_{i=1}^{I} a_{ji} * E_i \leq D_j^* \quad for\ every\ j=1,2,...,J \qquad (13.7)$$

Furthermore, the emission reductions are limited (non-negative and below the maximum feasible abatement).

The most important of the necessary (Kuhn-Tucker) conditions for a cost minimum are the following (Tietenberg, 1985):

$$(E_{i,\max} - E_i)\ [C_i' - \sum_{j=1}^{J} a_{ji}\ \mu_j] = 0\ for\ every\ i \qquad (13.8)$$

$$\mu_j\ [D_j^* - \sum_{i=1}^{I} a_{ji} * E_i] = 0\ for\ every\ j \qquad (13.9)$$

Furthermore, the deposition at each receptor has to be equal or less than the targets (condition 13.7). Equation 13.8 states that for a cost-effective solution either the emission reduction of the source has to be zero ($E_i = E_{i,max}$) and/or the marginal costs of emission reduction for each source have to equal a weighted sum of the shadow prices (μ_j) for each receptor. The shadow prices reflect the (marginal) costs of tightening the constraints. Weights are the transfer coefficients from source i to each affected receptor j. Equation 13.9 shows that either the required target load (D_j^*) is met exactly or the μ_j is zero. The latter means that the receptor is non-binding. The main message here is that in the cost-effective solution the marginal costs are a function of the transfer coefficients at the binding receptors and the prevailing shadow prices unless no emission reduction is required.

One possibility for selecting an exchange rate is to base the exchange rate on the ratio of the marginal costs in the cost minimum solution. To determine this "optimal" exchange rate we recall equation 13.5: $C'_x = W_{xy} * C'_y$. If the ratio of the marginal costs in the optimum is used to determine the exchange rate it would be:

$$W_{xy} = \frac{C_x'}{C_y'} = \frac{\sum_{j=1}^{J} a_{jx}\mu_j}{\sum_{j=1}^{J} a_{jy}\mu_j} = \frac{a_{1x}\mu_1 + a_{2x}\mu_2 + ... + a_{jx}\mu_j}{a_{1y}\mu_1 + a_{2y}\mu_2 + ... + a_{jy}\mu_j} \tag{13.10}$$

μ_j is the shadow price at the jth receptor. To explain how this exchange rate governs trading we first analyze the case of one, binding, receptor. Then we include more (binding) receptors.

Two countries and one receptor

To see how optimal exchange rate trading works, let us assume a transfer coefficient from source x to the receptor of 0.5 and from source y to the one receptor of 1. The following condition then applies in the optimum:

$$W_{xy} = \frac{C_x'}{C_y'} = \frac{a_{1x}\mu_1}{a_{1y}\mu_1} = \frac{0.5}{1} = 0.5 \tag{13.11}$$

The ratio of 0.5 is the exchange rate $W_{x,y}$. This implies that if source x would decrease its emissions with one unit (sells permits) source y would be allowed to increase its emissions with only half a unit. Clearly, in this case the deposition target load will not be violated during trading, provided it was not exceeded before trading (in this case deposition would remain constant at the receptor but it would remain too high).

Figure 13.1 illustrates the case for one receptor and two sources. The y-axis of figure 13.1 shows the emissions of source y and the x-axis emissions of source x. The line R1 shows combinations of emissions from both sources for which the deposition target at receptor 1 is exactly met. The curves C1 and C2 are isocost curves: combinations of emissions from both sources which lead to the same level of total costs. The closer these costs are to the origin, the higher the costs and the lower the emissions are. The figure shows that as long as emissions from both sources remain within the area OAB the deposition targets are met. The least-cost solution is point E. At this point the ratio of the marginal costs equals the ratio of the transfer coefficients for receptor 1, which at this point are both binding. This ratio is also the coefficient that determines the direction of line R1 as can be seen from (cost minimum) condition 13.7 if it is rewritten so that the emissions of y are a function of the emissions of x and the (binding) deposition target:

$$E_y = \frac{D_1}{a_{1y}} - \frac{a_{1x}}{a_{1y}} E_x \tag{13.12}$$

This implies that an exchange rate based on the ratio of the optimal marginal costs allows both sources to trade along a line parallel to the line R1. From condition 13.4c for exchange rate trading one can also express E_y as a function of other variables:

$$E_y = E_y^0 + W_{xy}E_x^0 - W_{xy}E_x \tag{13.13}$$

Clearly, if both equations lead to similar results for E_y, exchange rate trading will achieve the cost minimum solution. This is so if two conditions are fulfilled:

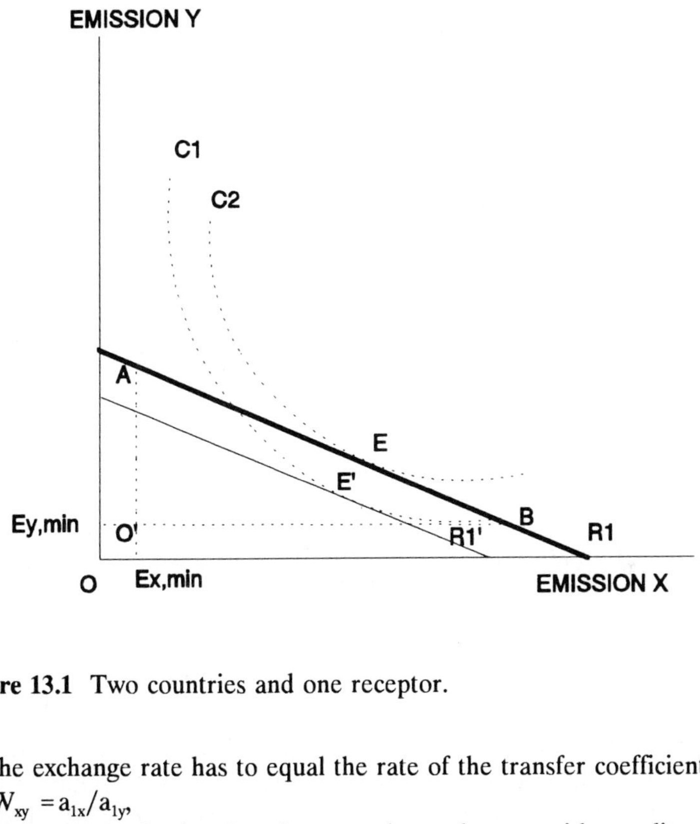

Figure 13.1 Two countries and one receptor.

- the exchange rate has to equal the rate of the transfer coefficients: $W_{xy} = a_{1x}/a_{1y}$,
- in the pre-trade situation the target has to be met with equality: $a_{1x}E^o{}_x + a_{1y}E^0{}_y = D_1$

The first condition is met if the exchange rate is based on the cost minimum solution. The second, however, is only met if initially the deposition target is exactly met.

In conclusion, with one receptor and two trading sources, the cost minimum is attained only if the one receptor is initially binding as well. If the deposition target is met initially (initial point inside the feasible area) it remains met, since trading will occur parallel to the constraint.

Two sources and two receptors

The question is whether these results are transferable when there are two (or more) binding receptors in the cost minimum. First, this will be illustrated graphically. Then we supply an analytical proof. Figure 13.2 is comparable to figure 13.1, the only difference being that there are now two receptors R1 and R2, both binding in the optimum. The line R1 depicts combinations of emissions of both emitting sources that exactly meet the deposition constraint at receptor 1. Line R2 does the same for receptor 2. The least-cost solution is point E. At this point the ratio of the marginal costs equals the weighted ratio of their transfer coefficients for receptors 2 and 1:

$$\frac{C_x'}{C_y'} = \frac{a_{1x}\mu_1 + a_{2x}\mu_2}{a_{1y}\mu_1 + a_{2y}\mu_2} \tag{13.14}$$

This ratio is also the coefficient that determines the direction of line R3. The direction coefficient of the line R3 can be seen as a weighted average of the direction coefficients of the lines R1 and R2. This implies that an exchange rate based on the ratio of the optimal marginal costs allows both sources to trade as long as they move (trade) along a line parallel to line R3. Again attainment of the optimum (E) is contingent on the initial solution. If the initial position is P, the trading ratio prevents reaching the optimum, although cost savings are possible. Starting from P, sources would only be allowed to trade along the dotted line R3' (parallel to R3). In this case, E' would be the least cost solution attainable from P with the given trade ratio. Obviously E' is not identical with E, and therefore costs (C2) are higher than in the cost minimum (C1). It appears that trade would end up in the cost minimum if it started from (any) point on the line R3. This is so because all other points on the line R3 would be more expensive, given the fact that isocost curves cannot intersect.

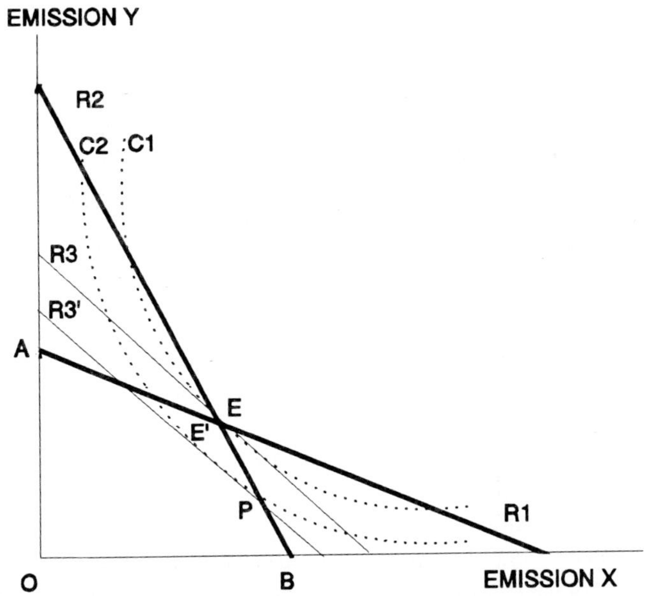

Figure 13.2 Two sources and two receptors.

Although figure 13.2 gives us some hints, an analytical examination is useful to figure out under which conditions the "optimal" exchange rate trading could end up in the cost minimum. For this purpose the conditions for a cost minimum solution will be compared with the conditions for an exchange rate optimum. It is assumed that there are two receptors and both of them are binding in the optimum. So both constraints are met exactly. This allows to write the problem as a minimization problem with equality constraints. The cost minimum formulation with two binding receptors is:

$$\begin{aligned} &\text{Minimize:} && C_x + C_y \\ &\text{Subject to:} && a_{1x}E_x + a_{1y}E_y = D_1^* \\ & && a_{2x}E_x + a_{2y}E_y = D_2^* \end{aligned} \qquad (13.15)$$

All volumes are non-negative. The following (first order) conditions for an optimum prevail:

$$C'_x = a_{1x}\mu_1 + a_{1x}\mu_2 \qquad C'_y = a_{1y}\mu_1 + a_{1y}\mu_2 \qquad (13.16)$$

$$a_{1x}E_x + a_{1y}E_y = D_1^* \qquad a_{2x}E_x + a_{2y}E_y = D_2^* \qquad (13.17)$$

From condition (13.17) y's emission can be formulated (by eliminating E_x) as:

$$E_y = \frac{a_{2x}D_1^* - a_{1x}D_2^*}{a_{1y}a_{2x} - a_{2y}a_{1x}} \tag{13.18}$$

For exchange rate trading, conditions 13.2 and 13.5 should be met for an exchange rate trading optimum. Setting the exchange rate at the rate of the marginal costs in the optimum (condition 13.14) and substituting this in condition 13.2 gives:

$$E_y = E_y^0 + \frac{a_{1x}\mu_1 + a_{2x}\mu_2}{a_{1y}\mu_1 + a_{2y}\mu_2} E_x^0 - \frac{a_{1x}\mu_1 + a_{2x}\mu_2}{a_{1y}\mu_1 + a_{2y}\mu_2} E_x \tag{13.19}$$

Clearly, this defines a line (see the coefficient of direction) parallel to the tangent of the isocost curves at point E. The question is under which conditions both cost minimum and exchange rate conditions for E_y imply the same optimum value for E_y. Obviously, as figure 13.2 suggested, the line has to go through the optimum solution to ensure that condition 13.18 equals 13.9. In this case pre-trade emissions of x could be equal to their cost minimum levels ($E_x^0 = E_x^*$) which immediately would imply (condition 13.19) that y's initial emissions would have to be optimal.

In conclusion, exchange rate trading achieves the cost minimum solution only under a very special condition on the starting point. The starting point has to be on a line that not only has the same coefficient of direction as the tangent of the isocost curves in the optimum (which is guaranteed if one uses the ratio of the marginal cost in the optimum as exchange rate) but, moreover, this line has to go through the optimum. The latter seems to be a rather restricting condition.

Assuming that the above stringent conditions on the starting point are not fulfilled, an important question now is whether exchange rate trading can result in violating the target loads or not. Turning to figure 13.2, this figure suggests that it might happen. Even if both conditions are initially met (trade starts from inside the area O'AEB), it could stay inside or they might go outside, dependent on the configuration of the isocost curves.

In conclusion, exchange rate trading with two sources and two binding receptors in the optimum can only achieve the cost minimum under special provisions for the pre-trade emissions. Even if exchange rate trading starts from inside the feasible region the analysis suggests that one deposition target might be violated, although the other is met.

Uncertainty on costs

If an environmental agency has incorrect information on costs, it is likely that the emission trading with the above exchange rate neither attains the cost minimum nor the deposition targets. However, there is a difference between the case of one receptor and more receptors. In case there is only one receptor, the environmental agency does not need any information on the costs to determine the "optimal" exchange rates. This is so since the exchange rates depend on the ratios of the transfer coefficients only. This is also sufficient to guarantee that the deposition targets are met if they were not exceeded before trading. In case of more than one receptor, the agency can not determine the exchange rates without knowledge of the cost functions, since in that case it would not know which receptors would be binding in the cost minimum and which not. Sources might then trade on the basis of their costs, but the exchange rates would be based on the environmental agency's (wrong) perception of the costs. The trading results will then differ from the expectations.

Conclusions

Clearly, the advantage of the sketched system of exchange rate trading of emission permits is that it is simple for the trading sources. Moreover, with one receptor, deposition targets will not be violated if the deposition constraint is initially met and no knowledge on the cost functions would be needed to set the exchange rates. However, even with one receptor, bilateral, sequential trading ends up in the cost minimum only if trading starts from a specific starting point. With two (or more) binding receptors there is neither a guarantee that the system will attain the cost minimum nor is there certainty that deposition targets will not be violated. This is the case even if the environmental agency has complete and correct knowledge of the costs necessary to determine the trading ratio. Generally, a simultaneous trade between all the sources would be needed to attain the cost minimum, unless the emissions of the non-trading sources are already optimal and trade starts at a specific point. Again, there is no guarantee that the deposition targets are met.

The introduction of such a trade ratio fulfills only one of the several necessary conditions (and only partly), stating that the ratio of marginal costs should be equal to the exchange rates for a cost minimum. One other condition for a cost minimum is that deposition is not allowed to exceed target loads. This, however, is not an explicit condition for the exchange rate trading. This was a deliberate choice, in order to first investigate unconstrained trade due to the complexities of the trading process of imposing target load constraints. Hence, the solutions to both problems are not identical. Mathematically there is neither a guarantee that the cost minimum will be attained by trading nor that the deposition targets will be met. The initial emissions of the sources and the trade ratios restrict the possible range of solutions that can be reached by trading. In summary, it is unlikely that the proposed exchange rate leads to the cost minimum since this depends on a special provision on the initial point.

Moreover, exchange rate trading does not guarantee that the deposition targets are met.

Although there is evidence that the optimum solution will only be achieved by chance, the question remains how close such a trading scheme could approach the optimum and to what extent deposition targets will be violated. As usual, the magnitude depends crucially on the actual problem specification (transfer coefficients, deposition target, cost curves, initial solutions). Therefore, the following sections describe a simulation of such trading processes. First, the optimization procedure and the data on used costs and atmospheric transport are described, since the trade simulation model makes extensive use of the optimization routine.

13.3 SIMULATION METHOD AND DATA

13.3.1 The model applied

The method developed to simulate bilateral, sequential trading using trade ratios as introduced above makes use of an adapted version of the optimization module in the RAINS (Regional Acidification INformation and Simulation) model (Alcamo et al., 1990). The model has the theoretical structure of Equations (13.6 - 13.7). The cost functions of emission reductions are expressed as piecewise linear curves denoting cost-minimal combination of measures within each country to achieve certain levels of national total emissions (see figure 13.3). The marginal costs are the slope of the total cost curve in each segment. $R_{i,s}$ is the purification in region i in segment s. The amount purified in each of the segments (s) is limited:

$$0 \leq R_{i,s} \leq R_{i,s}^{\max} \quad , i = 1,...,38 \; , \; s = 1,..,S \qquad (13.20)$$

The RAINS model simulates the flow of acidifying pollutants (sulfur and nitrogen) from source regions in Europe to environmental receptors. The current model (version 6.0) covers 38 source regions in Europe: 26 countries, 7 regions in the former USSR and 5 sea regions (ship emissions). Emissions can be controlled in 36 regions. Analysis of deposition is performed for 547 land-based receptor sites with a regular grid size of 150*150 km.

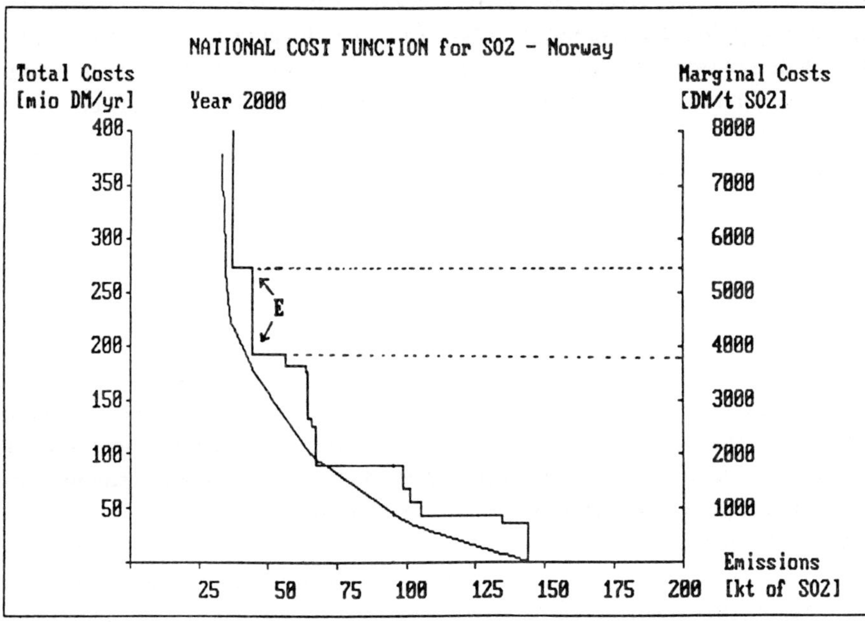

Figure 13.3 Cost function.

The optimization part of the RAINS model formulates possible strategies to minimize the costs of achieving deposition targets at certain receptors as a linear optimization problem (LP) that can be solved with LP packages (Amann, 1992; Batterman and Amann, 1991). The cost-effective solution requires that the total costs of emission reductions are minimized, subject to the constraint that the desired depositions are met at every receptor.

The piecewise linear cost function affects the practical definition of the exchange rate, being the ratio of the marginal costs of the optimum. Due the nature of linear optimization, the optimal status of most variables of an optimization problem are exactly located at the corner points of the solution space, on the intersections of the linear cost function segments (e.g. point E in figure 13.3). For these points a unique definition of marginal costs does not exist (for point E, marginal costs are either 3900 or 5500 DM/ton SO_2). The marginal costs for increasing emissions differ from the marginal costs for decreasing emissions. For this reason the simulation uses the higher marginal costs (the additional costs of further reducing emissions represented by the next step of the cost curve) by default. This also avoids problems with marginal costs of zero, for which the definition of the exchange rate is not applicable.

13.3.2 The method to simulate bilateral trading

Introduction

Until recently, many model studies simply assumed that the potential cost savings of emission trading would be identical to the results of optimization procedures; in other words, a perfectly working, competitive market where emission permits are simultaneously traded, was assumed. Practice as well as recent model studies (Atkinson and Tietenberg, 1991; Tietenberg, 1990) suggest that in reality trading takes place bilaterally and sequentially. With such restrictions, trading is not expected to capture the complete cost savings possible according to any LP cost minimization procedure. The trading algorithm presented here describes a process of repeated bilateral trading subject to an offset or exchange rate for every possible combination of bilateral trades. The procedure consists of the following steps:

1. Creation of a matrix of potential cost savings from each potential bilateral trade.
2. Specification of transaction costs and determination of a threshold level, below which trades will not take place.
3. Selection of the trading sequence, taking the most profitable one first.
4. Updating of emissions after implementation of the selected trade and recording cost savings.
5. Updating of the matrix of possible trades, accounting for the trades that took place (return to step 1).

Regarding the creation of the cost saving matrix, one should realize that cost functions estimated with the RAINS model are piecewise linear. As a result, RAINS does not work by equalizing marginal costs but sorts and ranks elements of two (or more) cost functions according to their marginal costs (in ascending order). If an exchange rate is introduced, the determination of the cost-optimal bilateral combination of emission reductions can be performed in a similar way, taking into account the exchange rate.

The two conditions for an optimum solution for each bilateral trade are condition (13.5) and the trade rule in (13.2), stating that the emission increase by source 2 should equal the emission decrease of source 2 multiplied by the exchange rate. Alternatively, the trade rule can be written as:

$$E_x + E_y/W_{xy} = E_1^0 + E_2^0/W_{xy} \tag{13.21}$$

Since emissions have different impacts on the receptors a simple summation is less relevant if one is interested in meeting certain deposition goals. In order to get the emissions on the same receptor-oriented unit of measurement, emissions from source y (arbitrary choice) are weighted with the exchange rate. On the left-hand side we have the emissions after trade, while on the right-hand side we have the pre-trade emissions, and on both sides the emissions are made comparable by weighing. We see clearly that it is only for the case of the exchange rate being 1 that pre-trade emission levels equal post-trade emissions in tons. Equations 13.5 and 13.21 have two unknowns, the post-trade emissions. For each pair of countries we can find the solutions and create a matrix of cost savings of all possible trades. The number of possible trades depens on the number of countries that can trade. In our example 33 countries can trade, giving 1056 (33 x (33-1)) possibilities.

Emission trading practice in the USA showed that transaction costs frequently prohibit trading. In the literature, transaction costs are estimated at 10-30% of the cost savings (Dwyer, 1991). Building in transaction costs thus gives a more realistic picture of the potential cost savings. The algorithm allows specification of a (exogenous) level of transaction costs for each trade. If cost savings of a potential bilateral trade are below the threshold the trade is not profitable and skipped from further selection.

After calculating the matrix of cost savings of all possible (and profitable) trades, the sequence of trading is determined. Currently, the algorithm ranks all possible trades according to their cost savings and selects the one with the highest cost savings. Cost savings are defined as the difference between the total cost increase of the source selling emission permits and the total cost decrease of the emission increasing source, between the pre-trade emission level (of every round) and the post-trade emission level. This is an optimistic assumption assuming perfect information and coordination of the selection of traders. An alternative assumption, currently not implemented, would be that, due to the imperfect information of traders and due to the competition to get the trade accepted that is best for two individual trade partners (not necessarily the best in terms of highest overall cost savings), the selection of trade takes place at random.

Updating the emission matrix and the matrix of possible trades is the final step. Upon completion of the trade(s), the following steps are implemented:

1. Update the (pre-trade) emission vector, accounting for the trade(s) that tookplace.
2. Record the cost savings of the trade (compared to the pre-trade situation) in a file.

Finally, the matrix of possible trades is updated. Sources that have already traded are allowed to trade again, but not with the same partner, and the cost savings of all the other trade relationships of these two countries are recalculated with the new emission levels.

13.3.3 Data on costs and atmospheric transport

The RAINS model contains a sub-module to assess the feasibility and costs of alternative emission abatement technologies. The evaluation is based on internationally reported performance and cost data of control devices (Amann, 1992). Cost estimates for specific technologies are extrapolated by the model to reflect country-specific conditions such as operating hours, boiler size, and fuel price. The following technical options are available:

1. Use of low sulfur fuels and fuel desulfurisation. This pertains to the use of fuels with a reduced sulfur content, such as fuels with a lower natural sulfur content or fuels that have undergone a desulfurisation process. For low sulfur hard coal, the sulfur content is set at 1%. Desulfurisation of gas oil and diesel oil can reduce the sulfur content in two steps: down to 0.3% and down to 0.15%. The desulfurisation of heavy fuel oil is assumed to be possible down to a level of 0.6%.
2. Desulfurisation of flue gases during or after combustion. This set of measures requires investments at the plant site. Three techniques are considered: desulfurisation during combustion with removal efficiencies of 50% at relatively low costs, flue gas desulfurisation with a removal efficiency of 95% at moderate costs, and the use of advanced flue gas purification with emission reduction of 98% at high costs.

Not all abatement technologies are applicable for all fuel types and energy sectors. Moreover, a distinction is made between new and existing plants to account for the additional costs of retrofitting existing plants.

For the optimization, RAINS creates 'national cost functions' for controlling emissions. National circumstances (such as sulfur content and operating hours) result in variations in the costs for applying the same technology in different countries in Europe. Another difference is the structural variation of energy systems, especially in the amount and structure of energy use, which determines the potential for application of individual control options. One way to combine these factors is to compile national cost functions. These functions display the lowest costs for achieving various emission levels by applying the cost optimal combination of abatement options. The cost curves used in this chapter are based on official energy use projections for the year 2000 provided by the UNECE.

Source-receptor transfer coefficients, which relate (country) emissions in the diffusion model to deposition at receptor points (for each grid), are based on the acid deposition model developed within the European Monitoring and Evaluation Program (EMEP) (Iversen et al., 1991; Sandnes and Styve, 1992). The model calculates transboundary fluxes of oxidized sulfur and nitrogen as well as reduced nitrogen (ammonia and its product ammonium). For the trade simulations presented in this chapter, EMEP model results have been applied that reflect the meteorological average of the years 1985, 1987, 1988, 1989 and 1990.

13.4 RESULTS FOR A 60 PERCENT GAP CLOSURE

13.4.1 Scenario setting

Current negotiations on a new sulfur Protocol (Amann et al., 1993) focus on sulfur deposition targets. Provisional (interim) deposition targets were proposed that reduce the difference (the gap) between the sulfur deposition in 1990 and the 5 percentile critical sulfur deposition values by 50 respectively 60%.

This section examines the cost-effectiveness and environmental impacts of allowing exchange trading. The instruments (or scenarios) for achieving the target loads are:

1. a cost-minimum allocation of (based on optimization) for attaining deposition targets based on closing the gap by 60% (scenario A6 of the TFIAM, Amman et al., 1993);
2. a cost-minimum allocation as scenario 1 but assuming that countries at least carry out their Current Reduction Plans (scenario A4 of the TFIAM, Amann et al., 1993);
3. emission trading with an exchange rate equal to the marginal costs in the optimum (scenario 1) but trade starting from the emissions of scenario 2.

The emission trading scheme simulates bilateral, sequential trading. The trading scheme (scenario 3) has a different ratio for each particular trade. In principle, emission trading can start from any initial distribution of emission reduction commitments. In this example, emission trading schemes start from scenario 2 since this is one of the scenarios considered as potential candidate for a new sulfur protocol. This has the additional advantage that initially the target loads are not violated and, therefore, it allows one to analyze whether or not emission trading results in violation of the targets.

13.4.2 Cost-effectiveness and environmental impacts of emission trading scenarios

Table 13.1 displays the major results of the various scenarios in terms of emission control costs, the total remaining emissions, ecosystem protection and the violation of target loads. The Table shows that Scenario 1 (optimized allocation) is the most cost-effective means to achieve the target loads. Costs of taking account for current plans (Scenario 2) are 7 percent higher than Scenario 1. Emission trading starts from the initial allocation of emission levels of Scenario 2. In this case, emission trading (Scenario 3) results in cost savings compared to Scenario 2 but does not achieve the cost minimum (Scenario 1). Table 13.1 also shows that Scenario 2 results in lower emissions than Scenario 1. Exchange rate trading increases emissions (starting from Scenario 2), from 22539 to 23323 kton (+3.5 percent). Consequently, the percentage of ecosystems protected after trading is smaller (92%) than before trading (Scenario 2, 93% protection). Hence, cost savings of trading are bought at the expense of environmental protection.

Furthermore, the violation of deposition targets is different for each scenario. Target loads are not violated, by definition, in Scenarios 1 and 2, but they are with exchange rate trading. The cost savings of Scenario 3 (exchange rate trading) are partly obtained by an increase in emissions. This increase results in exceedance of target loads after trading, due to the fact that pre-trade depositions in a number of locations are very close to or at the deposition targets. Figure 13.4 shows that exchange rate trading results in violation of the deposition targets at a number of locations (the hot spots). To a certain degree these violations are small (1 to 3% above the target), for a number of locations (i.e. in Finland) the deposition after trading is up to 40% higher than the deposition targets. This shows, as the analysis in section 13.2 already suggested, that exchange rate trading might violate the deposition constraints.

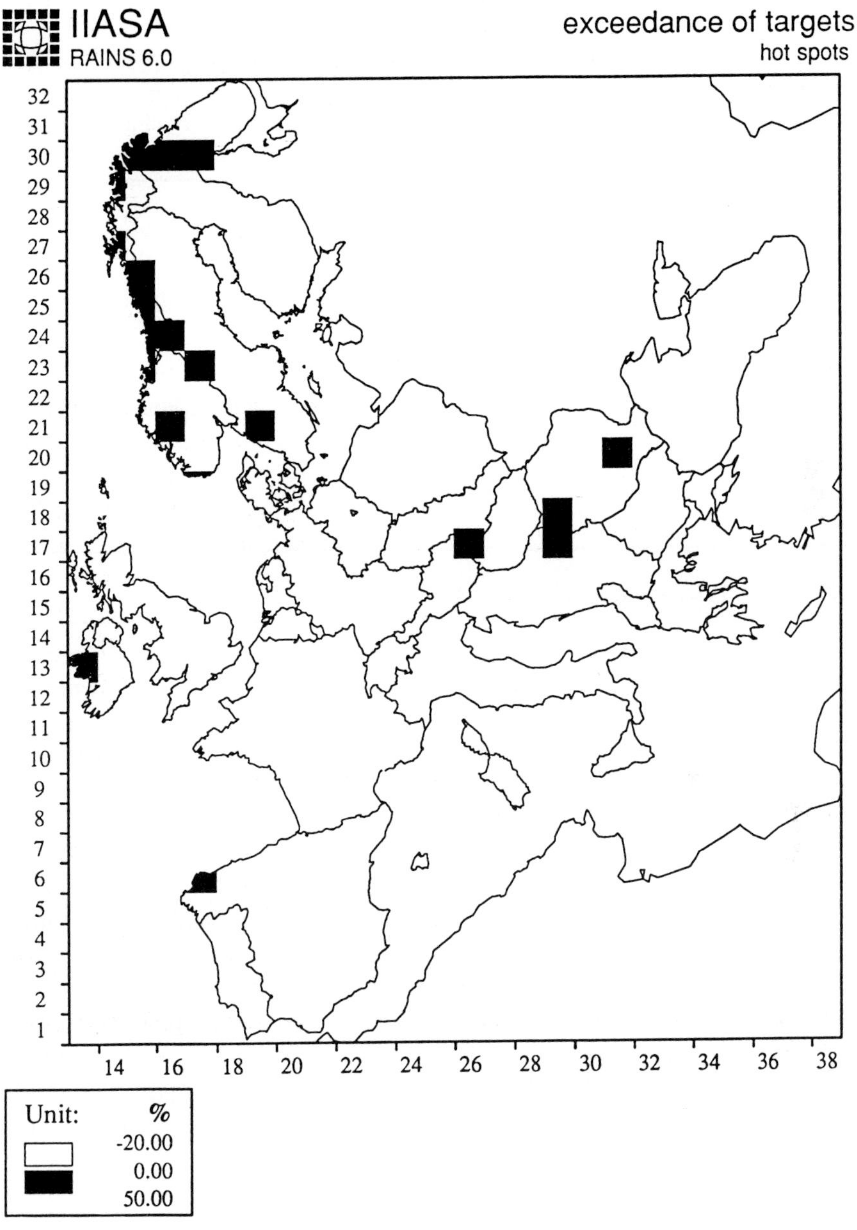

Figure 13.4 Hot spots - % excess over targets.

Table 13.1 Main results

	SCENARIO 1	2	3
SCENARIO:	Cost Minimum	Cost minimum plus CRP	Exchange rate trading
Annual costs (mio DM)	23969	25629	24298
Annual costs (as % of cost minimum)	100	107	101
Emissions (kt SO_2)	23516	22539	23323
Exceedance of target loads (%)	no	no	yes (0-39%)
Ecosystem protection (% of ecosystems)	92	93	92

Table 13.2 Trades implemented with exchange rate trading

Trade	Buyer	Amount (kt SO_2)	Seller	Amount (kt SO_2)	Exchange Rate	Cost Savings (mio DM/yr)
1	Germany-E	161	Netherlands	-40	.25	577
2	Bulgaria	296	France	-3	0.01	249
3	Austria	162	France	-169	1.04	243
4	Greece	311	Belarus	-5	.02	207
5	Norway	6	Belarus	-38	6.02	83
6	Kola-Karelia	248	Belarus	-144	0.58	83
7	Hungary	0	Moldavia	0	0.47	1
Total	All	1184	All	-399	0.34	1393

13.4.3 A closer look at exchange rate trading

Table 13.2 shows that under exchange rate trading seven trades are expected to be implemented. The first trade is between Germany-East and the Netherlands. The Netherlands decreases its emissions by 40 kilotons (dropping Dutch emissions to 66 kiloton) to enable East-Germany an increase in emissions of 161 kiloton (increasing emissions to 391 kiloton). Hence, the exchange rate is 0.25. The seven trades together result in an increase of emissions of 1184 kilotons (the buyer side) and a decrease of only 395 kilotons. This boils down to an average exchange rate of 0.34 and implies, unique in the history of emission trading in practice, that emissions would be allowed to increase.

Interestingly, Table 13.3 shows that some countries would benefit from emission trading, some would not be affected, whereas other (third) parties would experience no cost savings and less ecosystem protection. Table 13.3 compares for every scenario and region the annual costs and the percentage of ecosystem protection. The last column compares the results of scenario 2 (pre-trade) and scenario 3 (post-trade). + indicates that costs are less/ecosystem protection improves, 0 indicates no changes, - implies less ecosystem protection. For example, Austria, has +/- in the last column. This means that Austria will have less costs due to trading (these will drop from 514 million DM/year to 271-514 million DM, depending on the terms of trade) and the percentage of ecosystem protection decreases (from 88 to 82%). The latter is no surprise since Austria would increase emissions. The results of Table 13.3 can be summarized in Table 13.4, that summarizes the losers and the winners of trading. Table 13.3 and 4 allow drawing the following conclusions on the distribution of costs and benefits (ecosystem protection) before and after trading:

- total European costs will decline and ecosystem protection will decrease;
- some countries are not affected by trading since neither costs nor the percentage of ecosystem protection is altered: Albania, Denmark, Ireland, Italy, Poland, Portugal, Spain, Sweden, United Kingdom, St. Petersburg, the Baltic region, Ukraine and Russia;
- some countries have lower costs and more ecosystem protection (+/+ in the last column of Table 13.3): France and the Netherlands;
- other countries have lower costs and no change in ecosystem protection (+/0 in the last column of Table 13.3): Bulgaria, Hungary, Kola-Karelia, Belarus and Moldavia;
- for some countries costs are the same (they do not trade) but the environment improves: Luxembourg and Switzerland;

Table 13.3 Costs and environmental benefits

	COSTS (mio DM/yr)			ECOSYSTEM			Sum
Scenario	1	2	3	1	2	3	
Country	Cost minimum	Cost minimum with CRP	exchange rate trading	Cost mini-mum	Cost minimum with CRP	Exchange rate trading	Costs/ bene-fits
Albania	0	0	0	100	100	100	0/0
Austria	105	514	271/514	83	88	82	+/-
Belgium	974	816	816	76	63	76	0/-
Bulgaria	0	252	3/252	85	85	85	+/0
CSFR	1234	1009	1009	66	64	59	0/-
Denmark	656	656	656	93	93	93	0/0
Finland	788	864	864	95	95	94	0/-
France	442	317	-175/317	98	98	99	+/+
FRG, West	3259	3863	3863	59	64	61	0/-
FRG, East	1368	2247	1670/2247	59	64	61	0/-
Greece	1	213	6/213	99	100	99	+/-
Hungary	279	278/279	279	95	95	95	+/0
Ireland	99	109	109	96	96	96	0/0
Italy	1359	1349	1349	92	92	92	0/0
Luxembourg	10	7	7	87	80	87	0/+
Netherlands	856	719	142/719	43	39	43	+/+
Norway	232	295	212/295	78	77	76	+/-
Poland	2182	2182	2182	67	66	66	0/0
Portugal	68	68	68	100	100	100	0/0
Romania	841	790	790	91	100	91	0/-
Spain	514	516	516	98	98	98	0/0
Sweden	742	742	742	91	91	91	0/0
Switzerland	14	16	16	93	84	85	0/+
Turkey	0	0	0	99	100	99	0/-
UK	2457	1864	1864	91	91	91	0/0
Yugoslavia	795	780	780	93	93	82	0/-
Baltic Sea	101	101	101				
North Sea	238	238	238				
Atlantic Ocean	0	0	0				
Kola-Karelia	313	364	331/364	100	100	100	+/0
St. Petersburg	186	186	186	100	100	100	0/0
Baltic Region[2)]	642	642	642	99	99	99	0/0
Belarus	174	69	-254/69	100	100	100	+/0
Ukraine	1599	2065	2065	100	100	100	0/0
Moldavia	179	179	178/179	100	100	100	+/0
Russia[1)]	1256	1334	1334	100	100	100	0/0
EUROPE	23969	25629	24298	92	93	92	+/-

Notes: [1)] Russian Federation within the EMEP area, Kola-Karelia, St. Petersburg and Kaliningrad regions excluded

[2)] Includes Estonia, Latvia, Lithuania and Kaliningrad (part of Russian Federation).

Based on the IIASA RAINS model a routine has been developed to simulate bilateral, sequential emission trading for the specific situation of reducing SO_2 emissions in Europe. For a set of example target loads for the deposition, based on closing the gap between the sulfur deposition in 1990 and the 5 percentile critical loads by 60 percent, calculations identified the cost-minimal allocations of emission reductions as well as the cost minimum, given that countries at least carry out their Current Reduction Plans. Taking the latter as (slightly non-optimal) starting point, exchange rate emission trading achieved cost savings but could not attain the cost minimum. However, these cost savings are attained at the expense of an increase in emissions, less ecosystems protection and the violation of target loads. Moreover, although some countries have less costs and more ecosystem protection after trading, other countries have no cost savings and their level of ecosystem protection even decreases. In view of the above conclusions, it seems necessary to examine those possibilities for trading of emission reduction commitments between countries that ensure that third parties not involved in a trade have the same levels of deposition and ecosystem protection before as after trading.

REFERENCES

Alcamo, J., L. Hordijk and R. Shaw (eds.) (1990) *The RAINS model of acidification. Kluwer Academic Publishers*. Dordrecht/Boston/London.

Amann, M. (1992) *Zur effizienten multinationalen Allokation von Emissionsminderungsmassnahmen zur Verringerung der sauren Deposition - Anwendungsbeispiel Österreich* -. (Dissertation). Karlsruhe (Germany): University of Karlsruhe.

Amann, M., G. Klaassen and W. Schöpp (1993) *Closing the gap between the 1990 deposition and the critical sulfur deposition values*. Background paper prepared for the UN/ECE Task Force on Integrated Assessment Modelling, June 7-8, 1993, London. Laxenburg, Austria: International Institute for Applied Systems Analysis.

Atkinson, S. and T.H. Tietenberg (1991) Market failure in incentive based regulation: the case of emissions trading. *Journal of Environmental Economics and Management*, 21: 17-31.

Batterman, S.A. and M. Amann (1991) *Targeted acid rain strategies including uncertainty*. Journal of Environmental Management, 32; 57-72.

Borowski, A.R. and H.E. Ellis (1987) *Summary of the final federal emissions trading policy statement*. Journal of the Air Pollution Control Association, 37 (7):798-800.

Dwyer (1991) *California's tradeable emissions policy and its application to the control of greenhouse gases*. Paper presented at the OECD workshop "Tradeable permits for reducing greenhouse gases". June 26-27, 1991, Paris: OECD.

Ekelund, R.B. and R. Hébert (1983) *A history of economic theory and method*, Singapore: McGraw-Hill.

Iversen, T., N.E. Halvorsen, S. Mylona, and H. Sandnes (1991) *Calculated budgets for airborne acidifying components in Europe, 1985, 1987, 1988, 1989 and 1990*. Oslo: The Norwegian Meteorological Institute.

Klaassen, G. and M. Amann (1992) *Trading of emission reduction commitments for sulfur dioxide in Europe*. SR-92-03. Laxenburg, Austria: International Institute for Applied Systems

Kremeniouk, R. (1991) (ed.) *International negotiation*. San Francisco: Jossey-Bass.

Opschoor, J.B. and H.B. Vos (1989) *Economic instruments for environmental protection*. Paris: OECD.

Sandnes, H. and H. Styve (1992) *Calculated Budgets for airborne acidifying components in Europe, 1985, 1987, 1988, 1989, 1990 and 1991*. EMEP/MSC-W Report 1/92. Oslo, Norway: Meteorological Synthesizing Center-West, The Norwegian Meteorological Institute.

Tietenberg, T.H. (1985) *Emission trading: an exercise in reforming pollution policy*. Washington, D.C.: Resources for the Future, Washington.

Tietenberg, T.H. (1990) Economic instruments for environmental protection. *Oxford Review of Economic Policy*, 6 (1):17-33.

Chapter 14 TRADEABLE SO2 PERMITS: GUIDED BILATERAL TRADE IN EUROPE

Sonja Kruitwagen, Eligius Hendrix, Ekko van Ierland[1]

ABSTRACT

In this chapter a system of 'guided permit trading' is developed for pollutants that are non-uniformly mixed, taking into account that permit trading is a bilateral and sequential process. It is shown that in theory this system is able to achieve the cost effective allocation of emissions, required to reach given deposition targets. For this to take place, the number of permits traded has to be controlled by a trade institution. A simulation of the system of guided bilateral trading of SO_2 permits between European countries indicates that some non-profitable trade transactions should take place. This prevents the cost effective emission allocation being fully achieved. However, the calculations show that guided bilateral permit trading generates substantial cost savings and results in better environmental quality than reducing emissions by a fixed percentage across the board.

14.1. INTRODUCTION

Tradeable discharge permits that are among the economic instruments can help achieve environmental targets cost effectively. Their use to reduce emissions has been extensively discussed (e.g. Montgomery, 1972; Atkinson and Tietenberg, 1982; Krupnick et al., 1983; McGartland and Oates, 1985; Tietenberg, 1985). However, for non-uniformly mixed pollutants like sulphur dioxide (SO_2), there is no satisfactory permit trading system that can achieve the cost effective allocation of emission abatement, taking into account that permit trading is actually a bilateral sequential process (cf. Atkinson and Tietenberg, 1991). Given this nature of the trading process, in this chapter we will develop a permit system for abating sulphur dioxide emissions cost effectively to achieve deposition targets.

Sulphur dioxide, one of the main precursors of acid rain, causes substantial damage to sensitive ecosystems. After being emitted it is transported over long distances in the atmosphere. Due to this long distance transport, acid rain is a

[1] The authors are grateful to Prof. H. Folmer and Prof. L. Hordijk for their contribution in preparing this chapter.

transboundary environmental problem. Since SO_2 is a non-uniformly dispersed pollutant, its deposition depends not only on how much is emitted but also on the location of emitting countries. Accordingly, cost effective emission reduction is defined as the allocated emission which meets *deposition targets* for minimum abatement costs.

In order to take emission location into account in a system of tradeable discharge permits, the trade has to be constrained in some way. Atkinson and Tietenberg (1991) have outlined the need to regard permit trading as a bilateral and sequential process. They simulated bilateral permit trading by applying the 'non degradation' constraint, requiring every trade transaction to meet the deposition targets. However, this is a rather restrictive condition. If sources affect different receptors, a source selling permits might not be able to compensate at all receptors for increasing deposition caused by the increase in emissions from the source buying permits. In their simulations of Atkinson and Tietenberg found that the result of bilateral trade does not coincide with the cost-effective solution. It is therefore necessary to find other rules for trade, in order to be able to achieve the cost effective allocation. The alternative we develop in this chapter is called 'guided bilateral trade'. This system guides permit trading towards the cost effective allocation of emission abatement.

The central principle in the concept of guided bilateral trade is that not every bilateral trade transaction has to meet the deposition targets, because a subsequent bilateral trade transaction might compensate for the deposition exceedance at a receptor brought about by a previous trade transaction. Only after all bilateral trade transactions have been completed does deposition have to meet the targets. To ensure that deposition targets are met, the number[2] of permits traded in a bilateral trade transaction has to be controlled by a trade institution.

Below we apply the concept of guided bilateral trade is applied to SO_2 permit trading between European countries. The central question raised is: "can guided bilateral trade contribute to reaching a European cost effective allocation of emission abatement?"

After discussing the concept of guided bilateral trade (section 14.2), guided bilateral SO_2 permit trading between European countries is simulated (section 14.3).

14.2 GUIDED BILATERAL PERMIT TRADING

Guided bilateral trade aims at reaching the cost effective abatement allocation, taking deposition targets into account. As outlined in the introduction, permit trading has to

[2] Each permit will give the holder the right to a standard amount of emission (e.g. 1 kton SO_2).

be constrained to take account of these deposition targets. To control the trade transactions a trade institution has to be set up to limit the number of emission permits that countries are allowed to buy or sell.

As a first step in trading, countries have to receive a number of emission permits. Since permit trading does not affect the total number of permits, this implies to achieve the cost effective allocation of emissions, the sum of all permits over all countries has to correspond to the total emission in the cost effective emission allocation. According to the guided bilateral trade system, given an initial allocation of emission permits, the trade institution defines a 'trade vector' indicating how many permits countries are allowed to buy or sell. Numerically, the trade vector is the difference between the cost effective allocation and the initial permit allocation.

Figure 14.1 illustrates the concept of guided bilateral trade graphically. In this figure only two emitting countries are considered. The emissions of country 1 (e_1) are measured along the x-axis and those of country 2 (e_2) are measured along the y-axis. The shaded area represents the emission combinations of country 1 and country 2 that meet the deposition target (d^*) at all receptors. The abatement costs are represented by the iso-cost curves C1, C2 and C3. Since the cost of abatement increases concomitantly with a decrease in emissions, C1 represents a higher cost than C2 and C3 represents the least cost.

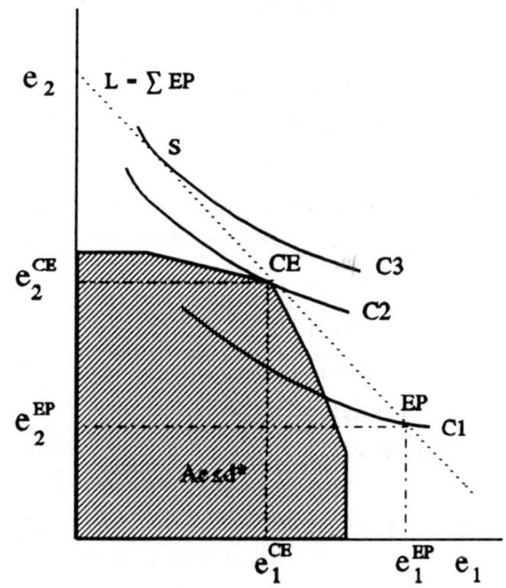

Figure 14.1 Bilateral Trading in Permits.

In figure 14.1 the cost effective emission allocation is represented by point CE, since this allocation meets the deposition target at all receptors at least possible cost. The dotted line L = ΣEP represents emission combinations of country 1 and country 2 that correspond with the total emission at CE. In order to reach allocation CE by permit trading, the initial permit allocation has to be somewhere on the dotted line L = ΣEP since permit trading keeps the total level of emissions constant.

Next, in figure 14.1 it is assumed that the initial allocation is given, for example at EP. Note that the initial permit allocation EP does not meet the deposition targets. Permit trading is necessary to meet these targets ultimately. Given this initial permit allocation and the cost effective allocation, the trade vector can be derived. In this graphical example the trade vector equals:[3]

$$\begin{bmatrix} (e_1^{CE} - e_1^{EP}) \\ (e_2^{CE} - e_2^{EP}) \end{bmatrix}$$

According to this vector country 1 is allowed to sell - $(e_1^{CE} - e_1^{EP})$ permits and country 2 is allowed to buy $(e_2^{CE} - e_2^{EP})$ permits. Since the initial permit allocation corresponds to a higher cost (C1) than the cost effective allocation (C2), it is profitable for both countries to trade. Permit trading corresponding with the trade vector results in the cost effective allocation of abatement.

Figure 14.1 also illustrates the allocation that would result if permit trading is not guided. Generally, countries will trade permits until the least abatement cost is reached. In this case, countries would trade permits up to point S, since the emission allocation in point S results in the least abatement costs. However, allocation S violates the deposition targets. This illustrates that permit trading has to be constrained to reach deposition targets in the case of non-uniformly mixed pollutants .

Guided bilateral trade only guarantees that deposition targets are met at lowest abatement costs if all the bilateral trade transactions allowed have taken place. Obviously, a bilateral trade transaction will only take place if two trading countries benefit from the trade, that is, if the total abatement costs decrease. This raises the question of whether all guided bilateral trades are profitable; the answer is not automatically an affirmative.

Since we are considering SO_2 emissions at a European level, the cost effective abatement allocation minimizes the total European abatement costs. But this does not mean that each individual country is confronted with minimum abatement costs. Although Europe as a whole is better off, it might not be in every country's interest to participate in a full cooperative agreement that aims at cost effective abatement, because an individual country might be disadvantaged by full cooperation. Correspondingly, there might be a discrepancy between what is rational for a certain

[3] $(e_1^{CE} - e_1^{EP})$ is a negative number whereas $(e_2^{CE} - e_2^{EP})$ is a positive number

country and what is rational for European as a whole. Two situations can be distinguished:

1. a situation in which every bilateral trade transaction is profitable for each trading partner and contributes to the cost-effective solution, i.e all sellers might trade with a buyer at a profit;
2. a situation in which a bilateral trade transaction is necessary to reach the cost effective abatement allocation, but which is not profitable for some individual countries.

Whether bilateral trade transactions actually are profitable depends upon the quantitative relationship between the abatement costs, the country-receptor matrix and the initial permit allocation. The empirical research described below casts light on the actual process of guided bilateral permit trading.

14.3 SIMULATION OF GUIDED BILATERAL TRADE

14.3.1 Method

The simulation of guided bilateral trade described below is intended to answer the question of how guided bilateral trading in SO_2 permits between European countries results in profitable trade transactions. The four steps that have to be taken to perform this simulation are represented schematically in Figure 14.2.

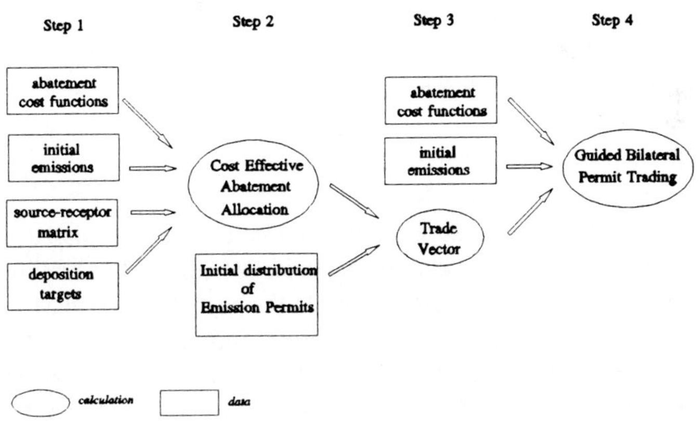

Figure 14.2 Schematic overview of steps to simulate guided bilateral permit trading.

First a cost effective emission abatement allocation is calculated, given the national abatement cost functions, the emissions, the source-receptor matrix and the deposition targets . The cost effective emission allocation that does not violate deposition targets follows from the optimization problem in which deposition targets (d_i^*) at every receptor i are met at minimum aggregate costs of abatement (ΣC_j) for the n countries. The deposition at a receptor depends upon the emission from the countries that affects that receptor, multiplied by the corresponding transfer coefficients (a_{ij}). A transfer coefficient represents the proportion of the emission of country j deposited at receptor i. The degree of abatement (r_j) depends upon the difference between initial emissions (E_j) and actual emissions (e_j). Mathematically the problem for calculating the cost effective solution is defined as follows:

$$Min \sum_{j=1}^{n} C_j(r_j) \tag{14.1}$$

$$s.t. \sum_{j=1}^{n} a_{ij}\, e_j \leq d_i^*, \quad for\ every\ i \tag{14.2}$$

$$r_j = E_j - e_j\,, \quad for\ every\ j \tag{14.3}$$

$$r_j \geq 0,\ r_j \leq E_j \tag{14.4}$$

The second step is to define an initial distribution of emission permits. To reach the cost effective abatement allocation by guided bilateral permit trading, the total emissions corresponding to the number of initially distributed emission permits must equal the total emissions in the cost effective abatement allocation. Since it is assumed that permits are traded on a one-to-one basis, which implies that the total emissions are fixed, this requirement is essential. The initial distribution can be based on several principles. One such principle could be equal percentage reduction as compared to emission levels in a base year.

Third, the trade vector is derived. In section 14.2 we stipulated that the trade vector indicates the number of permits countries are allowed to buy or sell. Numerically, the trade vector is the difference between the cost effective abatement allocation (CE in figure 14.1) and the initially distributed emission permit allocation (EP in figure 14.1). A positive number in the trade vector is the maximum number of emission permits a country is allowed to buy whereas a negative number in the trade vector is the maximum number of emission permits a country is allowed to sell. Obviously, the initial distribution of emission permits largely influences the trade vector.

Finally, having obtained the trade vector, the fourth step is to simulate guided bilateral trade for SO_2 emission permits, assuming that a trade coordinating institution restricts the amounts that countries are allowed to buy and sell in accordance with the maximum quantities indicated by the trade vector. To do this we developed an optimization model in OMP,[4] to simulate guided bilateral trade to clarify the sequence and the volume of profitable trade transactions between European countries. It is assumed that countries generating the largest cost savings trade first. The optimization model used for simulation is specified by equations (14.5) - (14.14)

$$Min \sum_S \sum_t C_{S,t}(r_{S,t}) + \sum_B \sum_t C_{B,t}(r_{B,t}) \tag{14.5}$$

$$s.t.\ e_{S,t} = e_{S,t-1} - \sum_B G_{S,B,t}\ ,\ \ for\ every\ S\ and\ t \tag{14.6}$$

$$e_{B,t} = e_{B,t-1} + \sum_S G_{S,B,t}\ ,\ \ for\ every\ B\ and\ t \tag{14.7}$$

$$\sum_t G_{S,B,t} = S_S\ ,\ \ for\ every\ S \tag{14.8}$$

$$\sum_t G_{S,B,t} = B_B\ ,\ \ for\ every\ B \tag{14.9}$$

$$r_{S,t} = E_S - e_{S,t}\ ,\ \ for\ every\ S\ and\ t \tag{14.10}$$

$$r_{B,t} = E_B - e_{B,t}\ ,\ \ for\ every\ B\ and\ t \tag{14.11}$$

$$G_{S,B,t} \leq Min\ \{S_S;\ B_B\}\ H_{S,B,t}\ ,\ \ for\ every\ S,\ B,\ t \tag{14.12}$$

$$\sum_S \sum_B H_{S,B,t} \leq 1,\ \ \ for\ every\ t \tag{14.13}$$

$$e_{S,t}\ ,\ e_{B,t}\ ,\ r_{S,t}\ ,\ r_{B,t}\ ,\ G_{S,B,t} \geq 0 \tag{14.14}$$

where:

$C_{S,t}(r_{S,t})$	= abatement cost function of a seller S
$C_{B,t}(r_{B,t})$	= abatement cost function of a buyer B
Variables:	
$r_{S,t}$	= emission reduction by a seller S at period t
$r_{B,t}$	= emission reduction by a buyer B at period t

[4] OMP is an LP modelling package.

$e_{S,t}$	= emissions of a seller S at period t
$e_{B,t}$	= emissions of a buyer B at period t
$G_{S,B,t}$	= traded permits between a seller S and a buyer B at period t
$H_{S,B,t}$	= binary variable for every buyer-seller combination at period t
Data:	
E_S	= initial emissions of country S
E_B	= initial emissions of country B
S_S	= maximum permits a seller S is allowed to sell
B_B	= maximum permits a buyer B is allowed to buy

Our aim is to minimize total abatement costs. By minimizing these costs over all periods t, the trade between countries S and B that generates the largest cost saving will take place first. As a result of a trade transaction, the emission of a country selling permits decreases, whereas the emission of a country buying permits increases. The emissions of countries not involved in trade remain constant. The number of permits sold equals the number of permits bought in every trade transaction. The total amount of permits countries are allowed to buy or sell is constrained according to the trade vector. The degree of emission reduction depends upon the initial emissions and the actual emission. To clarify the sequence of bilateral trade transactions, equations (14.12) and (14.13) stipulate that only one transaction takes place in a period t. Note that the parameter t is only introduced for a mathematical reason, to retrieve the sequence of trade transactions and does not imply that each period is the same length.

14.3.2 Data

Several assumptions have to be made when simulating guided bilateral trade between European countries. In this section the assumptions made to calculate the cost effective abatement allocation are discussed first. Second, the rule for initially distributing emission permits is discussed. For an overview of all data used for simulation, see Figure 14.2.

Cost effective abatement

We calculated the cost effective abatement allocation using the Regional Acidification INformation and Simulation (RAINS) model, developed at the International Institute of Applied Systems Analysis (Alcamo et al., 1990). The model links energy use with data on emission coefficients, costs and effects of abatement strategies, long range atmospheric transport and sensitivity of ecosystems. The model simulates the flow of acidifying pollutants from 38 countries/regions in Europe to land-based receptor sites with a regular size of 150 x 150 km.

The initial emissions are based on the Official Energy Pathway which is based on national projections of energy trends and description of energy policies up to the year 2000, as submitted by individual governments to international organisations (Amann, 1990). Given this energy scenario for the year 2000, RAINS calculates the corresponding emissions, assuming that no emission reduction takes place.

The atmospheric transport of acidifying emissions is based on the atmospheric model developed within the co-operative programme to monitor and evaluate the long range transmission of air pollutants in Europe (EMEP) (Iversen et al., 1991).

The abatement cost functions in the RAINS model are technology based and depend upon the energy scenario chosen. The functions are piece-wise linear. Each step in the abatement cost function implies another abatement technique. The step size differs for every technique, because the ranges of application of techniques differ. Moreover, the step sizes and the number of steps vary for every country, since the average installation capacity, the fuel type used etc. differs for every country.

The deposition target was chosen in accordance with the current negotiations concerning the 1994 SO_2 Protocol. The deposition target aimed at is based on a pan-European reduction of the difference (the 'gap') between the actual deposition level and the 5 percentile critical loads for sulphur. The 5 percentile critical load level protects 95% of the environment in a grid cell of 150 x 150 km against the damaging effect of sulphur deposition (Downing et al., 1993).

A 'gap closure' scenario would bring all European countries closer to fully achieving of the 5 percentile critical loads. An efficient 'gap closure' scenario takes into account regional differences in the environmental sensitivities, the dispersion of air pollutants, and differences between countries' economic and energy situation and some notion of economic efforts required to achieve these targets (Amann et al., 1992). Currently a gap-closure of 60% has been chosen as the policy target (Amann and Schöpp, 1993). In this scenario the difference between the 1990 deposition level and the 5 percentile critical load will be reduced by 60%. For grids where the 1990 deposition already meets the 5 percentile critical load, the deposition target equals these critical loads. This deposition target is applied to 483 receptors of 150 x 150 km each.

Apart from the necessary abatement efforts required to reach the specified deposition targets, some countries have already committed themselves to acieve a certain emission reduction by the year 2000. The calculated cost effective emission allocation takes into account these so-called Current Reduction Plans (CRP). Therefore, in the optimization problem a constraint is added that stipulates that the minimum emission abatement of a country equals the Current Reduction Plan. As a result the cost effective allocation does not represent the least possible costs, but the least possible costs given the Current Reduction Plans. Further it is assumed that ship emissions can be reduced. At the same time, it is assumed that ships will not participate in permit trading.[5]

The cost effective emission allocation taking into account the CRP for the year 2000 is represented in Table 14.1. The first column represents the emissions in the cost effective allocation. The second column shows the abatement required to reach these emissions and the third column shows the corresponding percentage of reduction compared to 1990. The fourth column shows the abatement costs.

Initial Distribution of Permits

As a first step in trading, countries should receive a number of emission permits. Since guided bilateral trade aims at reaching the cost effective emission allocation, the sum of all permits over all countries has to correspond to the total European emission in the cost effective allocation. Therefore the total number of emission permits has to allow emissions of 23296 kt SO_2.[6]

The distribution of the total number of permits between the countries is based upon historical emissions in 1990. The initial distribution of permits is equal to the emission for each country that would emerge if all countries reduced emissions by a uniform percentage. The reduction equivalent to a European emission level of 23296 kt SO_2 equals 42%.[7,8] The emission allocation resulting from this initial distribution of emission permits does not meet the deposition targets

[5] Ship emissions are the emissions in the Baltic Sea, the North Sea and the Atlantic Ocean.

[6] It is further assumed that 1 permit allows the emission of 1 kton SO_2.

[7] The political feasibility of this initial distribution is not taken into consideration, although this is a very important issue.

[8] Unfortunately the specified initial distribution requires Switzerland and Norway to exceed their maximum feasible abatement. Since this is not possible, the initial distribution has been adapted so that both countries can satisfy their abatement requirements. Therefore Switzerland and Norway are allocated emission permits for respectively 7 and 2 kton SO_2. Since the total number of emission permits cannot be changed, the permits initially distributed to the former GDR are reduced by 9. The former GDR was chosen because this is the only country that is able to reduce 9 kton SO_2 at zero costs because the number of initially distributed emission permits for the former GDR exceeds the amount of unabated emissions in the year 2000. In table 14.2 the adapted initial distribution and trade vector are represented.

Table 14.1 Overview of the cost effective emission allocation taking into account CRP (emissions and abatement in kton SO_2 ; abatement costs in Mio DM.).

Country/Region	Cost Effective Emissions	Emission Abatement	% Reduction (1990)	Abatement Costs
Albania	132	5	6	0
Austria	78	274	20	537
Belgium	200	521	52	712
Bulgaria	520	295	59	251
Czechoslovakia	1792	554	27	423
Denmark	51	318	72	790
Finland	116	487	55	916
France	675	634	46	461
Germany-West	520	1751	45	3914
Germany-East	230	2133	95	2437
Greece	595	310	20	212
Hungary	521	554	48	290
Ireland	53	181	68	234
Italy	1016	1984	53	1384
Luxembourg	10	5	38	27
Netherlands	106	318	49	716
Norway	34	110	37	379
Poland	1717	2021	47	1782
Portugal	294	95	-44	69
Romania	1036	1556	42	844
Spain	1486	1307	36	496
Sweden	88	324	48	979
Switzerland	60	14	3	16
Turkey	2887	0	-98	0
United Kingdom	949	2384	75	2533
former Yugoslavia	983	593	34	819
Baltic Sea	18	55	75	101
North Sea	42	131	76	238
Atlantic Ocean	317	0	0	0
Kola-Karelia	157	571	78	463
St. Petersburg	239	95	42	61
Baltic Region	266	363	59	317
Belarus	456	111	23	72
Ukraine	1696	3370	39	1988
Moldavia	231	147	26	89
Remaining Soviet Union	3725	2622	-11	1320
Europe	23296	26193	42	25870

Table 14.2 Calculation of the Trade Vector (kton SO_2).

Country	Emissions 1990	Initial Distribution of Permits	Cost Effective Emissions	Trade Vector (+: buyer -: seller)
Albania	140	81	132	51
Austria	98	57	78	21
Belgium	414	240	200	-40
Bulgaria	1266	735	520	-215
Czechoslovakia	2443	1419	1792	373
Denmark	184	107	51	-56
Finland	260	151	116	-35
France	1260	732	675	-57
Germany-West	940	546	520	-26
Germany-East	4800	2779	230	-2549
Greece	740	430	595	165
Hungary	1010	587	521	-66
Ireland	168	98	53	-45
Italy	2180	1266	1016	-250
Luxembourg	16	9	10	1
Netherlands	207	120	106	-14
Norway	54	33	34	1
Poland	3210	1865	1717	-148
Portugal	204	118	294	176
Romania	1800	1046	1036	-10
Spain	2316	1345	1486	141
Sweden	169	98	88	-10
Switzerland	62	43	60	17
Turkey	1459	847	2887	2039
United Kingdom	3774	2192	949	-1243
former Yugoslavia	1480	860	983	123
Baltic Sea	73	18	18	-
North Sea	173	42	42	-
Atlantic Ocean	316	317	317	-
Kola-Karelia	701	407	157	-250
St. Petersburg	409	238	239	1
Baltic Region	653	379	266	-113
Belarus	596	346	456	110
Ukraine	2782	1616	1696	80
Moldavia	313	182	231	49
Remaining Soviet Union	3350	1946	3725	1779
Europe	40020	23296	23296	0

Next, the trade vector follows from the difference between the initial distribution of permits and the cost effective emission allocation. The initial distribution of emission permits and the trade vector are presented in Table 14.2. For clarification the 1990 emissions and the cost effective emission allocation are also presented.

14.3.3 Simulation Results

Now we have discussed the pre-trade steps, we can start simulating guided bilateral permit trading to find out which trade transactions will actually occur. Countries are allowed to trade the initially distributed permits, while the maximum number of permits countries are allowed to buy or sell is defined by the trade vector. The sequence of profitable bilateral permit trading, assuming that the trade transaction with the greatest profit will take place first is represented in Table 13.3. The first column shows the trade sequence. The next two columns show the seller and the buyer in a trade transaction. The fourth column shows the number of permits traded. The fifth column shows the profit generated by the trade transaction. This profit depends both on the quantity of permits traded and on the difference in marginal abatement costs.

The last column in Table 14.3 shows the permit price assuming that both the seller and the buyer receive an equal profit share by permit trading. The benefit of permit trading by the buyer equals the decrease in abatement costs (ΔC_b) minus the permit price (P) that has to be paid, multiplied by the number of permits bought (Q). The benefit of permit trading by the seller equals the permit price (P) that is received, multiplied by the number of permits (Q) minus the increase in abatement costs (ΔC_s). The permit price can be calculated both from a seller's perspective and from a buyer's perspective. Here we illustrate the calculation from a seller's perspective. If both the seller and the buyer receive half the profit generated by trading, the following equation holds:

$$P * Q - \Delta C_s = 0.5 * Profit \tag{14.15}$$

The total profit equals the decrease in the buyer's abatement costs minus the increase in the seller's abatement costs:

$$Profit = \Delta C_b - \Delta C_s \tag{14.16}$$

Substitution of (14.16) in (14.15) and rewriting gives:

$$P = (\Delta C_b + \Delta C_s) \; / \; 2\,Q \tag{14.17}$$

Equation (14.17) defines that the permit price (P) equals the decrease in the buyer's abatement costs plus the increase in the seller's abatement costs, divided by twice the traded quantity of permits (Q). Analogously, (14.17) can be derived from the buyer's perspective.

Table 14.3 Sequence of Profitable Trade Transactions

No.	Seller	Buyer	Traded Amount (kton SO_2)	Profit (Mio.DM)	Permit Price (Mio.DM/ kton SO_2)
1	Germany-East	Turkey	1077	2700	1.63
2	Germany-East	Remaining Soviet Union	965	176	0.80
3	Hungary	Austria	21	155	4.17
4	Kola-Karelia	Switzerland	17	124	4.35
5	Germany-East	Portugal	176	122	1.06
6	United Kingdom	Yugoslavia	123	119	1.22
7	United Kingdom	Belarus	110	47	0.94
8	Hungary	Luxembourg	1	18	10
9	Kola-Karelia	Greece	124	17	0.90
10	Hungary	Norway	1	10	6
11	United Kingdom	Greece	41	9	0.84
12	Germany-East	Albania	26	4	0.81
13	United Kingdom	Albania	19	3	0.82
14	Hungary	Czechoslovakia	10	2	0.80
15	Italy	Czechoslovakia	12	1	0.79
16	United Kingdom	St.Petersburg	1	0.15	1.08

The simulation results indicate that the first trade transaction is between former Germany-East and Turkey, generating a profit of 2.7 billion DM. Since former Germany-East is able to sell a number of permits at zero costs, the profit from the first transaction is observably large. According to the cost effective emission allocation Turkey is allowed to increase emissions substantially in the year 2000, since its emissions only affect minor environmentally sensitive areas. As the initial allocation of emission permits is based on the emission level of 1990, the trade vector allowes Turkey to purchase a large number of permits.

The next trade transaction between former Germany-East and the Remaining Soviet Union, is about the same size in terms of amount of emission, but the profit is considerably smaller.

The calculations show that countries do not always trade the maximum number of permits allowed by the trade vector indicated in table 14.2. This can be explained by the intersection of marginal abatement cost curves. As long as marginal abatement cost curves do not intersect, countries will trade their maximum allowed number of permits, otherwise it is most profitable to trade until the point of intersection.

The variation in permit prices is considerable. It generally holds that the greater the profit, the higher the permit price, and the more permits are traded, the lower the permit price. Since the trade sequence is based solely on the profitability of a trade transaction, the prices do not follow a regular pattern.

After all profitable transactions have taken place, the cost effective emission allocation has still not been reached. The total profit generated by permit trading 3507.15 Mio. DM. However, the deposition targets are exceeded at 46 receptors (see Figure 14.3). A disproportionately large number of these receptors are located in Norway and in the Netherlands.

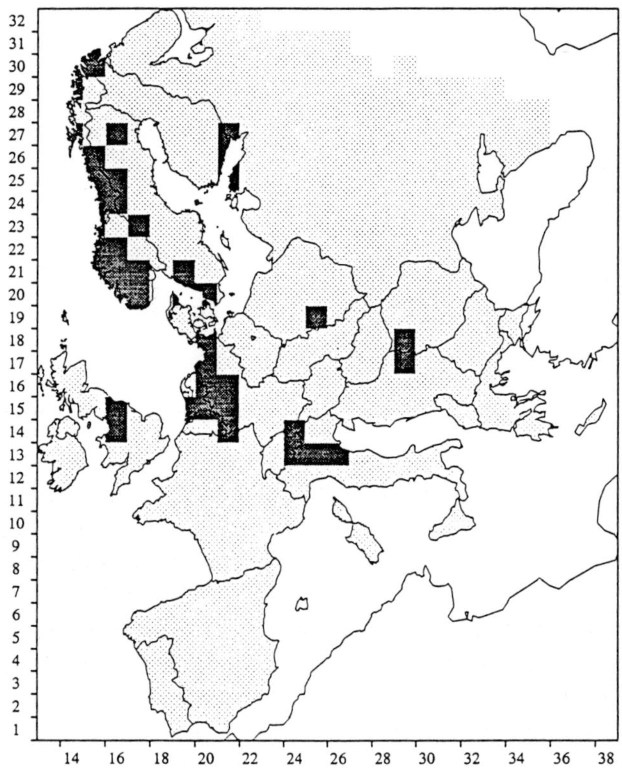

Figure 14.3 Location of receptors in Europe exceeding the deposition target in the Guided Trade Allocation.

The allocation reached by profitable guided permit trading can be compared with the pre-trade allocation in terms of environmental quality. In the pre-trade allocation the deposition target is exceeded at 88 receptors (of a total of 483 receptors), compared to 46 after profitable guided trade.

An additional point of interest concerns the relative exceedance of deposition targets. In general it can be stated that the smaller the relative exceedance the less environmental damage occurs. Figure 14.4 illustrates the relative exceedance of deposition targets in the pre-trade allocation and the guided trade allocation.

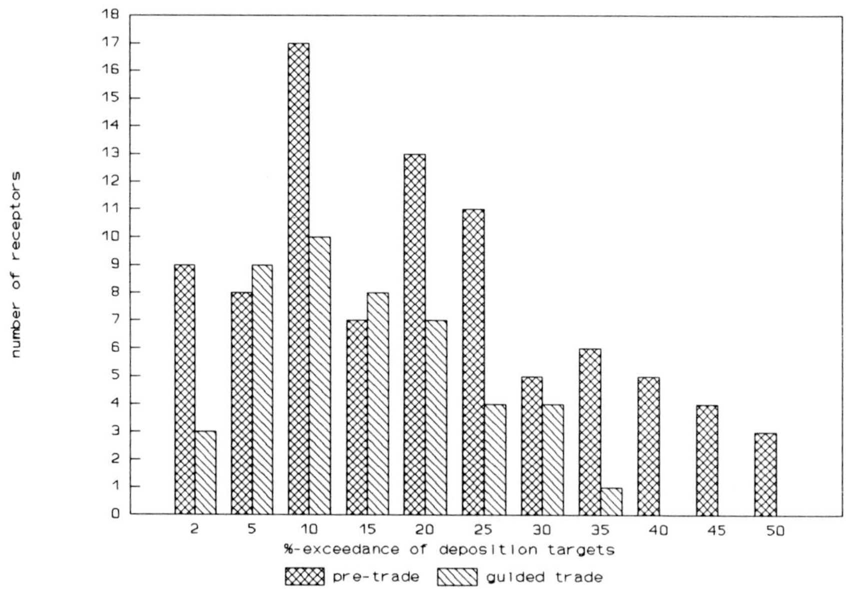

Figure 14.4 Relative exceedance of deposition targets for the pre-trade and the guided trade allocations.

From Figure 14.4 it can be concluded that the guided trade allocation clearly shows an environmental improvement compared to the pre-trade allocation. The pre-trade allocation results in relative exceedances of 50% at 3 receptors. The guided trade allocation only results in relative exceedances with a maximum of 35%.

14.4. CONCLUSION

In this chapter a system of guided permit trading for non uniformly mixed pollutants has been developed, taking into account that permit trading is a bilateral and sequential process. Theoretically, this system of guided bilateral trade has the potential of achieving the cost effective allocation of emissions. This property follows from the restrictions specified to permit trading. We have demonstrated that guided bilateral permit trading steers the trade process towards the cost effective allocation of emission abatement.

Although guided bilateral permit trading potentially results in cost effective abatement, in practice, this allocation might not be achieved because non-profitable trade transactions are required if the cost effective solution is to be reached. Whether this actually occurs depends upon the combination of abatement costs, country-receptor matrix and initial permit distribution.

The results of our simulation of the application of guided bilateral trade to SO_2 permit trading between European countries indicate that the cost effective allocation of emission abatement will not be fully reached by profitable trade transactions. However, guided bilateral permit trading generates substantial cost savings and improves environmental quality.

REFERENCES

Alcamo, J., R. Shaw and L. Hordijk (Eds.) (1990) *The RAINS Model of Acidification. Science and Strategies in Europe.* Kluwer Academic Publishers, Dordrecht.

Amann, M., (1990) *Energy Use, Emissions and Abatement Costs.* In: Alcamo, J., R. Shaw and L. Hordijk (Eds.) (1990) *The RAINS Model of Acidification. Science and Strategies in Europe.* Kluwer Academic Publishers, Dordrecht.

Amann, M, L. Hordijk, G. Klaassen, W. Schöpp (1992) *Reduction of SO_2 Emissions in Europe: A combination of the BAT and the Critical Loads Concept*, Report prepared for the 8th Session of the UN/ECE Working Group on Strategies, IIASA, Laxenburg, Austria.

Amann, M. and W. Schöpp (1993) *Reducing excess sulphur deposition in Europe by 60 percent. Further analysis of an international reduction scheme suggested as a basis for a new sulphur protocol in Europe.* Background paper prepared for the UN/ECE Working Group on Abatement Strategies August 30 - September 2, 1993, IIASA, Austria.

Atkinson, S. and T.H. Tietenberg (1982) The empirical Properties of Two Classes of Designs for Transferable Discharge Permit Markets. *Journal of Environmental Economics and Management, 9, pp. 101-121.*

Atkinson, S. and T.H. Tietenberg (1991) Market failure in incentive based regulation: the case of emission trading. *Journal of Environmental Economics and Management, 21, pp. 17-31.*

Downing, R.J., J-P Hettelingh and P.A.M. de Smet (1993) *Calculation and Mapping of Critical Loads in Europe.* Status Report 1993, RIVM report No. 259101003, Coordination Centre for Effects, National Institute of Public Health and Environmental Protection, Bilthoven, The Netherlands.

Iversen, T., N.E. Halvorsen, S. Mylona, H. Sandnes (1991) *Calculated Budgets for airborne acidifying compounds in Europe, 1985, 1987, 1988, 1989 and 1990.* The Norwegian Meteorological Institute, Oslo.

Krupnick, A.J., W.E. Oates and E. de Verg (1983) On marketable air pollution permits: the case for a system of pollution offsets. *Journal of Environmental Economics and Management, 10, pp. 233-247.*

McGartland, A.M. and W.E. Oates (1985) Marketable permits for the prevention of environmental deterioration. *Journal of Environmental Economics and Management, 12, pp. 207- 228.*

Montgomery, W.D. (1972) Markets in licenses and efficient pollution control programs. *Journal of Economic Theory*,5, pp. 395-418.

Tietenberg, T.H. (1985) *Emissions Trading; an exercise in reforming pollution policy*, Resources for the Future, Washington D.C.

Chapter 15 ACID RAIN IN EUROPE: A DYNAMIC PERSPECTIVE ON THE USE OF ECONOMIC INCENTIVES

Karl-Göran Mäler[1]

ABSTRACT

In this chapter, strategic interaction among countries negotiating on reducing sulphur emissions, is analyzed in terms of a simulation model based on simple game theory. Furthermore, different institutional structures are illustrated by the use of this model.

15.1 INTRODUCTION

During the last couple of years, the problems connected with emissions of sulphur- and nitrogen oxides and the resulting damages on ecosystems have come more and more in the focus. The forest deaths in central Europe and Scandinavia have been quite important news items and rightly so, because if the European ecosystems will change so drastically that there will be a substantial reduction in timber supply, that will effect not only nature conservationists but also everyday man because of the economic consequences in a conventional sense. Thus, the problems of acid rains seem to be a very important issue for economists to analyze. Moreover, it offers a fascinating multitude of intellectual challenges, one being that the information on causes and effects is very uncertain, another being that it concerns the use of a common property resource in a very asymmetric way, a third being that it is about a game with incomplete information and with many players so that problems with incentives compatibility will be at the heart, and finally the parties involved are different nations with no agreed rules of the game. In addition to these challenges, there are of course a multitude of equally challenging empirical problems.

However, the single factor that really characterizes the problem is its importance. We will later use illustrative estimates of the damage from acid rains and in spite of the fact that these estimates are seriously biased downward, they indicate quite huge economic effects from a continuation of the status quo. It is imperative that we can find solutions to the problems of international cooperation in the environmental field in order to secure continued welfare for the European people. This chapter is based on the Acid Rain Game (Mäler, 1989), which analyzed the incentive structures in

[1] This chapter is based on Mäler (1989).

international cooperation on reducing sulphur emissions in Europe. However, possible institutional setups were not discussed in that paper. Therefore, we will focus on possible institutional mechanisms here. Furthermore, we will touch upon what will happen when we leave the static framework that characterized the earlier paper and instead study either a repeated game framework or a differential game framework. However, in order to make the presentation selfcontained, parts of the earlier paper have been included. Furthermore, the transport of sulphur in the European atmosphere is based on data from the beginning of the 1980's and no attempt has been made to revise the calculations in view of changes in the transport patterns. The cost of abatement functions are also the ones used in the original paper, although improved estimates are now available. It is, however, not very likely that the main findings of the paper would change if more recent data had been used. The reason for this is that these findings are based on the general transportation pattern of sulphur, which has not changed from the early transport matrices in early 1980's. The abatement cost functions seem to play only an auxiliary role for the findings.

15.2 THE BASIC MODEL

Each country is emitting sulphur oxides. The expected control cost function for country i will be denoted $C_i(E_i)$ where E_i denotes the emission in country i. This function is, of course, decreasing in E_i, as the cost of abatement increases with abatement. We will also assume C_i to be strictly convex (although in some simulations it will be linear or piecewise linear). Moreover, we will assume it to be sufficiently smooth.

The sulphur emitted will be transported by winds in the atmosphere and will also be transformed by chemical processes from sulphur oxides to sulphates. Ultimately, the sulphates will be removed from the atmosphere by direct "dry" deposition or by rains - "wet" deposition. That process can be described by a linear system centred around a transfer matrix. The transfer matrix is denoted by A. If E is the vector of emission levels in the European countries and if Q is the vector of sulphur deposition in the countries, the steady state model for the transport of sulphur is simply $Q = A \times E$. Thus, the element a_{ij} of the matrix A denotes the portion of emission of one tonne of sulphur that is emitted in country j that is deposited in country i.

The deposition of sulphur results in environmental damage. Let the deposition in country i be Q_i. Then we will denote the expected damage cost function (or simply damage function) by $D_i(Q_i)$. For each country we thus have an expected net control cost function and an expected damage cost function. We will assume that these functions as well as the EMEP matrix that relates the emissions from one country to deposition in all other countries are known by all countries.

The net control cost functions, the EMEP matrix and the damage functions define a non-cooperative game, in which the strategy for a country consists of choosing an emission level and the payoff is equal to minus the sum of the net control cost and the damage cost (except for an uninteresting constant).

15.3 NON-COOPERATIVE EQUILIBRIA AND THE FULL COOPERATIVE SOLUTION

The purpose of the simulations to be presented in the next section is to get a feeling for the gains from cooperation the European countries could expect. Ideally, one would therefore be interested in the core and related equilibrium concepts of the acid rain game. In this section the interest is focused on the "full cooperative solution". This was defined in the previous section as simply that vector E of emission levels that minimizes the total cost, i.e.

$$\min \sum_i (C_i(E_i) + D_i(Q_i))$$

$$\text{subject to } Q = A \times E$$

This solution concept implicitly requires transferable utility, i.e. that gains in one country can be transferred to other countries in order to achieve another distribution of gains and losses. As both cost and damage functions are measured in monetary units we thus assume that utility is linear in income and transferable between countries (which incidentally requires that the current exchange rates are equilibrium rates).

The full cooperative solution is a special case of the "Pareto-efficient" outcomes. A vector $(E,Q) = \{E_1, E_2, ..., E_n, Q_1, ..., Q_n\}$ is said to be "Pareto-efficient" if there does not exist another feasible vector (E',Q') such that the total cost (control- and damage cost) for each country at these alternative emission and deposition levels is less than or equal to the total cost at the (E,Q). If the feasible set is convex (which it is with our assumptions on the control- and damage cost functions and the linearity of the transport model), then all Pareto-efficient outcomes can be characterized as vectors (E,Q) that minimize

$$\sum_i \alpha_i \{C_i(E_i) + D_i(Q_i)\}$$

for a certain selection of $\alpha_1, ..., \alpha_n$, with $\alpha_i > 0$, $i = 1, ..., n$.

If (E′,Q′) is any allocation of emissions and depositions that is feasible (i.e. $Q' = A \times E'$), then P(E′,Q′) denotes the set of feasible allocations of emissions and depositions that pareto dominates (E′,Q′), i.e.

$$P(E',Q') = \{(E,Q);\ C_i(E_i) + D_i(Q_i) \leq C_i(E_i') + D_i(Q_i'),\ Q = A \times E\}$$

A subset of the Pareto efficient allocations is thus a subset of the boundary of P(E′,Q′). Note that for a given (E′,Q′), the full cooperative solution need not be a member of P(E′,Q′). If that is not the case when (E′,Q′) is the initial situation, then the full cooperative solution will not be obtained unless there are side payments among the countries.

We can define (at least) two different non-cooperative equilibrium concepts.

i) Dominant equilibrium. $E_k^*,\ Q_k^* = A \times E_k^*,\ k=1, ..., n$ is a dominant equilibrium if for all E, Q such that $Q = A \times E$ it is true that for all $k=1, ..., n$

$$C_k(E_k^*) + D_k(Q_k^*) < C_k(E_k) + D_k(Q_k)$$

Thus, a dominant equilibrium is characterized by a set of strategies E_k^* such that irrespective of what other countries do, it would not be beneficial for country k to change its strategy.

It is very easy to see that in general no dominant equilibrium will exist and we will give a proof in connection with the discussion of Nash equilibrium. However, under one special condition which is the basis for the simulations discussed later, a dominant equilibrium will exist. This condition is that the damage function is linear so that the marginal damage is constant. If that is the case, the cost minimization emission in a country is determined by the condition that marginal abatement cost equals the marginal damage times the proportion of the emission that will be deposited in the own country. As the marginal damage is independent of the emissions in other countries, it follows that the cost minimizing is independent of the emissions in other countries and thus constitutes a dominant strategy.

ii) Nash equilibrium. A Nash equilibrium is defined as a pair (E^*,Q^*) such that

$$C_k(E_k^*) + D_k(Q_k^*) \leq C_k(E_k) + D_k(Q_k'),\ k=1, ..., n$$

where Q′ is defined by

$$Q' = A \times E'$$
$$E_l' = E_l^*,\ l=1, ..., n,\ l \neq k(E',Q')$$

Thus a Nash equilibrium is characterized by the condition that if all other countries are emitting their Nash equilibrium quantities, then it is optimal for the remaining country to do that too.

It is possible to prove that with the empirical transport matrix, there exists a unique Nash equilibrium. Furthermore, this equilibrium will be an equilibrium in dominant strategies if the damage function is linear.

15.4 FULL COOPERATIVE SOLUTION WITH SIDEPAYMENTS (Fcs)

We will start by looking at the "full cooperative solution" when sidepayments between the different countries are possible. This means that the net benefits from some countries may turn out to be negative but these countries can be compensated by cash payments if the total European benefits are positive. The results are given in Table 15.1.

The calculations show that the total emissions in Europe would be reduced by about 40 percent in the full cooperative solution compared with the present situation (emissions 1984), that almost all countries would gain from the full cooperative solution and that a few countries - UK, Italy and Spain - would lose from participating in the cooperation. (The losses that Finland and Luxembourg would experience are negligible.) Spain's loss is also almost negligible, while Italy would experience a moderate loss and the UK a substantial loss. Obviously, UK would have no incentives to participate in organized cooperation to reduce the sulphur emissions in line with the full cooperation.

These results do not depend crucially on the assumed control cost functions. If we have assumed too high a value for the marginal control cost for a particular country, this would to a certain extent be compensated by too high a value for the marginal damage cost, because of the way we calibrate the damage cost function. In fact, sensitivity analysis yield the expected result that it is the EMEP matrix that is crucial as long as we are willing to make the assumption that the damage cost function is linear.

A few things should be noted about the figures in this table (some of which also apply to later tables). First a few countries are required to abate their emissions of sulphur up to the maximum amount. In this group are the UK, Czechoslovakia, east and West Germany and a few others. In view of the rather arbitrary upper limits on emission reductions that have been imposed on the solutions, it can be concluded that these countries would have to reduce their emissions further given a more realistic cost of abatement function.

Table 15.1 Net benefits from the full cooperative solution

	Emission control	Percentage reduction	Benefits
	1000 ton SO_2	%	Mill. D-Mark
ALB	11	45	24
AUS	31	21	324
BEL	112	36	191
BUL	215	43	-7
CZE	1219	75	152
DEN	130	86	119
FIN	25	14	-2
FRA	104	10	879
GDR	1040	80	11
FRG	1183	86	328
GRE	303	86	53
HUN	635	77	5
IRE	27	38	71
ITA	634	33	-83
NET	105	62	565
NOR	3	6	272
POL	560	27	599
POR	15	19	10
ROM	83	83	422
SPA	231	14	-29
SWE	6	4	606
SWI	10	23	192
TUR	188	39	68
USS	107	2	1505
UK	1494	81	-336
YUG	465	79	344
TOTAL	8936	39	6290

Finally, a few countries will end up with net losses. It would not be rational for any country to sign a binding agreement by which they would expect losses. Therefore, we can conclude that even binding agreements between countries on emission reductions could be made, we should not expect a full cooperative solution to result, unless the strategy space for each country is expanded to include sidepayments. Without sidepayments, we would not expect UK, Italy and Spain to agree on an emission control plan that would result in non-negligible losses.

15.5 A DYNAMIC GAME

So far, we have used a very simple model for describing the interactions between the European countries in the area of acidification. Let us now use that simple basic model in order to incorporate one aspect of the importance of time. The game we have been studying has been a one shot game. No country has felt the need to look ahead at the future consequences of a decision taken today, simply because there were no such consequences in the basic model.

There are at least two ways by which decisions today may have consequences for tomorrow. The first way is that a decision today may affect other countries' decisions tomorrow and therby affect tomorrow's welfare. The other way has to do with the environmental consequences of deposition. The basic model was built upon the assumption that annual damage is related to the annual deposition of sulphur. If, instead, we make the assumption that it is the cumulated depositions that determine the damage, the model becomes quite different. Unfortunately, it has not been possible to develop that model in this chapter.

We will start by looking at the incentives to cooperate when the basic game model is assumed to be valid, but that the same game is played repeatedly in the future. This opens up the possibility for countries to "punish" countries that do not participate in that effort. The punishment will take the form of increases in the emission in the cooperating countries in order to increase the environmental damage in the non-cooperating countries. This can obviously only be achieved at substantial costs to the punishing countries themselves and the basic issue is whether these costs are too high to motivate that kind of strategy.

The basic theorem in this field is the Folktheorem,[2] which roughly says that if there is a game with a minimax solution v^* (i.e., such that player j is guaranteed at least v_j^* in payoff) and if the discount rate is sufficiently small (it is assumed that players' payoffs in the repeated game is the present value of future payoffs in the one-shot game), then any individual outcome v (i.e., with $v_j > 0$) is a Nash equilibrium in the infinitely repeated game.

[2] See Friedman (1986) and Fudenberg and Maskin (1986) for a presentation of the Folktheorem and various generalizations.

If NB > 0 for all i with corresponding emission vector E, and if the countries discount the future sufficiently little, then there exists a perfect equilibrium in the infinitely repeated game, such that the average payoff to country i is NB_i.[3]

Thus, if the financial transfers are used so as to guarantee that all countries will make positive net gains in all time periods, if the countries do not discount the future too much, and if they look sufficiently far into the future, cooperative outcomes will be equilibria, that is, they will be supported by strategies that are based on maximization of the own benefit in each country as a best response to what other countries do. Thus, it seems that by incorporating the time dimension, the incentive structure changes quite dramatically. Before accepting this conclusion there are a few issues that must be faced; first it is the issue of the interest rate, second the need for an infinite horizon, third the non-uniqueness of the Nash equilibirum in the repeated game, and finally the question whether the kind of strategies with punishments that are needed to support a cooperative outcome are to be expected in the real world.

Let us start discussing the issue of the discount rate. That will also lead us to the issue of non-uniqueness, and we will soon see that we have to restrict the possible transfers in order to establish the existence of an equilibrium supported by trigger strategies. The idea behind trigger strategies is that a deviation from cooperative behaviour on the part of one player in one time period will be punished in later time periods by the other players because they will then for the rest of the periods revert to their single period Nash equilibrium strategies. If the deviating player discounts the future sufficiently heavily, the reduction in the present value of the future net benefits because of the other player's punishment may be smaller than the gain from the deviation. Thus, the discount rate must not be too large for the punishment to have a bite.

Friedman (1986) has shown that the following condition is a necessary and sufficient condition for the existence of an equilibrium in trigger strategies:

$$\alpha \geq \frac{NB_i^{ch} - NB_i^{coop}}{NB_i^{ch} - NB_i^{Nash}}$$

here NB_i^{ch} is the net benefit to the cheating player in the period he deviates from cooperative behavior, NB_i^{coop} is the net benefit when all players behave cooperatively, NB_i^{Nash} is the net benefit in the periods when all players are using their non-cooperative strategies, and α is the discount factor $(1/(1+r)$ where r is the interest rate). In order to investigate whether this condition is satisfied or not, NB_i^{ch}

[3] See Fudenberg amd Maskin (1986), theorem C, or Friedman (1986), theorem 3.3, p.88.

has been calculated for all countries. The result is presented in the second column in Table 15.2.

Table 15.2 Net benefits to the cheating player and transfers.

Country	Defector's benefit (2)	Coop. benefit (3)	Gain from (4) deviation	Transfers (5)
ALB	27	24	3	-7
AUS	343	324	19	-104
BEL	309	191	118	8
BUL	90	-7	97	65
CZE	571	152	419	215
DEN	210	119	91	16
FIN	8	-2	10	7
FRA	934	879	55	-279
GDR	551	11	540	343
FRG	1106	328	778	383
GRE	53	53	0	-19
HUN	162	5	157	99
IRE	78	71	7	-21
ITA	157	-83	240	184
NET	647	565	82	-149
NOR	272	272	0	-97
POL	864	599	265	-44
POR	12	10	2	-2
ROM	425	422	3	-149
SPA	7	-29	36	33
SWE	606	606	0	-216
SWI	199	192	7	538
TUR	88	68	20	-11
USS	1505	1505	0	-583
UK	94	-336	430	396
YUG	457	344	113	-50
TOTAL	9775	6283	3492	0

The second column in this table shows the maximum net benefit a country can hope for by cheating in the full cooperative solution. It is thus assumed that there is an agreement on the the full cooperative solution. By cheating, i.e. increasing its emissions from the agreed level, a country can make gains and the larger these gains are the greater the incentive to cheat. By comparing the second and the third columns we get a picture of the gains from cheating in the full cooperative solution without transfers. This comparison is carried out in the next column with the title Gains from deviation. It is clear from these columns that a few countries would not gain anything at all from cheating. Neither Sweden, the Soviet Union nor Greece would stand to gain anything from cheating.[4] On the other hand it seems that Czechoslovakia, Federal Republic of Germany, German Democratic Republic, Poland and United Kingdom would stand to make substantial gains from cheating. For England, that is artificial, because UK would never sign an agreement unless it was guaranteed a positive gain from that.

Let us assume that the total net benefits are distributed equally among the participating countries by financial transfers according to the Nash cooperative solution. That would give each country about 240 million D mark. With this gain, UK would never find it profitable to cheat. In fact, even if UK gets as little in compensation as 336 million D mark, she would not find it profitable to cheat. The situation is slightly different if we look at the Federal Republic of Germany. She stands to make a net benefit of 1106 million D mark by cheating. Her gain would be 866 million D mark compared with the equal division scheme. If she discounts the future with more than 27 percent per annum cheating would be profitable if the other countries follow a trigger strategy (i.e., after the cheating been discovered, the one shot Nash equilibrium strategies will be followed). However, that is a very high interest rate, and thus it seems that even for West Germany, cheating would not pay in the infinitely repeated game.

At all interest rates interesting from a practical point of view there would exist an equilibrium in trigger strategies. Thus, the discount rate is in this respect really a non-issue.

So far we have assumed an infinite time horizon. With a finite horizon, the Nash equilibrium to the repeated game will be the unique Nash equilibrium in the one shot game. Thus, with a finite time horizon, even if it would be far away in the future, we would not expect cooperation. It is hardly reasonable to assume that governments have an infinite horizon, and thus we cannot apply the Folktheorem to the problem of cooperation among European countries on acidification.

[4] This is strictly not true as the figures in the table are truncated. Truncation is also the reason why the figures for the cooperative solution do not agree with the figures in Table 15.2.

There are at least two arguments that could be advanced for a cooperative outcome, even in a situation with finite horizon.[5] The two assumptions that make a Nash non-cooperative equilibrium inevitable are the assumption of complete rationality and the assumption of complete information. By relaxing these assumptions it is possible to show that cooperative behaviour can be expected.

Radner (1980)[6] has shown that if one weakens the assumption of rationality so as to assume only a certain kind of bounded rationality, trigger strategies will support ϵ-equilibria in repeated games with finite horizon. An ϵ-equilibrium is a set of strategies, one for each country, such that each country's average net benefit is within ϵ of the maximum of what it could obtain against the strategies of other countries. The trigger strategies are in this case slightly more complicated than in the infinite horizon case. If one country defects, the other countries will revert to their one shot Nash equilibrium strategies. However, irrespectively of what other countries do, during the last periods up to the horizon, all countries will play their one shot Nash strategies. Thus, the cooperation will break down completely when the time horizon comes close.

What is the reason that countries are satisfied with less than they could achieve? One reason may be the cost of gathering the information necessary for an optimal response to the strategies in other countries (which has not been modelled explicitly) becomes too high compared with the additional benefits from an optimal response. This seems to be quite relevant for governments which have in general quite weak information on emissions, abatement costs, and damages, not only in foreign countries but also in the own country. Thus, one should expect some kind of satisficing behaviour, which would lead to an ϵ-equilibrium.

The second factor that could lead to cooperative behaviour in finite horizon repeated games is the lack of information of other players' strategy sets and pay-off functions.[7] This argument is built around the idea that when there is incomplete information, it is possible for players to build up a reputation based on their behaviour. With complete information, each player would see through such a reputation, but with less information it is possible to go through with sending signals

[5] If the one period game would have more than one Nash non-cooperative equilibrium, it can be shown that there exist equilibrium strategies that would support a cooperative outcome in the repeated game with a finite horizon. Our assumptions imply the existence of a unique Nash equilibrium. See Friedman (1986), theorem 3.6, Chapter 3, for a discussion of an equilibrium in trigger strategies, and Benoit and Krishna (1985) for a more general discussion.

[6] The following discussion is based on Friedman (1986), in which a slightly more general presentation is given.

[7] The following argument is based on Kreps, Milgrom, Roberts, and Wilson (1982). Bounded complexity of the strategies to be used can also promote cooperation, see Neyman (1985). We will not, however, exploit this idea.

to the other players on the nature of the strategy one is playing. Although this seems to be quite relevant in the present context, any analysis of the basic acid rain game in these terms seems to be very complicated, and no attempts will be made here.

To sum up, in the infinitely repeated acid rain game with discounting, cooperation is an equilibrium strategy for all countries if the distribution of benefits is such that all countries gain in all time periods. Moreover, it is possible to find a transfer scheme such that the incentives to cheat are allocated among the countries in such a way that they will all have the same incentive to cheat.

The results that we have relied upon have been based on equilibrium trigger strategies. However, one could argue that the trigger strategies may be too costly to the countries that carry out the punishment, to be credible in the real world. As they are perfect equilibrium strategies they are of course credible in the game theoretic sense. However, Segerstrom[8] has recently shown that one can obtain essentially the same result with strategies that are not as costly to the punisher as the trigger strategies and therefore should be more credible.

What is the relevance for the real world of these results? Axelrod (1984) performed a computer tournament between invited programs exploring various strategies in the repeated prisoner's dilemma. Winner was the "tit-for-tat" strategy, a kind of very simple trigger strategy. Among the opponents to tit-for-tat we very sophisticated programs with Bayesian updating of the information on which strategy the opponent is using, with very strong punishments of defection from cooperative behaviour, with random choices, etc. It turned out that in order to win, a strategy should be tough in the sense that a defection from a cooperative solution should be punished immediately and severely, but should also be forgivable in order to establish new cooperation.

The analysis above has been carried out in terms of repeated games. However, it is doubtful whether this is an appropriate framework. The reason is that we have assumed that damage is determined by the current depositions only and not by the accumulated depositions of sulphur. According to present views on acidification, it should be the accumulated depositions that matter and we would be into a dynamic game that is not a repeated game. Let us look into this briefly. Let the stock of sulphur (or the stock of cations) in the environment in country i be denoted by Y_i. This change in the stock will depend on the rate of deposition and the natural depreciation of the stock. Assume then that the dynamics can be written

$$dY_i/dt = Q_i - L_i$$

[8] Segerstrom (1988).

where L_i is the natural removal of sulphur in country i. The pay off function of country i can now be written

$$\int_0^{\infty} (C_i(E_i) + D_i(Y_i))e^{-rt}dt,$$

where damage is now a function of the stock of pollutants. It can now be shown that the open loop non-cooperative equilibrium will converge in the long run to a state in which $Q_i = L_i$, i.e. a state in which deposition will be such that no further environmental damage will occur.[9] Such a level of deposition is usually called a critical load. Furthermore, in a full cooperative solution, the emissions will be such that depositions will in the long run tend towards the critical loads. If the model above is correct, then attempts to enhance cooperation should focus on the critical loads. This is also what is happening in the present discussions on a new protocol.

15.6 INSTITUTIONS

Suppose now that the incentive and information problems of cooperation have been solved in one way or another, i.e. the countries are willing to cooperate in order to reduce the problems of acid rains. What kind of agreements should the European countries aim at? Different structures of agreements may be possible and it seems to be of utmost importance to investigate these different structures to see what they would mean in terms of economic efficiency.

Present efforts to establish cooperation have been tied to the idea of a uniform emission reduction. As has been shown, this is a very inefficient way of reducing the costs of acidification. In the basic model, only about one third of the potential gains could be realised by that kind of scheme. Obviously, other schemes must be found. The previous analysis has shown that some kind of transfers between countries must be connected with these schemes, otherwise some countries would never participate. We have seen that these transfers need not be financial, but in order to simplify the analysis, let us assume that financial transfers are available and can be used in order to induce all countries to participate.

Three different kinds of schemes suggest themselves. First, agreements on financial transfers and emission reductions, calculated for each country in such a way that total costs are minimized. Second, a tax on "export" of sulphur, i.e., each country

[9] See Mäler (1992) and Mäler and de Zeeuw (forthcoming).

pays an international tax on the part of the domestic emissions of sulphur that is carried to other signing countries. The proceeds of the tax may then be distributed among the participating countries in such a way that there is a net gain for each country. Finally, one could think of a European market for emission permits, where the number of permits is determined in such a way that the total emissions are the desired emissions. We will discuss these schemes in turn.

15.6.1 Agreement on Emissions

The most natural scheme would be an agreement on the emissions in each country. However, such an agreement must be coupled to an agreement on transfers between the countries. If countries have perfect information on the costs of control of sulphur emissions and the damage from sulphur depositions in all participating countries *and* if these variables stay unchanged over time, the scheme seems to be optimal. By calculating the efficient allocation of emissions among the participating countries and then implementing these, an efficient outcome will be expected. If this allocation is then combined with a European "sulphur" fund, to which countries would pay agreed amounts which would then be redistributed, the incentive problem will have been solved.

However, in reality, governments do not know neither the cost nor the damage cost with any greater certainty. Moreover, these costs will change over time due to changes in income, relative prices, and new information. Countries must therefore be prepared to renegotiate the agreement quite often. As the agreement will be quite complicated, this cannot be expected. Once an agreement has been signed, this agreement will be in force (assuming that it bites) for a long period of time, during which external and internal factors may have changed in such a way that the agreement is completely out of date. It does not seem feasible to adapt the scheme in order to make it more flexible to unexpected changes. This inflexibility may work with or against the pure environment objectives. If, for example, new and less expensive technologies of reducing the sulphur emissions are developed, the inflexibility of the contract is such that a further reduction of the emissions would be called for.

Another difficulty with this approach has to do with the financial transfers. As the contract specifies the emissions in all participating countries, these transfers should not give incentives to further abatement or to reduce the abatement. They must therefore be independent of emissions and of damage, although they must be such that all countries will gain from the agreement. Finding mechanisms that satisfy these requirements may turn out to be impossible.

15.6.2 Taxing the Export of Sulphur

One possible scheme to achieve greater efficiency would be to tax the part of the sulphur emissions that is transport by the winds to other participating countries. Ideally, such a scheme should be built on taxes that are individually adjusted to each country, so that the rate one country has to pay equals to the total marginal damage these emissions create in other countries.[10] How would such a tax-structure look like? In the first best allocation, as defined in section 15.4, the full cooperative outcome is defined by minimizing the total European abatement- and damage cost (we now neglect any possible international weights α_i). The necessary conditions for this are

$$C_i'(E_i) = a_{ii}D_i'(Q_i) + \Sigma_{j\neq i}a_{ji}D_j(Q_j),\ i=1,\ ...,\ n$$

The marginal cost of abatement in each country should thus be equal to the sum of the marginal damage caused by emissions in that country. Thus, each country should behave as if they were taxed at a rate that correspond to the total marginal damage. For an international agreement on reduction in sulphur emissions, we can think of at least three different kind of *international* taxes: an emission tax based on the total marginal damage, including domestic damage; a tax on emissions based on foreign damage; a tax on export of sulphur by winds. In all three cases we assume that the tax revenues are returned to the countries in a lump-sum fashion so that their behaviour will not be affected by this (except for deciding to cooperate or not). The first tax system is, however, not going to result in an efficient reduction of emissions. Assume that the tax rate facing country i is t_i. If that country is rational, it would then minimize

$$C_i(E_i) + D_i(Q_i) + t_iE_i$$

which entails

$$C_i'(E_i) + a_{ii}D_i(Q_i) + \Sigma_j a_{ji}D'(Q_j) = 0$$

This is obviously not consistent with efficiency because the individual country would take its own damage account twice, first by the direct damage and second through the tax. We will therefore disregard this situation.

[10] See Mäler (1974) and Mäler (1985) for a discussion of the ideal tax.

In the second tax system, each country would pay a tax on export of sulphur where the tax rate would be

$$t_i = \Sigma_{j \neq i} a_{ji} D_j'(Q_j)$$

It is obvious that in this case we will achieve a first best optimum. Note, however, that the efficient tax rate will vary among countries depending on the amount of exports (that is the coefficients a_{ij}) and the damage functions in the other countries. It is clear that such a tax system would result in severe political problems. It would therefore be interesting to find a system with a uniform tax rate on the export of sulphur. Let the uniform tax rate be t. Each country i would then have to pay

$$t\Sigma_{j \neq i} a_{ji} E_i$$

If all countries are involved in the cooperation (that is not only European), we can write this as

$$t(1\text{-}a_{ii})E_i$$

(If the transport matrix explains where all of the sulphur emissions are going, the column sums will be unity). In this case, the single country could be assumed to behave as if it is minimizing

$$C_i(E_i) + D(Q_i) + t(1\text{-}a_{ii})E_i$$

with the necessary conditions

$$C_i'(E_i) + a_{ii}D_i'(Q_i) + t\Sigma_{j \neq i} a_{ji} = 0, \; i = 1, ..., n$$

The question is whether it is possible to determine t in such a way that the first best optimum is approximated? Obviously, t should be chosen such that

$$\Sigma_i(C(E_i) + D(Q_i))$$

is minimized over t subject to

$$C'(E_i) + a_{ii}D'(Q_i) + t\Sigma_{j \neq i} a_{ji} = 0.$$

This analysis can obviously also be applied to a different mechanism, the use of tradeable permits. The second tax system would correspond to export permits which could be traded internationally, however, not one to one to but only in ratios corresponding to the foreign marginal damage caused by the emissions in a country. The third tax system would correspond to trades one to one. While the problem with the tax system is to find the appropriate tax rate, the problem with tradeable export permits is to find the optimal permissible total export of sulphur. As the system with tradeable permits is dual to the tax system (given our assumption that there is no asymmetric information), there is no need to deal with tradeable permits separately.

As the losses from a second best system cannot be estimated theoretically, an attempt has been made below to illustrate the size of such a loss by using the simulation model from the Acid Rain Game (Mäler, 1989). By rather arbitrary imposing a uniform tax of 4 D mark per kg sulphur, the effect on the net benefits for the european countries could be calculated and compared with the full cooperative solution.

The total net benefit to the participating countries is almost as great as the benefit from the full cooperative solution. The total emission reduction is about 3 % lower than in the full cooperative solution and the distribution of the emission reductions among countries is approximately the same in the two cases. Poland is the only country that has to increase abatement substantially (by about 8%) in the tax solution. The distribution of net benefits is also the same in the two solutions. Thus, the case with export of sulphur seems to produce almost identical real effects on emission reductions and damages from sulphur depositions as the full cooperative solution. This result has been obtained within a quantitative model developed for illustrative purposes only, and one should therefore be very careful in drawing too far-reaching conclusions. However, this result seems so interesting and promising that a deeper study of the consequences of this scheme should be carried out. Moreover, if this result should be valid, it would have consequences for the discussion on markets for emission rights. We will for now assume that the result is valid. The total tax payments are substantial and are more than sufficient to compensate those countries that have a negative net benefit.

Table 15.3 Effects of a tax of 4 D mark/kg sulphur exported

Country	Emiss Red	Perc Red	Net Ben	Tax Paym	Tax-Ben
ALB	13	52	24	8	16
AUS	33	22	313	150	163
BEL	74	24	235	325	-91
BUL	248	50	-37	262	-300
CZE	1219	75	147	585	-438
DEN	104	69	142	53	89
FIN	23	13	-1	126	-127
FRA	78	8	830	787	43
GDR	1040	80	-11	374	-385
FRG	1107	81	363	332	31
GRE	303	86	58	20	38
HUN	635	77	5	273	-269
IRE	22	31	71	21	50
ITA	567	30	-39	907	-946
NET	74	43	552	108	444
NOR	2	4	255	29	227
POL	612	30	530	1495	-965
POR	29	37	4	18	-14
ROM	83	83	430	18	411
SPA	180	11	-14	350	-364
SWE	6	4	580	114	466
SWI	9	20	175	43	133
TUR	216	45	67	64	3
USS	142	2	1508	234	1274
UK	1451	79	-326	221	-546
YUG	465	79	350	133	217
TOTAL	8733	38	6213	7067	-856

15.7 SUMMARY AND CONCLUSIONS

In this chapter we have developed a conceptual model of the issues connected to cooperative efforts in Europe to solve acid rains problems. The basic model is a simple game theoretic model with a unique equilibrium in dominant strategies. In order to illustrate the arguments, the model has been calibrated by assuming that i)

1984 corresponded to a Nash non-cooperative equilibrium (i.e. each country chose their best strategy, that is, the emission level, taking the emission levels in other countries as given), and ii) the damage cost function is linear so that the marginal damage cost is constant. This latter assumption means that the Nash non-cooperative equilibrium is really an equilibrium in dominant strategies. This assumption of a constant marginal damage cost has not been made because it is realistic but simply because it gives an opportunity to calibrate the model. The calibration is made possible by the fact that in a Nash non-cooperative equilibrium, the marginal abatement cost will be equal to the marginal damage cost times the part of the domestic sulphur emissions that is deposited within the country. By using information on the abatement costs, the marginal damage cost and therefore the whole damage cost function can be calibrated.

This calibration can be seen as a way of applying the idea of revealed preferences on the governments, and the damage costs so calibrated will obviously only reflect the damage as perceived by the governments and revealed in their decisions on permitted emissions. The damage function calibrated does not need to be related to any conventional concept of damage, except those concepts that have influenced the government's policies.

However, if the time dimension is taken into account, the situation changes. By an appeal to the Folktheorem, it is shown that if the basic acid rain game is repeated an infinite number of times, cooperative outcomes may be supported by equilibrium strategies. Even if one limits oneself to finite horizons, these cooperative outcomes are likely to emerge as equilibria. The basic mechanism is the punishment of countries defecting from cooperation. In the repeated game, these punishments take the form of increases in the emissions and therefore an increase in environmental damage in the defecting country. It is argued in this chapter that the punishments may very well take the form of changes in trade policies, changes in military concessions, etc. If the strategy sets are made larger and richer in this way, cooperation is more likely to emerge.

What is important is that countries make public the strategies they will follow and that these strategies are both based on credible threats and on tolerance and forgiveness so that cooperation can be restored, if it has once been broken.

If the countries have incentives to cooperate, appropriate institutions must be built up. One such institution is the European Acidification Fund, EAF. EAF could follow one of two strategies. According to the first strategy, it would tax each country for its export of sulphur to other countries. It is shown that with the proper tax rate, the outcome will approximately resemble the full cooperative solution. The tax revenues can then be used for compensating countries that would lose from the cooperation, to give countries incentives not to break off from cooperation and to reveal true estimates of abatement costs and damage costs to EAF. Moreover, the revenues could also be used for buying emission reductions in countries where the markets for capital and foreign exchange are severely distorted.

REFERENCES

Air Pollution across National Boundaries (1972) The Impact on the Environment of Sulphur in Air and Precipitation. Sweden's case study for the United Nations Conference on the Human Environment. Stockholm

Aivazian, V.A., J.L. Callen and I. Lipnowski (1987) The Coase Theorem and Coalitional Stability. *Economica*, Vol. 54, pp. 517-520.

Amann, M. and G. Kornai (1987) *Cost Functions for Controlling SO2 Emissions in Europe*. Working Paper, WP-87-065, IIASA, A-2361 Laxenburg, Austria.

Andreasson, I.M. (1988) *Costs of Controls on Farmers' Use of Nitrogen*. Stockholm School of Economics, Stockholm.

Auman, R. (1959) Acceptable Points in General Cooperative n-person Games, in *Contributions to the Theory of Games IV*. Princeton, N.J.

Axelrod, R. (1984) *The Evolution of Cooperation*. Basic Books, New York.

Benoit, J-P., and V. Krishna (1985) Finitely Repeated Games. *Econometrica*, Vol. 53, No. 4.

Bernheim, D., B. Pegel and M.D. Whinston (1987) Coalition-Proof Nash Equilibria. I. Concepts, *Journal of Economic Theory*, 42, 1, pp. 1-12.

Binmore, K. and P. Dasgupta, eds. (1986) *Economic Orgnizations as Games*. Basil Blackwell Ltd..

Binmore, K., A. Rubinstein and A. Wolinsky (1985) *The Nash Bargaining Solution in Economic Modeling*. ICERD, London School of Economics, Discussion Paper Series, no. 112.

Brooke, A., D. Kendrick and A. Meeraus (1988) *GAMS*. The Scientific Press, Redwood City.

Dasgupta, P. (1982) *Control of Resources*. Basil Blackwell, Oxford.

Dasgupta, P. and G. Heal (1981) *Economic Theory and Exhaustible Resources*. Cambridge.

Eliasson, A., and J. Saltbones (1983) Modeling of Long-range Transport of Sulphur over Europe: a Two-year Model Run and Some Model Experiments. *Atmos. Environ.*, 17, pp. 1457-1473.

Friedman, J. (1986) *Game Theory with Applications to Economics*. New York.

Fudenberg, D., and E. Maskin (1986) The Folk Theorem in Repeated Games with Discounting or with Incomplete Information. *Econometrica*, Vol. 54, pp. 533-554.

Green, J.R., and J.-J. Laffont (1980) *Incentives in Public Decision-Making*. North Holland, Amsterdam.

Groves, T. (1982) On Theories of Incentive Compatible Choice with Compensation, in *Advances in Economic Theory*. W. Hildenbrand (ed.), Cambridge University Press, Cambridge.

Harsanyi, J.C. (1986) *Rational Behavior and Bargaining Equilibrium in Games and Social Situations*. New York.

Hinrichson, D. (1986) Multiple Pollutants and Forests Decline. *AMBIO*, Vol. XV, No. 5.

Kneese, A. (1988) Environmental Stress and Political Conflicts: Salinity in the Colorado River, Paper presented at the conference *Environmental Stress and Political Conflicts*. The Royal Academy of Sciences, Stockholm.

Kreps, D.M., P. Milgrom, J. Roberts, and R. Wilson (1982) Rational Cooperation in the Finitely Repeated Prisoners' Dilemma. *Journal of Economic Theory*, Vol. 27, pp. 245-252.

Krutilla, J. (1966) The International Columbia River Treaty: An Economic Evaluation. in *Water Research*, eds. A. Kneese and S. Smith, The Johns Hopkins Press, Baltimore.

Krutilla, J. (1968) The Columbia River Treaty: *A Study of International River Basin Development*. Baltimore, Johns Hopkins Press.

Laffont, J.-J., and E. Maskin (1982) The Theory of Incentives: an Overview. in *Advances in Economic Theory*. ed. W. Hildenbrand, Cambridge University Press.

Lehmhaus, J., J. Saltbones and A. Eliasson (1986) *A Modified Sulphur Budget for Europe 1980*. EMEP/MSC-W Report 1/86, Norwegian Metereological Institute, Oslo.

Luce, D., and H. Raiffa (1964) *Games and Decisions*. New York.

Mäler, K.-G. (1989) The Acid Rain Game. in *Valuation Methods and Policy Making in Environmental Economics*. Eds: H. Folmer and E.C. van Ierland, Elsevier Science Publishers, Amsterdam.

Mäler, K.-G., Critical Loads and International Environmental Cooperation. in *Strategy and Conflict in Environmental Policy*, ed. R. Pethig.

Mäler, K.-G. (1974) *Environmental Economics - A Theoretical Inquiry*. Johns Hopkins University Press, Baltimore.

Mäler, K.-G. (1985) Welfare Economics and the Environment. in *Handbook of Natural Resources and Energy Economics*, Vol. I, eds. A. Kneese and J.L. Sweeney, Elsevier Science Publishers, Amsterdam.

Neyman, A. (1985) Bounded Complexity Justifies Cooperation in the Finitely Repeated Prisoners' Dilemma. *Economics Letters*, 19, pp. 227-229.

Nilsson, J. (ed.) (1986) *Critical Loads for Nitrogen and Sulphur*. The Nordic Council of Ministers, Report 1986:11, Copenhagen.

OECD (1975) *Polluter Pays Principle*. Paris.

OECD (1981) *Transfrontier Pollution and the Role of States*. Paris.

Olsson, C. (1988) *The Cost-Effectiveness of Different Strategies Aimed at Reducing the Amount of Sulphur Deposition in Europe*. Research Report 261, EFI, Stockholm School of Economics.

Radner, R. (1980) Collusive Behaviour in Noncooperative Epsilon-Equilibria of Oligopolies with Long but Finite Lives. *Journal of Economic Theory*, Vol. 22, pp. 136-154.

Segerstrom, P. (1988) Demons and Repentance. *Journal of Economic Theory*, 45, 1, pp. 32-52.

Shapley, L.S,, and M. Shubik (1969) On the Core of an Economic System with Externalities. *The American Economic Review*, Vol. 59, pp. 578-584.

Shapley, L.S., and M. Shubik (1966) Quasi-cores in a Monetary Economy with Nonconvex Preferences. *Econometrica*, Vol. 34, no. 4.

Starret, D.A. (1973) A Note on Externalities and the Core. *Econometrica*, Vol. 41.

The Norwegian Meteorological Institute (1987) *Sulphur Budgets for Europe for 1979, 1980, 1981, 1982, 1983, 1984 and 1985*. EMEP/MSC-W Not 4/87.

Theutenberg, B.J. (1984) *The Evolution of the Law of the Sea*. Tycooly International Publishing Limited, Dublin.

Tietenberg, T.H. (1985) *Emission Trading - an Exercise in Reforming Pollution Policy*. Resources for the Future, Inc. Washington D.C..

UNECE (1988) Workshop Report (Unpublished Draft) from the Workshop "Critical Loads for the Effect on Soils and Groundwater of Long Term Deposition of Nitrogen and Sulphur Compounds", held at Skokloster, Sweden, 20-24.

Weitzman, M.L. (1974) Prices vs. Quantities. *Review of Economic Studies*, 41, pp. 477-491.

Wijkman, P.M. (1982) Managing the Global Commons. *International Organization*, 36, 3, pp. 511-536.

Zeeuw, Aart de and K.-G. Mäler (forthcoming) *Critical Loads in Games of Transboundary Pollution Control*.

INDEX

C

D

E

F

G

H

I

M

N

O

P

Q

R

S

T

U

V

W

X

Y

Z